M 2106832

Irrigation Management

Principles and Practices

Mixed Sources
Product group from well-managed forests, controlled sources and recycled wood or fibre
FSC
www.fsc.org Cert no. SGS-COC-002953
© 1996 Forest Stewardship Council

Irrigation Management

Principles and Practices

Martin Burton

CABI is a trading name of CAB International

CABI Head Office
Nosworthy Way
Wallingford
Oxfordshire OX10 8DE
UK

Tel: +44 (0)1491 832111
Fax: +44 (0)1491 833508
E-mail: cabi@cabi.org
Website:www.cabi.org

CABI North American Office
875 Massachusetts Avenue
7th Floor
Cambridge, MA 02139
USA

Tel: +1 617 395 4056
Fax: +1 617 354 6875
E-mail: cabi-nao@cabi.org

© M.A. Burton 2010. All rights reserved. No part of this publication may be reproduced in any form or by any means, electronically, mechanically, by photocopying, recording or otherwise, without the prior permission of the copyright owners.

A catalogue record for this book is available from the British Library, London, UK.

Library of Congress Cataloging-in-Publication Data

Burton, Martin, Dr.
Irrigation management: principles and practices/M.A. Burton.
p. cm.
Includes bibliographical references and index.
ISBN: 978-1-84593-516-0 (alk. paper)
1. Irrigation–Management. 2. Drainage–Management. I. Title.

TC812.B87 2010
333.91′3–dc22

2009033245

ISBN: 978 1 84593 516 0

Commissioning editor: Nigel Farrar
Production editor: Shankari Wilford

Typeset by SPi, Pondicherry, India.
Printed and bound in the UK by CPI Antony Rowe Ltd.

Contents

About the Author

My interest in irrigation management stems back to an assignment in 1977 as a junior professional on the East Java Design Team, Indonesia, working as part of a team helping to modernize the operation and maintenance (O&M) procedures for the East Java Irrigation Service. For 18 months I worked with an experienced Indonesian colleague, Arief Effendi, in the Mojokerto office inspecting all irrigation and drainage systems in the 32,000 ha command area, and then worked with the O&M engineers and technicians to introduce updated O&M procedures for these systems. At the same time we worked with the *Juru Pengairan* (Irrigation Service water master), village leaders and the *jogotirto* (village water master) of the 108 ha Blendren tertiary unit on measures to improve on-farm operation and maintenance. I am indebted to Arief Effendi and our colleagues in Mojokerto for sharing their knowledge and experience with me, and hope that in some small way this book repays the debt I own them.

This interest in irrigation and drainage management developed over the years with various assignments as an O&M Engineer and a return to East Java as the Training Officer on the World Bank-funded East Java Irrigation Project, where I again worked with Arief Effendi and two colleagues, Bin Yali and Satrio, on organizing training courses for Irrigation Service sub-section office staff and water masters in one irrigation district of 140,000 ha command area. This training programme was novel at the time in its focus on practical classroom exercises and practical fieldwork, with the trainers travelling to work with the staff in their offices and on their systems rather than the trainees travelling to a central training centre. The concept was considered to be successful and expanded under subsequent World Bank projects to other regions in Indonesia.

In 1986 I joined the staff of the Institute of Irrigation Studies at the University of Southampton to lecture in management, operation and maintenance of irrigation and drainage schemes on the MSc Irrigation Engineering course. I quickly learnt that not everyone shared my enthusiasm for irrigation and drainage management, and that designing and building schemes was considered more interesting and challenging. It was, however, noticeable over the 14 years I spent at the University how this attitude changed, and how those attending the MSc course and associated short courses had a growing concern and interest in improving the management of irrigation and drainage schemes.

While teaching at Southampton I was aware that I needed to better understand general management, and therefore studied for an MBA at Henley Management College. This developed my awareness and understanding of management and administration systems, and led to work in restructuring of government-run irrigation and drainage agencies. I am sure that this is an area where we will see significant changes in the coming years, as government agencies modernize to meet the challenges we are facing in irrigation and water resources management.

Acknowledgements are due to many people over the years. To Robert Chambers in the initial instance for his work in the 1980s on irrigation management, and the identification of 'blind spots', which included main system management, night irrigation, and incentives and motivation for managers. Also thanks to my many professional colleagues in consulting engineers Mott MacDonald and later at the University of Southampton in the Institute of Irrigation and Development Studies. I am grateful to Alan Beadle, Mike Snell, Melvyn Kay and Tim Jackson for comments on initial drafts of this book, and to Masood Khan, Ian Smout, Don Brown, Mark Svendsen, Ian Carruthers, Rien Bos, Hector Malano, Charles Abernethy, Flip Wester, Laurence Smith, Jerry Neville, David Molden, Hammond Murray-Rust, Ian Anderson, Sam Johnston III and Joop Stoutjesdijk for their contribution over the years to my understanding of irrigation management. I am indebted to Dr Safwat Abdel-Dayem for never letting me forget that it is irrigation *and* drainage, and that for many schemes drainage is sometimes the central issue for sustainable irrigated agriculture.

Martin Burton
Itchen Stoke
July 2009

Preface

There is increasing pressure worldwide on available water resources. These pressures arise from a number of factors, including growing populations, increased wealth and urbanization, increased industrialization, and demands from society and environmental groups for safeguards to protect water resources and the aquatic environment. In many locations climate change is adding to these pressures.

In many countries irrigated agriculture consumes a large proportion of the available water resources, often over 70% of the total. There is considerable pressure to release water for other uses, and as a sector, irrigated agriculture will have to increase its efficiency and productivity of water use. A new era is dawning for water management in the irrigated agriculture sector, where the management effort and returns to management are required, recognized and rewarded.

This book draws on the author's experience and work over 30 years and in some 28 countries in the management, operation and maintenance of irrigation and drainage schemes. The book provides knowledge for management of irrigation and drainage systems in the 21st century, covering the traditional technical areas related to system operation and maintenance and expanding managerial, institutional and organizational aspects related to the changing political, social and economic environment. It lays emphasis on the management of irrigation as a business enterprise, moving management thinking out of traditional public sector mindsets to more customer-focused, performance-oriented service delivery.

A significant proportion of the irrigation and drainage systems worldwide are manually operated gravity systems managed by government agencies with large numbers of water users farming relatively small landholdings. The total area worldwide in this category is over 165 million ha, which is over 60% of the total area irrigated worldwide. It is in such systems where improvements in management are most required, and in which the most substantial benefits can be obtained.

The book seeks to provide practical guidelines to improve the three key processes of *management*, *operation* and *maintenance* of such systems. In the management context it deals with institutional issues, such as water law, and management structures and management processes, including establishing and working with water users associations, restructuring irrigation and drainage agencies, fee setting and cost recovery. In the operation context the book provides practical guidance on key operation processes, including irrigation scheduling at main system and on-farm level, and performance management tools. In the maintenance context it covers maintenance management processes, including maintenance identification, planning, budgeting, implementation, supervision and recording. Asset management is increasingly used as a tool for maintenance management, and is covered in some detail.

1

Introduction

This chapter looks at the historical development of irrigation and the pressure that this development has placed on the world's water resources. The issues facing irrigation and the associated development of water resources are discussed and the role that irrigation management can play in addressing these is outlined.

Historical Development

Irrigation and drainage development

The irrigation area worldwide has increased threefold over the last 50 years, from 94 million ha in 1950 to over 287 million ha in 2007 (Fig. 1.1). Despite this massive increase the irrigated area per member of the world's population has varied relatively little, from 37.3 ha/thousand people in 1950 to 43.0 ha/thousand people in 2007, with a peak in the late 1970s of 47.6 ha/thousand people.[1]

Table 1.1 shows the irrigated area, population and irrigated area per thousand people in a number of countries. The total irrigated area of these 42 countries represents 86% of the total area irrigated worldwide. The countries with the largest areas include India (57.3 million ha), China (53.8 million ha), the USA (21.4 million ha) and Pakistan (17.8 million ha). There are four countries with a significant irrigated area in the range of 5 to 10 million ha and a further 21 with irrigated areas in the range of 1 to 5 million ha. The irrigated area per thousand people ranges from 2 ha/thousand people in Nigeria to 232 ha/thousand people in Kazakhstan. The generally low level of irrigation development in some sub-Saharan countries in Africa can be seen from the data for Nigeria, Kenya, Mozambique and Senegal.

A valuable assessment of the current situation related to irrigated agriculture and water resources development has been published by the International Water Management Institute (IWMI). The Comprehensive Assessment of Water Management in Agriculture (Earthscan/IWMI, 2007) was a multi-agency study coordinated by IWMI in association with a number of other organizations, including the Consultative Group on International, Agricultural Research (CGIAR) and the Food and Agricultural Organization of the United Nations (FAO).

The Assessment found that agriculture continues to be the largest consumer of water, taking 71% of all withdrawals, compared with 18% for industry and 8% for domestic/municipal use. In total, in 2000, some 3800 km^3 of water were withdrawn from surface and groundwater resources, with approximately 2700 km^3 being abstracted for irrigated agriculture and 20% of the total abstraction being from groundwater. The dramatic change in the

©Martin Burton 2010. *Irrigation Management: Principles and Practices* (Martin Burton)

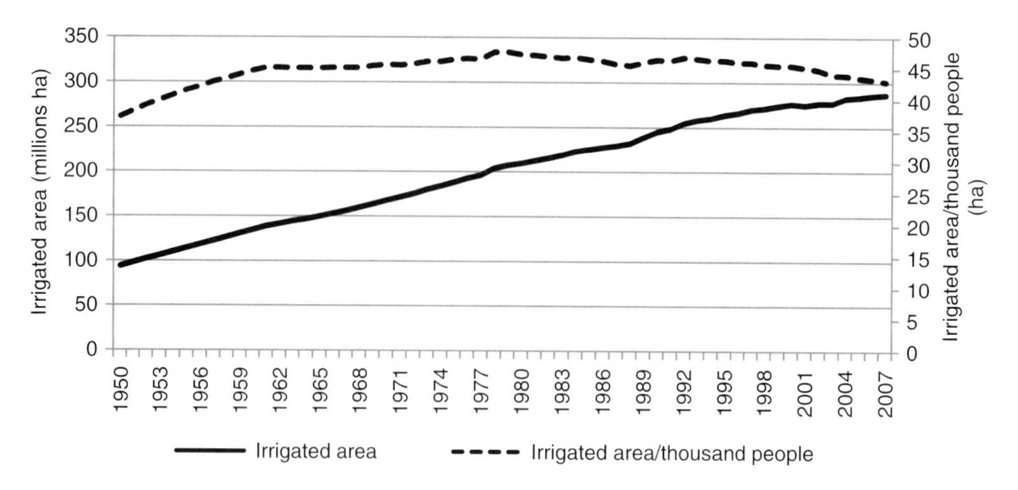

Fig. 1.1. Growth of irrigated area worldwide, 1950–2007. (Data from EPI, 2009.)

amount of water abstracted for various uses is shown in Fig. 1.2. The total quantity abstracted has increased from around 1400 km^3/year in 1950 to around 3800 km^3/year in 2000. As can be seen, the proportion abstracted for municipal and industrial use has changed as the population balance shifts from rural to urban. This shift in the rural–urban population balance has significant consequences for irrigated agriculture. As economic development takes place, the urban population grows and the proportion of the population deriving its livelihood from agriculture (irrigated and rainfed) declines, as does the proportion of the Gross Domestic Product derived from agriculture. The domestic water demands of the urban population increase, as does the demand from the industries[2] that provide work for these populations. The political balance shifts, with a focus on protecting the municipal and industrial demands for these growing urban populations.

The Assessment found a number of promising and disturbing trends and forces, which are summarized in Box 1.1. Though there are some positive trends, there is much of concern in these findings, with increasing levels of pollution and desiccation of rivers, over-committed river basins, increasing demands from urban populations, and rapidly depleting groundwater reserves.

Water resources development and the changing role of management

In many countries irrigation is the main user of water, with over 70% of all abstracted water being used for irrigated agriculture. Water, rather than land, has become the limiting constraint on development, with many basins being closed or approaching closure.[3] Figure 1.3 is helpful in understanding how the development of irrigation in many countries has led to this pressure on water resources and how institutional arrangements have adapted to cope with this development. Based on the work of Keller *et al.* (1998), Molden *et al.* (2001) identified four broad phases:

- development;
- utilization;
- allocation;
- restoration.

In each of these phases, different needs and therefore different institutional structures exist. In the *development* phase the amount of naturally occurring water is not constrained and expansion of demand drives the need for construction of new infrastructure, with institutions heavily involved in planning, design and construction of water resources projects. Civil engineers dominate the development process, and as water becomes scarce due to growing demand, additional spare capacity

Table 1.1. Populations and irrigated areas in selected countries. (From FAO Aquastat website,[a] http://www.fao.org/nr/water/aquastat/data/query/index.html)

Country	Population (000s)	Average precipitation (mm/year)	Irrigated area (000 ha)	Irrigated area per thousand people (ha)
India	1,151,751	1,083	57,286	50
China	1,328,474	n/a	53,820	41
USA	302,841	715	21,400	71
Pakistan	160,943	494	17,820	111
Iran	70,270	228	8,132	116
Mexico	105,342	752	6,256	59
Russian Federation	143,221	460	5,158	36
Thailand	63,444	1,622	5,004	79
Turkey	73,922	593	4,983	67
Indonesia	228,864	2,702	4,428	19
Uzbekistan	26,981	206	4,223	157
Italy	58,779	832	3,973	68
Bangladesh	155,991	2,666	3,751	24
Kazakhstan	15,314	250	3,556	232
Egypt	74,166	51	3,422	46
Afghanistan	26,088	327	3,199	123
Japan	127,953	1,668	3,128	24
Viet Nam	86,206	1,821	3,000	35
Brazil	189,323	1,782	2,870	15
Ukraine	46,557	565	2,605	56
Australia	20,530	534	2,545	124
Chile	16,465	1,522	1,900	115
Sudan	37,707	416	1,863	49
Greece	11,123	652	1,594	143
Philippines	86,264	2,348	1,550	18
South Africa	48,282	495	1,498	31
Morocco	30,853	346	1,484	48
Nepal	27,641	1,500	1,134	41
Kyrgyzstan	5,259	533	1,077	205
Republic of Korea	48,050	1,274	889	18
Romania	21,532	637	808	38
Portugal	10,579	854	617	58
Sri Lanka	19,207	1,712	570	30
Venezuela	27,191	1,875	570	21
Algeria	33,351	89	569	17
Malaysia	26,114	2,875	363	14
Nigeria	144,720	1,150	293	2
Israel	6,810	435	225	33
Senegal	12,072	686	120	10
Mozambique	20,971	1,032	118	6
Kenya	36,553	630	103	3

n/a, data not available.
[a]The database provides information on population and irrigated areas in each country during the period 1993–2007.

is created through the construction of more infrastructure, particularly dams, resulting in step changes in the amount of water available for use.

In the *utilization* phase the infrastructure is established and the broad goal is to make the most out of these facilities. Creation of additional supplies through further construction

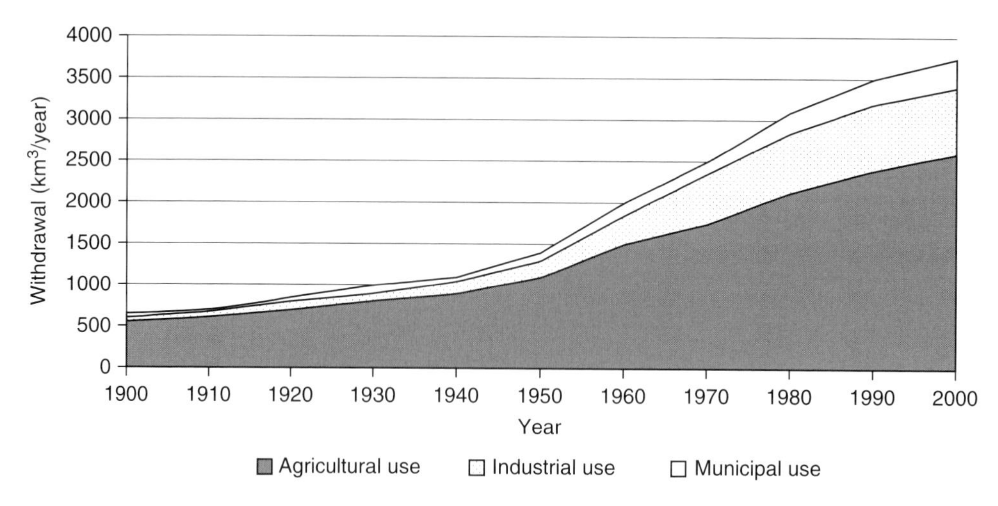

Fig. 1.2. Change in the water abstracted for human use, 1900–2000. (From IWMI, 2006 after Shiklomanov, 2000 with permission.)

Box 1.1. Emerging Trends (Earthscan/IWMI, 2007)

Promising trends

- A steady increase in the consumption of food, leading to better nutrition for many and a decrease in famines. The average global energy intake increased from 2400 kcal/capita/day in 1970 to 2800 kcal/capita/day in 2000, enough to feed the world in spite of a growing population.
- A steady increase in land and water productivity, with average grain yields rising from 1.4 t/ha to 2.7 t/ha during the past four decades and significant gains in water productivity.
- New investments in irrigation and agricultural water management have the potential to support economic growth within agriculture and other areas.
- An increase in global trade in food products and consequent flows of virtual water,[a] offering prospects for better national food security and the possibility to relieve water stress.

Very disturbing trends

- The number of malnourished people worldwide remains about 850 million.
- The average daily per capita food supply in South Asia (2400 kcal) and sub-Saharan Africa (2200 kcal) remains far below the world average (2800 kcal) in 2000.
- Pollution and river desiccation are increasing because of greater agricultural production and water consumption. Fisheries, important for the livelihoods of the rural poor, have been damaged or threatened.
- Land and water resources are being degraded through erosion, pollution, salinization, nutrient depletion and the intrusion of seawater.
- Pastoralists, many relying on livestock as their savings, are putting grazing lands under pressure.
- In several river basins water resources are over-committed and poorly managed, with insufficient water to match all demands.
- Groundwater levels are declining rapidly in densely populated areas of north China, India, North Africa and Mexico because of over-exploitation.
- Water management institutions have been slow to adapt to new issues and conditions.

Double-edged trends

- Increasing withdrawals for irrigation in developing countries have been good for economic growth and poverty alleviation, but bad for the environment.
- Subsidies, if applied judiciously, can be beneficial to support income generation for the rural, but can distort water and agricultural practices.

- The growing demand of cities and industries for water offers possibilities for employment and income, but it also shifts water out of agriculture, puts extra strain on rural communities and pollutes water.
- Fish and meat consumption is rising, increasing the reliance on aquaculture and industrial livestock production, with some positive well-being and income benefits but greater pressure on water resources and the environment.

And emerging forces

- The climate is changing, which will affect existing temperatures and patterns of precipitation. Agriculture nearer the equator – where most poor countries are situated – will be affected most.
- Globalization continues over the long run, providing opportunities for commercial and high-value agriculture but posing challenges for rural development.
- Urbanization increases the demand for water, generates more wastewater and changes patterns of demand for agricultural products, all affecting agricultural practices.
- Higher energy prices increase the costs of pumping water, applying fertilizers and transporting products. Greater reliance on bioenergy is affecting food crop prices and water used by agriculture.
- Perceptions and thinking about water are changing, with more attention to green water[b] resources (in the soil), not just to blue water[c] resources (in lakes, rivers and aquifers).
- More attention is also being given to ecosystem and integrated approaches and to understanding how forces outside water for agriculture influence both water and agriculture.

[a]Virtual water is the water used to produce food products. If 1t of grain requires 2t of water to grow, importing 1t of grain is equivalent to importing 2t of water.
[b]Green water is the term applied to water provided by rainfall, stored in the root zone and consumed by natural vegetation and rainfed agriculture.
[c]Blue water refers to the runoff from rainfall, which is stored in lakes, wetlands and aquifers.

activities is constrained, and thus increased attention is paid to water management to conserve water and optimize productivity of available water. In this phase institutions are primarily concerned with management within discrete units for irrigation, water supply, industry, hydroelectric power, etc.

In the *allocation* phase, when closure starts as depletion approaches the potential available water, there is limited scope for further development. Various measures are taken to maximize the productivity of water and managing demand becomes an issue. With little opportunity for making real water savings, reallocation of the available water from lower- to higher-value uses takes place. Institutions are primarily involved in allocation, conflict resolution and regulation, with several management and regulatory functions gaining prominence, such as inter-sectoral allocation and water trading. Coordination between the different, competing interests becomes an issue and moves are made for coordinating river basin management forums to resolve conflict and facilitate management.

In the *restoration* phase efforts are made to restore the river basin to a balance with its renewable resources. In many cases water is abstracted beyond the renewable resource; this is particularly the case with groundwater where the resource is mined and groundwater levels fall year-on-year. Measures here may include taking irrigated areas out of production and limiting further population growth and industrial development in the river basin. Some technical interventions may be possible in this phase, such as inter-basin transfers, but regulation (particularly enforcement) and management are most prominent. Political involvement is also required where tough decisions are needed to return the basin to a balanced situation.

Thus in the early stages of river basin development the focus is on planning and construction of infrastructure to increase the quantity of the renewable resource made available for use. Over time the focus changes to management rather than construction, initially with measures to match supply with the increasing demand (supply management) and later with measures

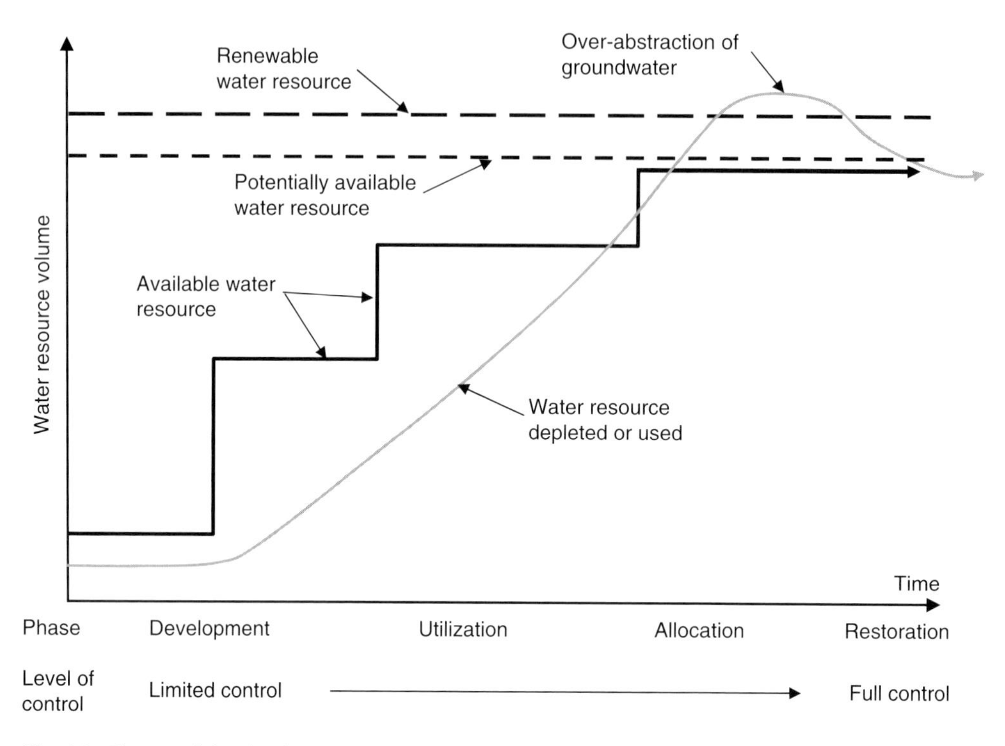

Fig. 1.3. Phases of river basin development. (Modified from Molden *et al.*, 2001.)

to limit demand to match the available supply.

As noted above, the institutional arrangements and management processes change over time as the pressure on the renewable resources increases. As one approaches the limit of the renewable water resource there is reduced room for manoeuvre and increased risk and consequences associated with periods of drought (Fig. 1.4). The management decisions change from generally top-down development of new infrastructure to more bottom-up approaches, incorporating dialogue with and empowerment of water users. Areas to focus on at this stage in the irrigation sector include education, training and capacity building (of both service providers and water users), development of information systems, institutional reform and organizational restructuring, operations management and performance management in order to keep abstractions to a minimum, reduce wastage, minimize pollution in return flows and increase productivity per unit of water abstracted.

With many rivers approaching closure, management of the water resources, including irrigation, has to improve (see Svendsen, 2005 for a more detailed discussion). Good management is dependent on reliable data, and increasing investment is being made in many countries in strengthening information management systems, particularly in relation to water sharing and allocation and operational management (Table 1.2).

Future Scenarios

In addition to assessing the current situation, as summarized in Box 1.1, the Comprehensive Assessment looked at future scenarios, as summarized in Box 1.2. The scenario is relatively bleak from the environmental perspective: with increasing pressure for food some governments will be looking to develop new irrigation areas, possibly on more marginal lands, thereby increasing water abstraction. In addition,

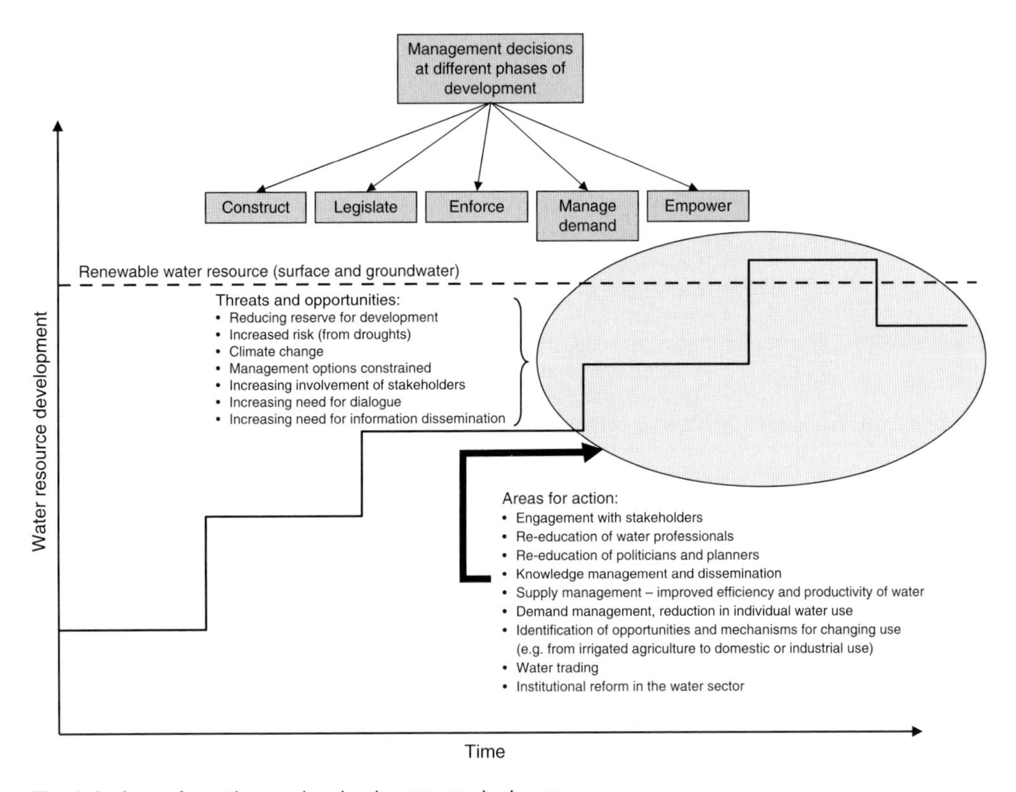

Fig. 1.4. Areas for action as river basins approach closure.

any increase in irrigated cropping will result in increased evapotranspiration, raising the water demand by an estimated 60–90% of the current levels.

The message is clear and stark: far greater attention and resources will need to be paid to management of the Earth's finite water resources if we are to both feed the Earth's population and sustain the aquatic ecosystems.

The Way Forward

Those involved with water resources and irrigation development will need to think more broadly, they will need to understand and consider the multiple uses of water and integrate the management and use of these resources. Managers of irrigation schemes will no longer be able to dismiss the impacts of their management on the natural environment; account will have to be taken of the levels of water abstraction from rivers and groundwater, the impact of agricultural pollutants in drainage wastewater on natural ecosystems. With understanding and knowledge comes an ability to manage better the resource for all uses and users. This understanding will need to be applied to greater appreciation of rainfed agriculture as well, particularly in Africa where farmers often cultivate both irrigated and rainfed crops as part of a mixed farming system.

Scheme managers will require greater understanding of and commitment to the needs of vulnerable groups and the poor. Ensuring secure and reliable irrigation water supplies to tail-ends, where these groups are often found, can make a significant difference.

Efficiency and productivity need to become keywords for irrigation managers. Efficiency in all operations: such that water abstraction is kept to a minimum, thereby leaving water in the river or groundwater

Table 1.2. Typical water uses, information uses and users within a river basin. (From Burton and Molden, 2005.)

Water uses	
Watershed water uses	• Lakes/reservoirs • Forests • Natural vegetation
Instream water uses	• Hydropower • Recreation • Navigation • Fisheries
Extractive water uses	• Irrigation (surface/groundwater) • Potable water (surface/groundwater) • Industrial water, including mining (surface/groundwater)
Environmental water uses	• Aquatic, wetlands and flood plain environment and ecology • Drainage disposal • Waste dilution and disposal • Repelling salinity intrusions • Erosion control
Information uses	
Development and master planning	• Planning and forecasting • Decision making in relation to resource development and protection
Water sharing and allocation	• Resource management and allocation • Allocation of water rights • Rule formulation • Pricing • Dialogue with, and among, users
Operational management	• Flow control and regulation • Flood control, protection and warning • Effluent control • Monitoring and evaluation (abstractions, effluent levels, environment, etc.) • Infrastructure asset management • Conflict resolution
Research	• Water resources, irrigation, environment, ecology, etc.
Information users	
Government	• Ministries of: Water Resources, Irrigation, Agriculture and Livestock, Energy, Hydrology and Meteorology, Health, Environment and Natural Resources, Fisheries, Forestry, Navigation and Marine Transport, Planning and Development • Legislatures • State, regional or local government • Municipalities
Regulatory and management authorities	• River boards, river basin councils, drainage boards • Regulatory bodies (rivers, groundwater, environment, etc.) • Courts
Companies, groups and associations	• Industry (manufacturing, services, mines, forestry, etc.) • Associations (irrigation, rural water supply, environmental lobbies, etc.) • Universities, research centres and training centres • Development agencies and agents • Non-government organizations
Individuals	• Domestic household users • Irrigation farmers • Livestock owners • Recreators

Box 1.2. Assessment of the Future (Earthscan/IWMI, 2007)

Demand for food

- Food demand will rise dramatically in the next 50 years, to almost twice present-day levels.
- This demand will be as a result of rising population, but also changing dietary habits with economic development. Diets will change to consumption of cereals, but also to more livestock and fish products. An estimated 25% of the increase will be for grains for livestock feed.

Availability of water

- The amount of evapotranspiration will increase by 60–90% depending on population growth and the change in dietary habits, increasing from 7200 km^3/year today to 11,000–13,500 km^3/year in 50 years.
- To meet the increasing demand will require:
  - using more blue water from rivers and aquifers for irrigation;
  - using more marginal-quality water for agriculture;
  - using more green water by upgrading rainfed agriculture;
  - increasing the productivity of blue and green water to reduce the abstraction;
  - managing demand for agricultural water by changing diets and reducing postharvest losses;
  - reducing water use in water-scarce regions through trade (importing virtual water).

for the environment, and water is delivered where, when and in the quantity required. Improving scheduling procedures to make better use of rainfall is one way of reducing river or groundwater abstraction, particularly in the humid tropics. Waste needs to be reduced in all parts of the supply chain, whether it is caused by over-irrigation of the farm plot by the farmer, a failure to reduce flows at night in the main systems when water is not required by farmers, or poor storage and loss of harvested crops. The focus needs to change from a narrow perspective on, say canal conveyance efficiencies, to a broader perspective such that as much as possible of the water abstracted for irrigation is converted into useable product at the point of use. In this context proper irrigation scheduling coupled with adequate control and measuring structures may be more relevant than focusing simply on conveyance losses.

Productivity will need to improve, though irrigation scheme managers will need to focus on more than just the physical productivity of water (the more 'crop per drop' approach). While the physical productivity of water is important, and generally within the control of scheme managers and farmers, consideration needs to be taken of the economic water productivity (the value of agricultural water production per unit of water) and the agricultural water productivity (the net gains from all uses of water for agriculture, including crops, fisheries, livestock, forestry, firewood, etc.). Improving these facets of productivity will involve improving the support given for inputs and supporting processes other than irrigation water and drainage water removal, such as credit, input provision, agricultural machinery and marketing.

With the increasing pressure on land and water resources to produce increasing quantities of agricultural produce, the need for better educated, informed and motivated managers in the irrigation sector is evident, whether they are the head of an irrigation district, the manager of an irrigation scheme, the executive director of a water users association or a farmer. Reforms are taking place in many countries through the process of irrigation management transfer, giving more rights and responsibilities to water users for the management, operation and maintenance of all, or parts of, their irrigation and drainage systems. In many countries these changes need to be matched by reforms to state agencies responsible for water resources and irrigation development and management, and correspondingly in the education and training institutions that feed young professionals into these agencies and the sector in general. As Robert Chambers advocated in the 1980s (Chambers, 1988; Box 1.3), a new cadre of irrigation managers is required, with enhanced

Box 1.3. On Irrigation Managers

None of these measures for bureaucratic reorientation, as it has been called, could in itself reform canal irrigation management. Nor would it make sense to recommend implementing them all simultaneously. As always, the best mix and sequence depends on conditions. As more is known, through more research and writing, about the real world of managers, so also it will become clearer what best to do. Exhortation or moralising are unlikely to make much difference. The conditions, motivation and incentives of the managers are the key. And among these, professional methodologies and behaviour present one promising point of attack. If it is clear what managers should be doing, it will be easier for them to do it; which brings us to what they and others can do – diagnostic analysis of existing canal irrigation systems, and selected actions to improve performance such as making and implementing operational plans.

(Chambers, 1988)

knowledge, skills, status and remuneration for the important work that they do. The irrigation manager in the 21st century will need to:

- understand the wider agricultural and water resources issues associated in using water for agriculture;
- understand the importance and value the natural ecosystems within which irrigated agriculture lies and upon which it depends;
- manage irrigation systems to provide reliable, adequate, timely and equitable irrigation water supply, and associated drainage water removal;
- manage water, in all its forms, to the benefit of all communities and the natural ecosystem;
- understand and accept the concept of service provision, and the need to liaise and work with water users in the provision of a responsible and fair service in return for timely and adequate payment of the service fee;
- understand the wider dimensions of irrigated agriculture, from catchment management through to marketing, and work towards enhancing efficiency and productivity in all parts of the supply chain;
- understand and treat irrigated farming as a business, to which the supply of irrigation water and removal of drainage water in a reliable, timely and adequate manner makes a significant contribution to the success or failure of the enterprise.

It is hoped that the information provided in this book will assist in addressing these issues and contribute to developing the necessary understanding, knowledge and skills required for effective and productive management, operation and maintenance of irrigation and drainage schemes.

Endnotes

[1] Data compiled by the Earth Policy Institute (http://www.earth-policy.org/index.php?/datacenter/xls/book_pb4_ch2_8.xls) using data from the Worldwatch Institute, the FAO and the United Nations Population Division.

[2] These demands are not just the abstracted water quantities; they will include increased minimum flow requirements to dilute the wastewater being discharged by these industries.

[3] A basin is considered closed when all available water has been used. A basin remains open when there are water resources remaining to be developed or used.

2

Components of Irrigation and Drainage Systems

Irrigation and drainage is a complex mixture of technical, institutional, economic, social and environmental processes. This chapter outlines the different elements of these processes and how they interact.

Overview

Chambers (1988) identified irrigation and drainage schemes (I&D schemes)[1] as a complex mixture of *physical*, *human* and *bioeconomic* domains. Figure 2.1 outlines these domains, the activities involved and their interactions. In the physical domain we are dealing with the climate, soils and physical infrastructure. In the human domain we are dealing with the irrigation agency personnel and with farmers, their families and other stakeholders. In the bioeconomic domain we are dealing with the crops, livestock and markets. Overlying these three domains are the political, economic and legal domains.

Those involved with irrigation development need to be aware of, and understand, all these domains, and know which factors they can control and which are out of their control and influence. Politicians and government officials, for example, have control over the political, economic and legal domains, and have the ability to set the direction in terms of political input, legislation and economic policy to support the irrigated agriculture sector. Managers of irrigation systems have control over the physical domain, and can organize the capture and distribution of water. Within the human domain the irrigation staff are under the control of the system manager, while the farm households and labour are under the control of the water users themselves. The bioeconomic domain is influenced by the market and government policies, such as for subsidies and food pricing.

Using a systems approach, Small and Svendsen (1992) separated the various components involved in irrigated agriculture into a series of nested systems (Fig. 2.2). The systems approach focuses on the *inputs*, *processes*, *outputs* and *impacts* at different levels. The nested system begins with the outputs of the irrigation and drainage system (I&D system) (supply of water to crops) feeding in as one of the inputs (along with the land, labour, seed, fertilizer, etc.) to the irrigated agriculture system. The outputs in the form of agricultural production from this system feed into the agricultural economic system and with other inputs (traders, market price, etc.) provide inputs (incomes) into the rural economic system. Together with other income, such as off-farm labour, and expenditure, the rural economic system feeds into

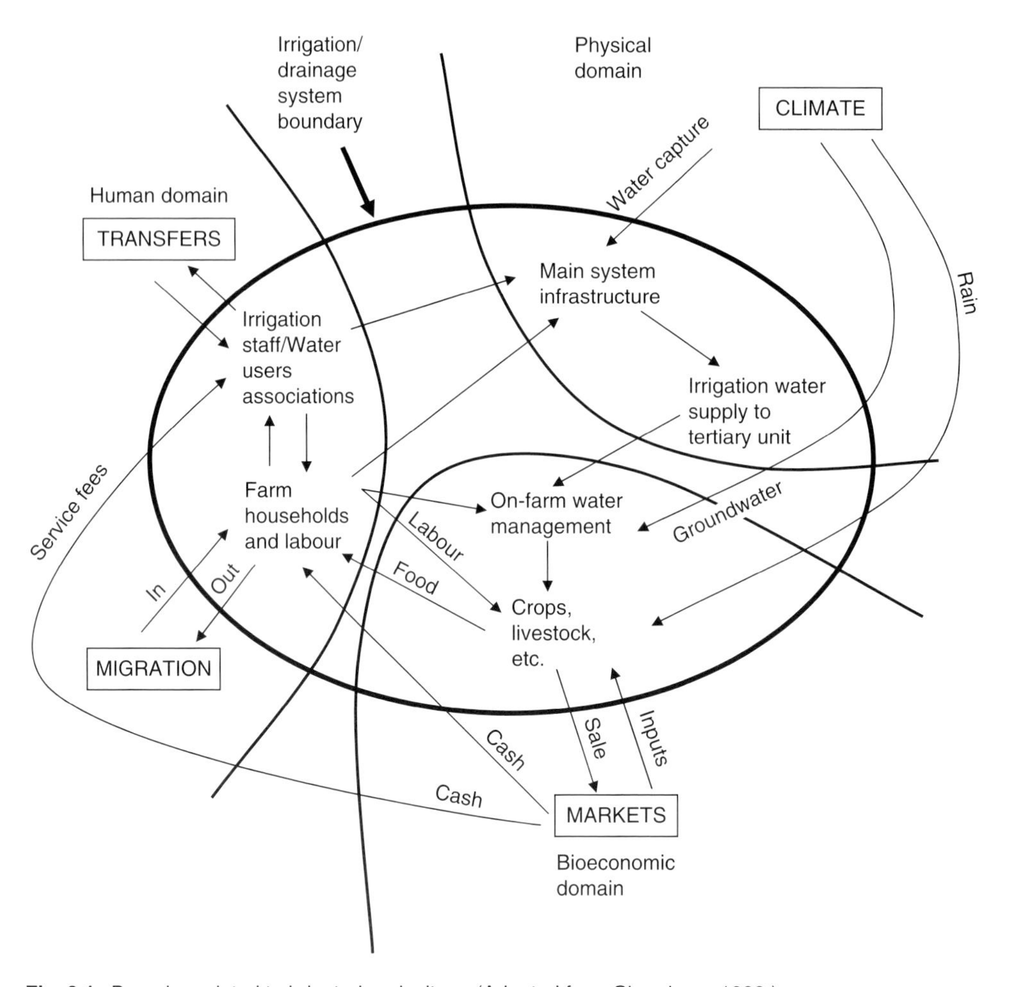

Fig. 2.1. Domains related to irrigated agriculture. (Adapted from Chambers, 1988.)

the politico-economic system to deliver national development objectives.

Building on the above and other work, a useful categorization of domains in relation to irrigated agriculture is:

- technical;
- institutional;
- economic;
- social;
- environmental.

Technical covers the physical infrastructure related to I&D systems, the canals, drains, roads, field layouts, etc., and includes analysis of the physical environment to facilitate the design, construction and implementation of the I&D system. *Institutional* covers the political, legal and organizational frameworks influencing irrigated agriculture, while *social* covers the interaction of people within the irrigation schemes and the ways that they live and work together. *Economic* covers the financial and economic aspects of irrigated agriculture, the cost and value of inputs, resources and outputs. Finally, *environmental* covers the physical environment impacted by the scheme and the health issues related to I&D systems. Table 2.1 summarizes some of the components in each of these domains, while more detail is given in the following sections.

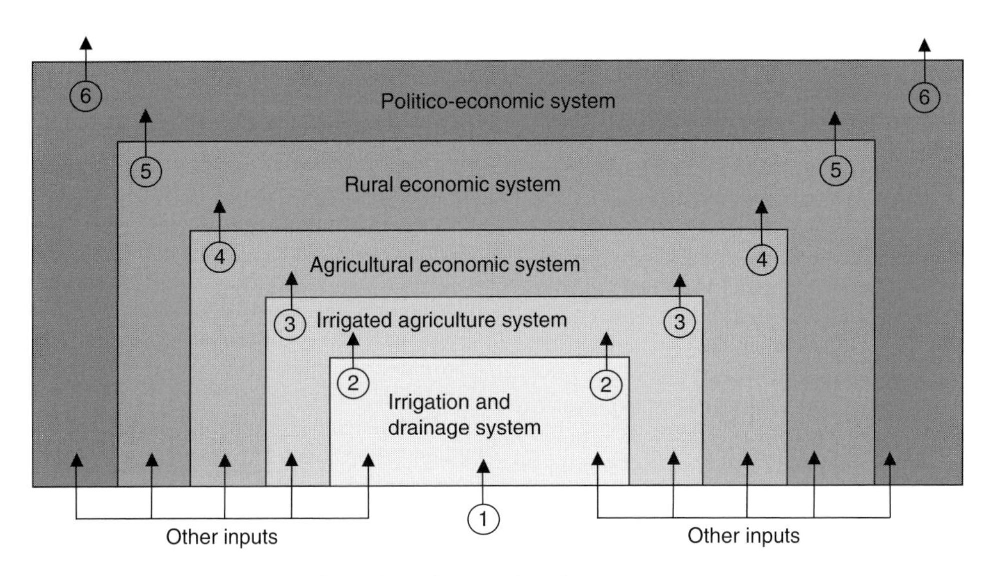

Inputs/outputs to each system

(1) Operation of irrigation facilities
(2) Supply of water to crops
(3) Agricultural production
(4) Incomes in rural sector
(5) Rural economic development
(6) National development

Fig. 2.2. Irrigation and drainage functions in the context of nested systems. (From Small and Svendsen, 1992 with permission.)

Table 2.1. Outline of components within each domain.

Domain	Component
Technical	• Physical conditions related to design and operation (climate, topography, soil, water resources) • Physical infrastructure (irrigation and drainage systems, roads, housing, etc.)
Institutional	• Political system • Legal framework • Organizations and organizational structures and functions
Economic	• Markets • Market price • Development and operational costs • Economic development and livelihoods • Employment opportunities
Social	• People • Communities and social structures • Social norms (religion, attitudes, patterns of behaviour)
Environmental	• Impact on the water environment (downstream flows, water quality) • Waterlogging and salinity • Health issues

Technical Domain

The technical domain for I&D schemes covers both physical aspects and the technical processes involved in developing the scheme.

The physical components of an I&D scheme are shown in Fig. 2.3. The main components from the water management perspective, starting from the field level and working up to the water source, are summarized in Table 2.2.

A more detailed breakdown of the physical infrastructure components for I&D systems is outlined in Table 2.3. The various structures that exist in an I&D system can broadly be divided into *conveyance*, *control* and *ancillary*. Conveyance structures include siphon underpasses, culverts and aqueducts, while control structures include cross and head regulators, and measuring structures. Ancillary structures comprise bridges, access points, bathing points, etc. Control structures form an important part of the operation of the irrigation system as they are the means by which water is managed and distributed in a controlled and regulated manner according to need.

Institutional Domain

The institutional domain covers the political, legal and organizational aspects related to irrigated agriculture.

Politics and politicians play a large part in irrigation development and irrigated agriculture, as agricultural production and rural livelihoods are key areas of political interest. This interest can be either beneficial or harmful depending on the context. Political support for the irrigated agriculture sector can result in measures to improve availability of inputs, access to markets and market prices. It can also be beneficial in allocating funds, either for capital investment for new schemes or rehabilitation of existing schemes, or for scheme management, operation and maintenance. Strong political support is also required to introduce, revise or update legislation, particularly in relation to the transfer of the management of I&D systems to water users associations (WUAs). In this context political influence can be harmful where politicians interfere in the setting and levying of irrigation and drainage service fees, either by setting an unreasonable cap on the service fee that can be levied, or by suggesting during election periods that water users need not pay such service fees. In essence, a strong irrigation sector depends on strong and consistent political support; where this support is missing or muted it is likely that the irrigation sector will have difficulties, principally in adequately funding the management, operation and maintenance of I&D schemes.

The irrigation sector is usually covered by legislation in the form of all or some of the following: Water Resources Law, Irrigation and Drainage Law, Water Users Association Law, Public Health Law, Environment Law, Tax Code, Civil Code and Employment Law. The water resources legislation covers the abstraction and use of water resources (Fig. 2.4), while the irrigation and drainage legislation covers the use of these water resources for irrigation and the drainage of agricultural lands. WUA legislation is more recent and covers the establishment of water users associations, and will involve changes in the water resources, irrigation and drainage legislation and the tax code. Public health and environment legislation generally relates to the impacts of irrigation and drainage, and will look to control its adverse impacts through regulation on wastewater discharge, pollutants, limiting of standing water, etc. The tax code will detail the tax regulations related to the irrigation and drainage sector, and is of particular interest to water users associations in terms of whether they have to charge sales tax (VAT) on the services they provide, and property tax on the infrastructure assets. The civil code and employment legislation are generally applicable within society.

The number of organizations involved in irrigated agriculture, their structure and functions have significant bearing on irrigated agriculture. The types and functions of organizations involved vary from country to country, but generally include the following:

- Ministry (or Department)[4] of Agriculture;
- Ministry (or Department) of Water Resources;

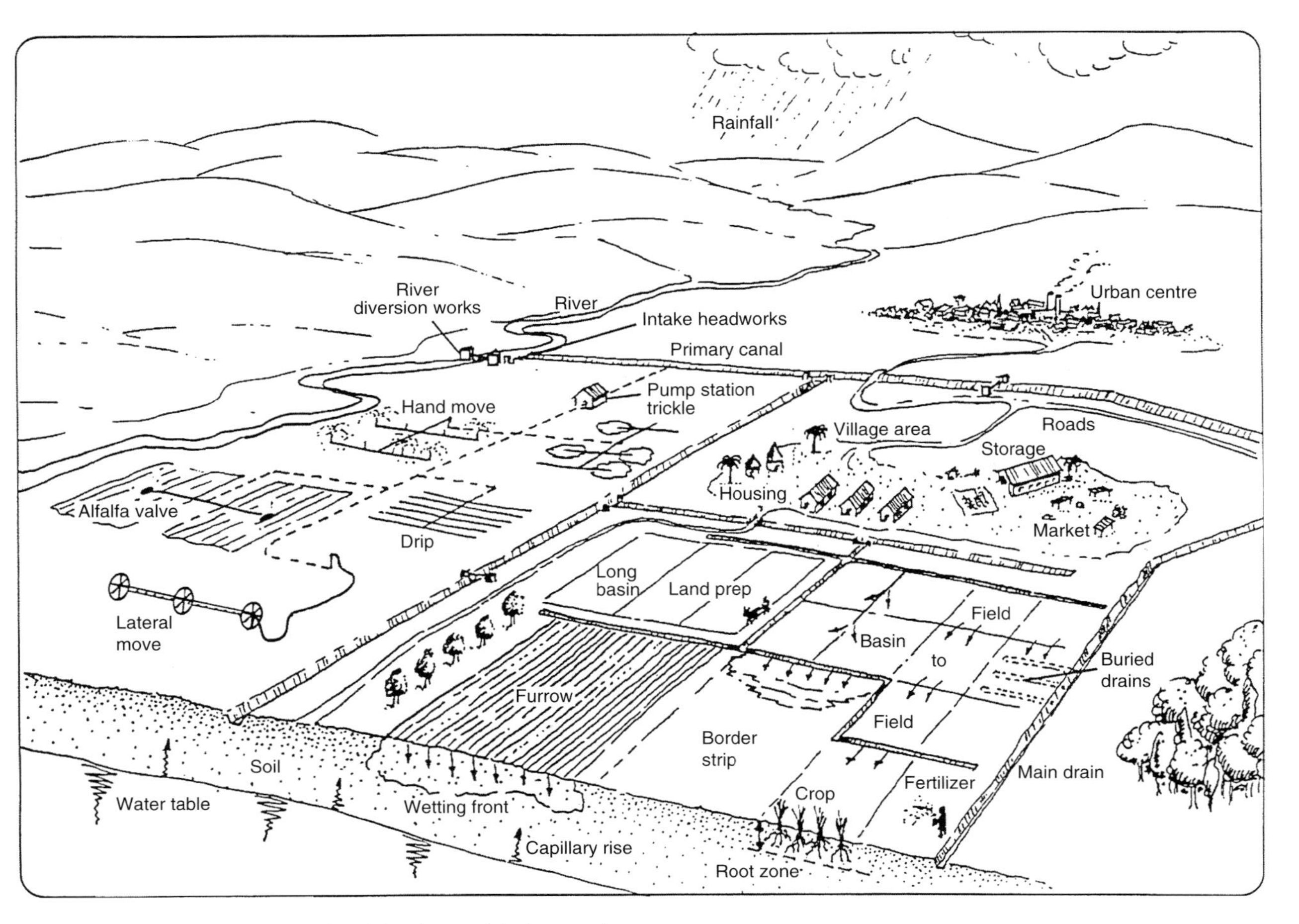

Fig. 2.3. Physical components of an irrigation and drainage scheme.

Table 2.2. Physical components of an irrigation and drainage scheme.

Component	Explanation of role
Crop	The end product of the farming and water management activity. Determines the quantity and timing of irrigation water requirements
Root zone	The storage reservoir for irrigation water. The greater the storage capacity, the greater the interval that is possible between irrigations. The depth of the root zone is determined by the type of crop and its rooting characteristics
Soil	The soil texture determines the water-holding capacity of the soil and its permeability. Heavier soils, such as clay, hold more water than lighter soils, such as sand. As a result, in general, lighter soils need irrigating more frequently than heavier soils
Groundwater	The depth to groundwater can influence the irrigation needs of the crop. If the groundwater is within 1–2 m of the soil surface there may be a contribution to the crop's water needs from groundwater, thus reducing the demand for irrigation. If the groundwater level is too high (<1 m) it will cause waterlogging and salinization, and reduction or loss of crop yield
Climate (sunshine, temperature, rainfall, etc.)	The climatic conditions drive the evaporation of water from the soil surface and the transpiration of water from the crop. Evapotranspiration increases with temperature, wind speed, lowering of the relative humidity and sunshine hours (day length). Irrigation is required to replace the water lost to evapotranspiration
Field layout and irrigation method	The field layout and irrigation method influences the efficiency and uniformity of irrigation. Various factors influence the choice of irrigation method, including the crop type, soil type, streamflow rates and topographic conditions, including land slope and topographic uniformity. Cost is also an important factor in determining the irrigation method
Field size	The field size influences irrigation management. Smaller field sizes generally mean more variation in the pattern of irrigation demand and increase the planning and management that is required to schedule and supply irrigation water
Tertiary unit[2] irrigation channels	Irrigation water is conveyed to the fields by the tertiary and quaternary or field canals. The type and condition of these canals (lined, unlined, piped, well/poorly maintained) influences the losses from the canals
Tertiary unit drainage channels	Open (surface) drainage channels remove excess irrigation and rainfall from the fields and ensure that the soils are adequately drained. Subsurface (closed pipe) drains discharging into open drains are used to regulate groundwater levels below the crop root zone
Tertiary unit control structures	Control structures are required to divide and regulate the discharges entering each canal. Lack of control structures within the tertiary unit restricts the ability to manage the supply of irrigation water to match demands
Tertiary unit discharge measurement	Measurement of irrigation supplies is required to know how much water is being delivered and to assess if the supply is adequate or excessive in matching the crop needs
Main system[3] canals	Main system canals, comprising primary and secondary canals, convey irrigation water to the tertiary units
Main system drains	Main system drains (primary and secondary) collect drainage water from the tertiary unit drains and dispose of it into natural drainage channels (streams, rivers, lakes and the sea)

Continued

Table 2.2. Continued

Component	Explanation of role
Main system control structures	Control structures, such as gated cross and head regulators, are required to regulate the flow entering each canal, while measuring structures are required to determine the amounts delivered. Without functioning control structures it is not possible to closely match irrigation supply with irrigation demand
Main system conveyance structures	Conveyance structures, such as inverted siphons and culverts, are required to pass the irrigation canal over or under natural or man-made obstructions
Water source	The nature of the water source has a significant influence on water management. The pattern of flow in the river controls the cropping pattern within the irrigated area. The quantity of water can influence the water management activity; in water-short systems water management is generally more carefully performed than in systems with adequate water supplies
Storage reservoirs	In the river, main system, tertiary unit or field, providing storage for a variety of durations: over-year, within-year, within-season, weekly, daily, overnight, etc.

Table 2.3. Physical infrastructure of irrigation and drainage systems.

Component	Level(s)	Purpose
Reservoir	River Main canal	To store water for the irrigation scheme, either on the river or in the system. The river reservoir may also be built to help alleviate flooding in the scheme
Flood bunds	River	To protect the scheme from flooding
Canals	Primary Secondary Tertiary Quaternary	To convey water. They may be open channels or closed pipes
Drains (open and closed)	Primary Secondary On-farm	To remove surplus water from the field. Surface flow protection drains may be required to protect infrastructure from upslope runoff
River weir	Main canal	To divert and control irrigation supplies
Pump station	Main canal Main drain	To lift water to command level for irrigation. To remove water from drainage channels that are below river level
Headworks	Main canal	To control the inflow to the main canal. Comprises a gate and usually a pipe to throttle the flow and thus limit inflow in flood periods
Sediment excluder or sediment trap	Main canal	To exclude or trap silt, which is then returned to the river by flushing or mechanical removal
Cross regulator	Primary and secondary canals	To raise and maintain the water surface at design elevation
Head regulator	Primary, secondary and tertiary canals	To regulate discharge entering a canal
Measuring structure	Primary, secondary and tertiary canals	To measure discharge for operational purposes
Aqueduct	All levels of canal	To pass canal over an obstruction (another canal, a drainage channel, etc.)
Culvert	All levels of canal or drain	To pass the canal or drain under an obstruction (road, drainage channel, etc.) or to pass cross-drainage flow under a canal

Continued

Table 2.3. Continued

Component	Level(s)	Purpose
Super-passage	Main canals	Used to pass flood flows and associated sediment over main canals
Drop structure	All levels of canal or drain	To lower the canal or drain bed level in a safe manner. Used to reduce canal or drain gradients on steep land
Escape structure	All levels of canals	Used to escape water from a canal into the drainage network in the event of oversupply or underutilization, or for emergencies (e.g. a breach in the canal downstream)
Inverted siphon underpass	All levels of canals	Used to pass the canal below an obstruction, such as a road or drainage channel
Distribution box	Tertiary canal Quaternary canal	Simple distribution structure to distribute the water between tertiary and quaternary canals
Night-storage reservoir	Main canal or on-farm	Reservoir to store irrigation water during the night. Main canals can operate 24 h/day, while lower-order canals can be operated during the daytime only. On-farm storage reservoirs allow the farmer to take water as delivered and then use it when needed
Tubewell, open well	On-farm	Abstraction of groundwater for irrigation. Often used in conjunction with surface water system
Bridges and culverts	Road bridges Foot bridges	To allow human and animal traffic over the canal or drain
Access points	Main canals	Access points into the canal for human and animal traffic for obtaining water, washing, etc.
Roads	Inspection roads Access roads	To gain access to the irrigation system, fields and villages. Also for inspection and maintenance alongside canals and drains
Fields	Within tertiary unit	Prepared land to cultivate the crop. Laid out for different methods of irrigation (basin, furrow, sprinkler, etc.)
Villages	Throughout the scheme	Living space for the farming community. The distance from the village to the fields is important

- Ministry (or Department) of Irrigation and Drainage;
- Ministry (or Department) of Environment and Natural Resources;
- Ministry (or Department) of Public Works;
- Ministry of Finance;
- Regional and Local Government.

The roles of these different organizations in relation to river basin and irrigation management in Mexico are illustrated in Fig. 2.5 in terms of the functions that they each perform. Such organizational mapping can be extremely helpful in understanding the different functions and linkages.

Irrigation management depends on both physical and human resources. The two are tightly interrelated; good infrastructure with poor management will not deliver reliable, adequate and timely water supplies. The human resource is a key aspect of good irrigation management; a good irrigation manager and his/her team can achieve a considerable amount, even though the physical condition of the irrigation and drainage network might not be so good. The key actors involved in irrigation and drainage management, from the field to the water source, are summarized in Table 2.4.

Economic Domain

Irrigation and drainage development in general contributes to improved livelihoods and economic development. The greater reliability of water supply provided by irrigation systems over rainfed agriculture results in higher

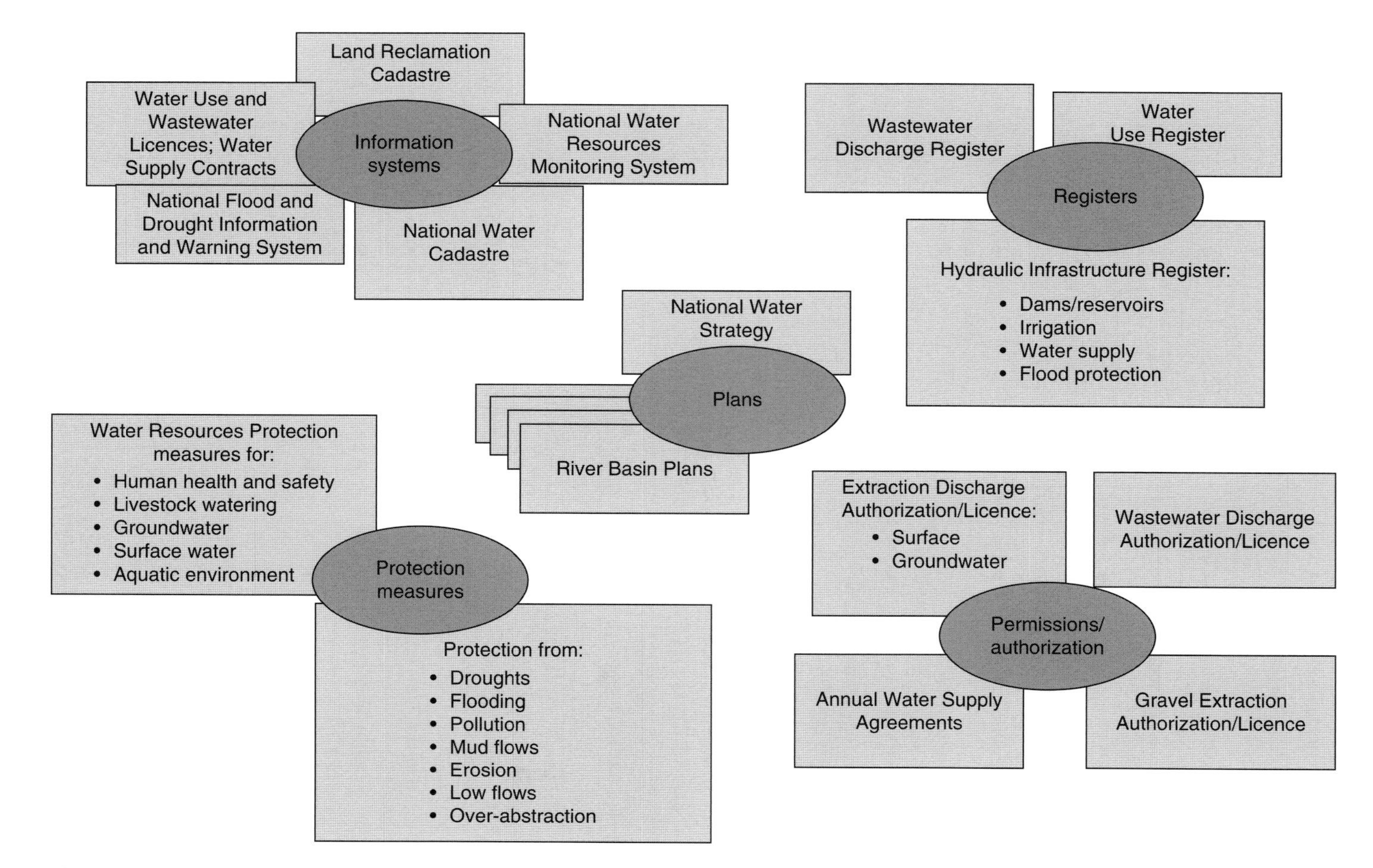

Fig. 2.4. Typical elements of water resources legislation.

Key players	Surface water									Groundwater							Wastewater					
	Plan (basin-level)	Allocate water	Distribute water	Construct facilities	Maintain facilities	Monitor quality	Ensure quality	Protect against flooding	Protect environment	Plan (basin-level)	Allocate water	Withdraw/distribute water	Construct facilities	Maintain facilities	Monitor quality	Ensure quality	Plan (basin-level)	Allocate/distribute	Construct facilities	Operate/maintain facilities	Monitor quality	Enforce quality
Ministry of Environment	●			●			●		●		●										●	●
CNA National Headquarters	●	●		●	●	●	●	●	●	●	●		●		●		●		●		●	●
CNA Regional Office	●	●	●	●	●	●	●	●	●	●	●		●		●		●		●	●	●	●
River Basin Council	○	○		○					○		○						○			○		
CNA State Office	○	●	●	●	●	●	●	●			●	●	●		●				●		●	●
State Water Commissions	●	○	○	●	●			●	○	○	○		●	○	●		○		●	○		
CNA Irrigation District Office			●	●	●																	
WUAs Irrigation Districts	○		●	○	●							●						●				
WUAs Irrigation Units			●		●							●										
Aquifer Management Councils (COTAS)	○								○	○	○		○									
Municipal Water Supply Utilities			●	●	●			●				●	●	●	●			●	●	●	●	
Industries												●	●	●					●	●		
NGOs	○								○													
Irrigators			●									●	●	●				●				

Fig. 2.5. Essential basin management functions and key actors in the Lerma Chapala River Basin, Mexico. ● indicates activity; ○ indicates limited activity; CNA, National Water Commission; WUA, water users association; NGO, non-government organization. (Modified from Burton *et al.*, 2002; Wester *et al.*, 2005.)

Table 2.4. Management components of irrigated farming systems.

Farmer	The capability of the farmer has a significant influence on irrigation water management. A good farmer will know when and how often to irrigate and will apply the correct quantity of water to match the crop and soil needs
Water Master[5] (within the tertiary unit/on-farm)	The water master is a central figure in water management within the tertiary unit, organizing and overseeing the distribution of water between farmers. A good water master can have a very beneficial impact on the productivity and efficiency of water use
WUA O&M Engineer/ Technician	The WUA O&M engineer/technician will work with the water master to plan and schedule irrigation water, and monitor its distribution. The better the planning and monitoring, the better the implementation
WUA management	The WUA management can influence performance in relation to irrigation water management by monitoring the planning, allocation and use of irrigation water, and in setting the standards required
Water Master (main system)	As at the tertiary unit/on-farm level, the main system water master is a key figure, responsible for the day-to-day distribution of water to tertiary units. A good water master can have a significant impact on system performance; it is important that they are well trained and well motivated
Main system service provider staff	As with the water master, the staff of the main system service provider can strongly influence the performance on an I&D system. The more professional and motivated the staff, the better the water delivery performance
O&M procedures	The O&M procedures, both within the tertiary unit and at the main system level, influence the performance of irrigation water management. Well-organized and systematic procedures lead to more efficient and productive water management
Setting and collecting service fees	With an increasing focus on I&D systems being financially self-reliant, the setting and collection of the service fees is growing in importance. Management systems for setting, collecting and utilizing the service fee need to be transparent and accountable

WUA, water users association; O&M, operation and management; I&D, irrigation and drainage.

crop yields, greater levels of production and increased income when the produce is sold. In poor areas irrigation can make a significant contribution to poverty reduction and provision of basic foodstuffs for families. In general, irrigated agriculture requires more labour than rainfed agriculture, resulting in increased employment opportunities for landless labour. As a result of the increased productivity, secondary industries develop, including traders, shopkeepers, agricultural machinery repair workshops and the like.

It is important to distinguish between the *economic costs and benefits* of irrigation and drainage development and the *financial costs and benefits*. The economic analysis looks at the benefit to society and government in general, while the financial analysis looks at the actual monetary transactions that take place. Economic analysis will take account of the benefits of creating employment, of developing secondary industries, of increasing demand for processed goods and the like. The financial analysis can be carried out for the operating authority and for farmers. For the operating authority it will look at the costs required to run the organization and provide its services, and the sources of income to cover these costs. For farmers it will look at the income generated for the farmers and their families as a result of the scheme. In general terms the government and any financial backers of the development will be interested in the economic costs and benefits, while the operating authority and farmers will be interested in the financial costs and benefits and whether the scheme can pay its way and improve the farming family's standard of living and food security.[6]

Allied to the potential financial benefits arising from irrigated agriculture are increased costs to cover the expenditure required to manage, operate and maintain the I&D system. Unfortunately governments and farmers in many countries are still reluctant to cover the real costs of managing, operating and maintaining these systems (MOM costs) despite the obvious financial and social benefits arising from them. For the last 20 years, significant sums of money have been invested in rehabilitating I&D systems, only for them to fail again several years later due to inadequate investment in maintenance. In an effort to overcome the difficulties of adequately funding management, operation and maintenance, some governments have transferred the management of the schemes or parts of the schemes to the water users, in the belief that as the direct beneficiaries they will be willing to cover the real MOM costs. However, despite being prepared to cover the costs of seeds, machinery hire, fuel and the like, farmers in some countries still have difficulty in accepting that they should pay the real cost for providing irrigation water and drainage water removal. This is one of the biggest challenges facing the irrigation and drainage sector.

Social Domain

Farming within I&D schemes requires more social cohesion, cooperation and discipline than rainfed farming. Water users need to collaborate to share the available water supplies; in many systems fairness or equity of allocation and distribution is valued more highly than adequacy of supply.[7] In general, where the social cohesion is strong irrigation is productive, while where the social cohesion is weak irrigation suffers. Successful traditional hydraulic societies, such as the Chagga on Mount Kilimanjaro, some hill tribes in Nepal and the Balinese with their *Subak* irrigation, have long-standing social norms and rules for irrigation, to which all users subscribe.

This ability to work together and enforce compliance with a set of agreed rules is not apparent in all I&D schemes, and becomes more difficult to engender the larger the scheme becomes. In the larger schemes the social domain will encompass the management and staff of the organization responsible for managing the main system as well as the farming community. The nature of the relationship between these two social groups will have a direct bearing on the nature and quality of the service provided. Where there is corruption and/or lack of transparency the service delivery may be good to some farmers but poor to others, with high levels of distrust between the two groups. Where there is accountability, openness, trust and communication between the two groups service delivery will be measurably better, and productivity consequently higher.

Environmental Domain

Interest in the environmental domain focuses on the impacts of irrigation and drainage development on the natural environment, as well as the potential impact of external factors on the scheme. The impacts of the scheme on the environment are generally most extreme when the scheme is first developed, though poor operation can be very harmful, particularly in the excessive abstraction and application of irrigation water and/or excessive use of pesticides and fertilizers.

The main environmental impacts include (Table 2.5):

- land degradation within the scheme;
- degradation of water quality, both in surface and groundwater;
- groundwater depletion;
- ecological degradation.

As can be seen from Table 2.5, responsible management can help to significantly reduce the adverse environmental impacts of I&D schemes. In summary such measures include:

- keeping river and groundwater abstraction to an absolute minimum by efficient operation and effective maintenance of the system;
- keeping drainage water flow to an absolute minimum through controlled application of irrigation water;

Table 2.5. Impact of irrigation and drainage development on the environment. (Modified from ICID, 1993; Neville, 1996.)

Impact	Explanation	Mitigation measures
Hydrological impacts		
Low flows in river	Abstraction of water for irrigation can have negative impacts on the downstream aquatic environment and downstream users	Minimize abstraction through efficient scheme operation. Adjust cropping pattern to minimize abstraction during low-flow periods
Flood regime	Uncontrolled flooding can cause damage to the human environment, but can be a central part of natural flood plain ecosystems, especially in recharging wetlands. Cutting off the natural floods can also adversely affect fish migration	Operate dams and river weirs to allow passage of (controlled) flood flow at critical times. Maintain connectivity of the river with the flood plain and wetlands
Operation of dams	The manner in which dams are operated has a significant impact on the downstream environment. Reservoirs can have adverse impacts in relation to human health (malaria, schistosomiasis, etc.)	Operate the dam in consideration of the environmental impacts. Allow (controlled) flood flows, variation in water levels, minimum low-flow regimes
Lowering of groundwater table	Depletion of groundwater can have serious implications for the environment, leading to saltwater intrusion, land subsidence, drying out of wetlands, acidification of (sulfate) soils and exposure of toxic layers (arsenic). In extreme cases, such as with fossil aquifers, the aquifer can dry up completely if over-abstracted	Limit groundwater abstraction through licensing of boreholes and regulation and control of abstraction quantities
Raising of groundwater table	One of the most common long-term impacts of irrigation, caused by over-irrigation. A rising groundwater table leads to waterlogging and salinization, loss of crops, loss of productive land and health risks from standing water	Good irrigation management, especially at the field level, to minimize water application losses. Provision of adequate drainage system
Water quality impacts		
Solute dispersion	Reduced flows in a river can reduce the river's ability to dilute and treat water-soluble pollution. Reduced flood flows in wetlands can lead to a build-up of pollutants and salts	Minimize water abstractions. Allow (controlled) flood flows at critical periods
Toxic substances	Irrigation can flush out toxic substances in the soils, leading to adverse downstream impacts. Pesticides are particularly dangerous for the environment	Implement responsible regimes for pesticide use. Minimize irrigation application losses. Implement flushing procedures for salts at high-flow periods
Agrochemical pollution	Though beneficial to irrigated crops, natural and chemical fertilizers can be harmful to the environment. Nitrates are soluble, while phosphates can fix to soil particles and be transported by erosion, resulting in harm to aquatic life and algae growth. Nitrates in water are harmful to human health	Implement responsible and informed fertilizer application regimes. Educate farmers in the proper use of fertilizers and pesticides. Minimize application losses from fields

Continued

Table 2.5. Continued

Impact	Explanation	Mitigation measures
Anaerobic degradation	Anaerobic conditions can be brought about by high levels of nutrients or decomposition of organic matter (such as in reservoirs), leading to emission of greenhouse gases (methane, hydrogen sulfide, ammonia)	Minimize nutrient runoff from irrigation schemes. Clear new reservoir areas of vegetation prior to filling
Soil quality impacts		
Soil salinity	Build-up of salts in soils may be caused by salts brought in by irrigation water, capillary rise from saline groundwater, or solutes leached from natural or artificial fertilizers	Manage the salt balance in the irrigated soils and provide adequate drainage so that salts do not build up. Implement responsible leaching regimes
Soil properties	Soil properties can change with irrigation or drainage. Drained peat soils can degrade rapidly, clay soils with high sodium-ion content (alkaline soils) can lead to collapse of the soil structure	Understand the behaviour of soils and manage them responsibly. Apply gypsum to alleviate problems with alkaline soils
Saline groundwater	Over-irrigation and leaching of salts into the groundwater is a common problem in some localities	Manage irrigation water application to minimize losses. Provide drainage to avoid build-up of salinity levels
Saline drainage water	Drainage systems pick up drainage water, which can have five to ten times the salt concentration of the irrigation water	Dispose of the drainage water responsibly, to the sea or a sink if high levels. Manage the salt balance to avoid build-up of salts over time
Saline intrusion	Reduced river flows and/or reduced groundwater flows can lead to saline intrusion in coastal areas, adversely affecting groundwater-based drinking water and the coastal/estuarine aquatic environment, particularly mangrove swamps	Maintain minimum-flow regimes in rivers and limit groundwater abstraction
Erosion and sedimentation impacts		
Local soil erosion	Soil erosion on fields can lead to the loss of good topsoil and have adverse impacts on the downstream ecology	Design I&D system appropriately; align furrows, borders, etc. to minimize soil erosion. Manage the land appropriately; keep vegetative cover during rainy periods. Design drains to avoid gullying
Hinterland degradation	The increased population supported by an irrigation scheme may lead to increased erosion as a result of tree-cutting for fuel wood, livestock grazing, etc.	Identify possible impacts during design and implementation. Build fuel wood plots on the scheme, allow for growing fodder for livestock, etc.
River morphology	Altering the natural flow regime of the river by construction of dams and weirs can lead to sediment deposition upstream and aggradation of river beds downstream of structures	Where possible, design structures to pass sediment loads downstream

Sedimentation	Sedimentation in reservoirs can significantly shorten their lifespan, leading to the need for more storage elsewhere and thus impacting the environment. Sediment in irrigation canals reduces capacity	Implement soil erosion measures upstream of reservoirs to maintain vegetative cover. Keep sediment out of irrigation systems with sediment basins, vortex tube silt extractors, etc. Design and operate canals to maintain flow velocities and carry sediment to fields
Estuary degradation	Loss of natural sediment in rivers can lead to degradation of estuaries and associated aquatic ecosystems dependent on the sediment	Design and operate structures to pass sediment load downstream
Biological and ecological impacts		
Scheme area land use	Development of an irrigation scheme will radically alter the human and ecological environment of the area	Assess the local environment at the planning and design stage and allow for mitigation measures to protect wildlife, migratory routes, traditional grazing, etc.
Water bodies	Existing water bodies will be affected by irrigation and drainage development (rivers, wetlands, flood plains, etc.) and new bodies created (reservoirs, canals, drains, storage ponds, etc.). Waterborne disease may increase	Identify mitigation measures at planning stage and apply at design and implementation stage. A variety of measures will be required
Degradation of surrounding area	The increased population supported by an irrigation scheme can have adverse impacts on the local environment, through tree-cutting for fuel wood, livestock grazing, etc.	Consider possible impacts during design and implementation. Build fuel wood plots on the scheme, allow for growing fodder for livestock, etc.
Valleys and shores	Traditional flood plain and estuarine ecosystems may be adversely affected by flood attenuation and water abstraction	Identify mitigation measures at planning stage and apply at design and implementation stages. Minimize abstraction, maintain partial flood flow regimes and minimum low flows
Wetlands and plains	Includes marsh, fen or peat land with areas of standing fresh or brackish water. Wetlands (including mangrove swamps) are very vibrant and productive ecosystems, in supporting both aquatic ecosystems/wildlife and humans (fishing, etc.)	Identify wetlands and associated ecosystems at planning stage, assess impact and formulate and apply mitigation measures. Drainage of wetlands is nowadays not considered to be justified on environmental grounds
Pests and weeds	Some pests and weeds benefit from development of I&D schemes. Natural predators such as snakes, birds and insects may be reduced by pesticides and land-use changes	Adopt measures to avoid certain types of pests and diseases, such as avoiding monocropping
Animal diseases	Animals are subject to waterborne disease, and may be adversely affected in an I&D scheme	Adopt measures to mitigate contact of livestock with potentially harmful water bodies
Aquatic weeds	An I&D scheme can provide an ideal habitat for aquatic weeds. The capacity of reservoirs and canals can be seriously impeded by aquatic weeds such as water hyacinth, taifa reed, etc. Vegetation in canals/drains is a habitat for disease vectors such as snails and mosquitoes	Can be a major problem in some areas. Need to adopt management and maintenance measures to control the spread of aquatic weeds. Use of herbicides can be very harmful to the aquatic environment

Continued

Table 2.5. Continued

Impact	Explanation	Mitigation measures
Human health	I&D schemes can have serious adverse impacts on human health. Malaria, Japanese encephalitis, schistosomiasis (bilharzia), lymphatic filariasis and onchoceriasis (river blindness) are carried by mosquitoes and other waterborne vectors	An extensive public health programme can help considerably in reducing the incidence and impact of these diseases. A variety of measures is required, such as use of mosquito nets at night, public hygiene measures, and operation and maintenance measures such as maintaining flow rates in canals and preventing the occurrence of standing or stagnant water
Socio-economic impacts		
Population growth	Irrigation schemes generally increase the population density in an area and can stimulate other economic development and growth. This can have a variety of impacts on the locality	Not always possible to predict the nature of the changes that may take place. Design of settlements and associated facilities, such as sewage treatment, can help mitigate impacts
Income and amenity	Development of irrigation schemes can change the levels of income and the balance within the local community, for example for fishers and pastoralists	It is important to identify all the stakeholders in irrigation and drainage development and to allow for the needs of each. For example, where fishers lose the previous ability to fish in the flood plain, they may benefit from fishing in the reservoir
Human migration	Human migration into an area as a result of work (e.g. as cotton pickers, farm labour) can alter the balance of the local society	A social impact assessment can be a useful part of the planning process
Resettlement	Resettlement of communities living in areas to be flooded by reservoirs and in lands to be irrigated can be traumatic for those affected	Adequate planning for resettlement and consultation with those affected can mitigate the impacts
Gender issues	In some cases women have been adversely impacted by irrigation and drainage development, with increased workloads from more intensive farming	A social impact assessment is required to understand the potential impact and mitigation measures for all stakeholders, but particularly for women and vulnerable sections of society
Vulnerable groups	Pastoralists often lose out as a result of irrigation and drainage development, as do fishers	As above, identify the groups and the issues they face and formulate mitigation measures
Regional impacts	If irrigation and drainage development is successful if may affect the local economy, increasing the price of labour and reducing local prices due to greater availability of agricultural produce	The socio-economic analysis of the project should identify possible impacts, and the nature of the costs and benefits
Recreation	Irrigation and drainage development can increase the opportunity for recreation, particularly bathing in reservoirs and canals. In some locations there may be significant health risks, for example with schistosomiasis	Where waterborne diseases are a potential risk, settlements should be located away from canals and night-storage reservoirs to prevent children using them as swimming pools

I&D, irrigation and drainage.

- minimizing the level of pesticide and fertilizer in drainage water though controlled application and controlled irrigation;
- controlling application of irrigation water at field level to avoid over-irrigation contributing to rising groundwater levels and subsequent salinization of the land;
- reducing losses from irrigation canals through measures such as canal lining and timely maintenance;
- reducing health risks through a knowledge and application of measures to alleviate health risks, such as avoiding areas of standing water (for mosquito control) and health education for farmers and their families.

Phases of Development

As noted above there are different phases involved in the development of I&D schemes. It is useful to look at the activities involved in these different phases as they can have a significant bearing on how the scheme is managed, operated and maintained.

Six relatively distinct phases can be identified:

- planning;
- design;
- construction;
- operation;
- maintenance and asset management;
- rehabilitation.

Overlying these phases is a support network, generally comprising government agency personnel for large schemes and community leaders for smaller schemes.

Support network

The support network encompasses a variety of activities that permit the execution of the above processes. Support encompasses the organizational structure that is present through which the scheme is identified, planned, designed, implemented and operated. Support activities include resource acquisition, personnel management, financing and general management, including contracting-out of services.

In order for an I&D scheme to be developed some form of organization needs to exist to conceive of the idea and follow it through to completion. This can be a group of farmers or a government agency, a combination of the two, or a private enterprise. The essential feature is that some element of cooperation and organization is required to realize the project.[8]

Planning

Planning is the process of identification of the potential for irrigation and selection of the best approach for its development. Planning will look at the feasibility of the development in technical, economic, physical, social, institutional and environmental aspects. Questions to be asked will include the following.

- Can it be done?
- How will it be done?
- What are the objectives for the development?
- What will it cost, who will pay and is it economic?
- What are the likely consequences and impacts of the development?
- What will be the benefits, and how will they be distributed among the various stakeholders?
- How will it be managed, operated and maintained?
- Will it be sustainable?

Though the planning stage is crucial to the long-term success of the scheme, it is often the case that insufficient time and resources are spent on it. It is also the stage at which the least is known about the scheme, its people and the environment in which it will have to function.

It is important at this stage to be clear about the objectives for the development, whether it is for political, economic, financial or social purposes. Some schemes are established for political or social purposes (to settle areas and/or to resettle farmers from overpopulated areas). While such schemes might be uneconomic, they are considered a political necessity. In general I&D

systems are developed for economic purposes, often with a focus on poverty alleviation or ensuring security of livelihoods in drought-prone areas. Other schemes, such as those in the private sector (e.g. sugar estates), are developed for the financial benefits they can generate.

In the planning stage a feasibility study will be carried out to ascertain the feasibility and likely cost and benefits of the development. The feasibility study will include:

- semi-detailed topographic surveys;
- soil surveys;
- soil mechanics survey and analysis at locations of significant structures;
- mapping of the project area;
- data collection and analysis of the climate and water resources (quantity and quality);
- determination of suitable crop types, availability of markets, costs of inputs and prices of outputs;
- social surveys to ascertain the pre-project conditions and needs of the rural community;
- institutional development studies to ascertain the current and required institutional framework;
- environmental impact study;
- outline designs with cost estimates;
- implementation development options and timeframes;
- economic and financial analysis.

An important part of irrigation development, especially in areas with existing agriculture, is the active participation of the intended beneficiaries and other stakeholders in the development process. Failure to involve beneficiaries and other stakeholders at the planning stage has been found to have serious and detrimental effects on the subsequent stages of development and the scheme's long-term sustainability.

Land consolidation is an important issue in some locations, which should be considered and investigated at the planning stage. Though considered preferable from a technical viewpoint, bringing a farmer's individual plots together in one location may require considerable resources (surveying, consulting and negotiating with farmers, etc.) and may face strong resistance from farmers.

Design

Once the development has been planned full designs will be prepared. These may require further data collection. The design stage may include the following:

- detailed topographic surveys;
- design of scheme layout, including headworks, canals, drains, control and measurement structures, flood control measures, villages, water supply and roads;
- determination of the cropping pattern;
- selection of irrigation method (surface, sprinkler, drip);
- estimation of crop and irrigation water requirements, leading to canal sizing and hydraulic design of all structures;
- estimation of surface runoff leading to drain sizing;
- costing;
- preparation of tender documents (specification, bills of quantities and album of drawings);
- implementation work planning (with realistic timeframe);
- specification of the scheme's organization, management, operation and maintenance, and preparation of a manual for such.

The design work is often carried out by consultants procured through a process of competitive bidding, generally based on quality and experience of previous work, and cost. The consultant prepares the designs together with the bill of quantities, an estimate of the likely cost (termed the Engineer's Estimate) and the specification. The work is put out to tender and similar criteria (quality, experience and cost) used to select a contractor to carry out the work. Water users should be involved in the design and procurement process, particularly for schemes that are to be managed by farmers, or projects where rehabilitation is being carried out within the tertiary unit.

Construction

Once finances have been secured, designs completed and contracts tendered and awarded, construction can commence. Procurement of suitable contractors is important. Construction may include the following processes:

- establishing effective construction supervision personnel and procedures;
- establishing a construction camp;
- site clearance;
- setting out of the works;
- agreement with farmers on construction timing and methods (to avoid disruption to farming activities, where possible);
- construction of infrastructure (canals, drains, structures, roads, villages, etc.);
- supervision and checking that the works comply with the specification;
- measurement of work done;
- preparation of as-built drawings, operation and maintenance (O&M) manuals and training of O&M staff;
- certification for payment (monthly and final);
- commissioning (the contractor is responsible for rectifying defects arising during the commissioning period, usually 12 months);
- handing over of the completed scheme.

Different procedures will be followed depending on whether the scheme is constructed by a contractor or by the developer with assistance from the beneficiaries. Generally speaking, large-scale irrigation schemes are constructed by a contractor, small-scale irrigation schemes through beneficiary participation.

Operation

Operation of the system can be by the beneficiaries, a government agency or a private enterprise. Again size often determines who operates the system; small systems are easier for farmers to run, government often runs the larger-scale systems. In the large-scale systems the government may operate the primary and secondary canals (the main system), while farmers operate the system 'below the outlet' within the tertiary unit. In a private development the management company will manage the whole enterprise.

For operation, a set of procedures, rules and regulations will be required if the I&D system is to operate efficiently and conflict is to be avoided. Procedures will be required to plan and manage the water distribution as the irrigation water demand is constantly changing. Operation activities will include:

- planning cropping patterns;
- determining crop and irrigation water demands;
- estimating available irrigation supply;
- making adjustments to match supply and demand;
- making water allocations;
- reporting and record-keeping;
- monitoring and evaluating performance;
- liaising with water users;
- conflict resolution.

Maintenance and asset management

Maintenance and asset management are an integral part of scheme operation, without which the system will deteriorate and productivity decline. Despite the very close relationship between performance and the physical condition of the system, it is often the case that inadequate funds are allocated for maintenance. Maintenance activities will include:

- identification and reporting of maintenance needs;
- prioritizing, planning and budgeting for maintenance;
- carrying out maintenance;
- monitoring and evaluation of work done;
- payment;
- liaising with farmers on maintenance;
- reporting on work carried out.

As the terms suggest, asset management relates to the management over time of the system's assets. It looks at the short-, medium- and long-term maintenance, repair and replacement of the system's physical assets and the income stream required to sustain the system at the required service level. Asset management includes:

- identifying and quantifying a system's assets (lengths of canals/drains, numbers, types and sizes of structures, etc.);
- assessing the condition and performance of the assets and their component parts;
- creating an asset database;
- discussing with water users and agreeing on target standards, levels of service and costs;
- formulating an asset management plan with details of operational and capital expenditure over time (typically 20–25 years, in 5-year packages) and associated service fee charges;
- implementing, over time, the asset management plan;
- monitoring and evaluation of implementation.

Rehabilitation

A further process, which has become all too common, is the rehabilitation of I&D systems, arising from the failure to properly operate and maintain schemes. A distinction needs to be drawn between *rehabilitation projects* and *modernization projects* or programmes. The focus of rehabilitation projects is to repair the system, and in most aspects will return the system to its original designed state, while the focus of modernization projects is primarily to upgrade components of the system, such as providing automatic control structures, automated flow measurement, changes in field irrigation methods or operating procedures. Rehabilitation projects often give reasonable economic returns, with possible improvements in agricultural production being achieved through relatively small investments as existing infrastructure is taken as 'sunk costs' from which benefits arise but no charge is made to the rehabilitation project.

Rehabilitation projects may appear economically favourable, but often this is because the 'without-project' case is taken to be continued poor performance – to have properly maintained the system in the first place would have provided better economic returns as: (i) production would not have been lost over time due to the decline in the physical condition of the system and difficulties in supplying irrigation water or removing excess drainage water; and (ii) it costs more to repair and rehabilitate physical systems than it does to maintain them ('a stitch in time saves nine'). There are also the social implications associated with allowing a system to deteriorate, with tail-enders suffering the most from inadequate water supplies due to poor physical condition of canals and structures.

Despite covenants in loan agreements that government will provide adequate funds for MOM following completion of the rehabilitation project, and preparation of detailed O&M manuals and maintenance costings, it is common to find that rehabilitated schemes need further rehabilitation after some years due to inadequate funds for system maintenance. In developed countries asset management planning has been developed and is now an established process for ensuring timely maintenance and replacement of physical infrastructure, thereby avoiding the need for expensive rehabilitation. It is to be hoped that such approaches will be used in developing countries, so that the depressing cycle of construction, deterioration, rehabilitation, deterioration and further rehabilitation is finally ended. Asset management planning is discussed in some detail in Chapter 6.

Irrigation Methods

The method used to apply the irrigation water to the crop has an important bearing on the management and performance of an I&D scheme. There are four principal methods, which can be subdivided into a number of variations and modifications. The basic classification is given below.

1. Surface irrigation:
    - uncontrolled flooding, wild flooding;
    - controlled flooding;

- ○ basin,
- ○ contour levee,
- ○ border,
- ○ furrow,
- ○ corrugation.

2. Sprinkler irrigation.
3. Trickle (drip) irrigation.
4. Subsurface irrigation.

Each of these irrigation methods has its own advantages and disadvantages, and is suited to particular physical conditions such as crop type, soils, land slope, water availability, etc. and also to other factors such as availability of funds, labour costs and labour availability.

Surface irrigation

Surface irrigation methods are the oldest and most widely used methods of water application. According to the FAO (1989) 95% of the irrigated area in the world uses surface irrigation methods. Surface irrigation methods can be used for most crops, but are not recommended for highly permeable soils or steep slopes. In their basic form they are the least expensive of the possible systems, though costs rise if land-forming measures such as land levelling are required.

The main advantages of surface irrigation methods are that land preparation is relatively straightforward, they are relatively easy to operate and maintain, are not affected by wind conditions, generally have low energy costs, and can be highly efficient under skilled management. They are less efficient than sprinkler and drip irrigation, and require more skilled operation if the water is to be applied uniformly to the land surface without undue losses.

The efficiency of the water application is highly dependent on the knowledge and skill of the farmer. It is often thought that farmers are very experienced in surface irrigation methods simply because they have been practising them for years. However, in many countries it is rare for farmers to evaluate their irrigation application by assessing the soil moisture status in the root zone before and after irrigation. It is therefore difficult to know if an excessive quantity of water has been applied and lost to deep percolation below the root zone; a farmer may well have been over-irrigating for many years without knowing it. Experience in using surface irrigation methods and skill in their use are therefore different; significant improvements in a scheme's water-use efficiency and productivity can be gained through assessment of farmers' actual application practices followed by training should their practices be found to be poor or inefficient.

Uncontrolled flooding

Uncontrolled flooding is the oldest and simplest irrigation practice, with water being diverted from natural streams and distributed over the land without any land-forming measures. The method is very basic, easy to set up and operate, and low-cost in terms of time and resources consumed. As a consequence of this simplicity and low level of inputs crop yields are generally poor, but may be suited to some situations, such as occur in parts of Africa where farmers plant in the moist soil as flood waters recede.

Controlled flooding

BASIN IRRIGATION Basin irrigation is the most common form of irrigation due to its simplicity. The land is divided by ridged earth bunds to form basins and water turned into the basin from a quaternary or field channel. The land within each basin is then levelled to allow uniform irrigation application.

The method is well suited to smallholder irrigation where farmers have small landholdings and grow a variety of crops, but is equally suited to large mechanized farms where laser-controlled land grading and large stream sizes are possible. Row crops can be grown in the basin on ridges; this is often termed furrow-in-basin irrigation.

For dry foot crops the water is turned into the basin and the water ponded before the supply is closed off and the next basin filled. To avoid deep percolation losses the basin should be filled quickly, large stream sizes are therefore required. In rice systems the water is often turned into one basin and then flows through a series of basins before discharging into the field drain. In this case land preparation and puddling[9] of the soil are required to reduce the deep percolation losses. Puddling of the soil can reduce deep percolation losses from 10 mm/day down to 1–2 mm/day.

CONTOUR LEVEES This method is similar to basin irrigation but is used on hillsides and steeply sloping land. The basins are cut into the slope and run along the contour to form a terrace. For dry foot crops water is applied and held in the basin until the required depth has been applied. For paddy rice, water usually flows from basin to basin, though care has to be taken to avoid gullying and cutting back into the terraces.

BORDER STRIP IRRIGATION With border strip irrigation,[10] land with a gentle and uniform slope is divided into parallel strips of land separated by earth ridges. Water is let into the border strip and flows down the slope. The end of the strip can be closed (borders without runoff) or open (borders with runoff). To achieve uniform irrigation application the border should not have a cross slope. The border strip can vary in size from 60 to 800 m in length and from 3 to 30 m in width depending on the soil type, the slope, the available stream size and farming practices.

The method can be used for irrigation of close-growing crops on most soils, and water application efficiency can be good under skilled management.

FURROW IRRIGATION Furrow irrigation is the most common method for irrigating row crops such as maize, sugarbeet, potatoes, sunflower, vegetables, orchards and vineyards. In this method water flows down or across the slope of the field and infiltrates into the soil. Unlike border strip irrigation furrow irrigation can be practised where there is a cross slope on the land.

The method is applicable to most soil types, except those which are highly permeable or easily erodible. Erosion is a key concern with this method; slopes are limited to 2% in arid climates and only 0.3% in humid climates with intense rainfalls. Care should be taken where salinity is a problem as salts tend to accumulate on the tops of the ridges. In these cases the plants are positioned on the sides of the ridges.

The furrow size depends on the method used to form the ridges but varies from 0.25 to 0.40 m in depth and from 0.15 to 0.30 m ridge width, with spacing between furrows of 0.75–1.0 m. The spacing of the furrows is governed by the soil type; on lighter (coarser) soils the furrows should be closer together as the lateral movement of water in the soil is significantly less than the vertical movement. On heavier soils water moves laterally as well as horizontally and furrows can be more widely spaced. The length of the furrow depends on the soil type, stream size, land slope and required irrigation depth, and can range from 60 to 300 m, with the shorter lengths being used on light (coarse-textured) soils and longer furrows on heavier (fine-textured) soils.

With good land preparation and skilled water application the water application efficiency can be high with furrow irrigation. Labour requirements can be greater than for other irrigation methods.

CORRUGATION IRRIGATION With this method water flows in small corrugations (channels) pressed into the soil in a similar manner to furrow irrigation. The method is suitable for irrigation of close-growing crops such as barley and wheat.

Sprinkler irrigation

Sprinkler irrigation is used on about 5% of the irrigated land worldwide. The method is flexible and can be adapted to suit most soil types and terrains, though it does not function well under windy conditions. It can also be used for frost protection, and application of fertilizers and pesticides.

A major drawback is the initial cost of the equipment and the energy costs required for pumping. A further constraint is the need for relatively good-quality water, particularly in relation to sodium and chlorite. Measured against this is the relative efficiency of application (about 75% compared with surface irrigation's estimated 60%) and generally lower water requirements than surface irrigation methods.

The sprinkler system consists of sprinklers or other sprinkler devices operating under pressure, a pump unit and a water distribution network, often comprising buried pipes. There are different sprinkler devices, including revolving head sprinklers (single- and multiple-nozzle sprinklers), fixed head sprinklers, nozzle lines and perforated pipes, and a number of different sprinkler systems classified according to their mobility. These include the following (Kay, 1983).

- Portable systems:
  - hand-moved systems;
  - lateral-move systems.
- Solid set or permanent systems.
- Semi-permanent systems.
- Mobile raingun systems:
  - hose-pull system;
  - hose-reel system;
  - rainguns.
- Perforated pipes or spraylines:
  - stationary spraylines;
  - oscillating spraylines;
  - rotating spraylines.
- Mobile lateral systems:
  - centre-pivot systems;
  - side-move systems.

Trickle (drip) irrigation

Trickle irrigation was developed initially for irrigation of glasshouse crops but has been adapted and extended for use with field crops. The trickle/drip irrigation system usually consists of a pump, filter, flowmeter and pressure gauge, fertilizer injector, valves, pipe networks (main, sub-main and laterals) and emitters.

It is the least used system worldwide, used on an estimated 0.1% of irrigated land. As the name suggests, the method comprises trickling or dripping small quantities of water from a pipe onto the soil surface next to the plant. Because the rate of application is low almost all the water is absorbed into the soil, there is little or no runoff. If the application rate and frequency of irrigation are matched to a crop's needs the application efficiency can be very high, potentially around 90%. In addition the system can be used to apply fertilizers direct to the crop's root zone.

The system is not without its drawbacks, however. The equipment and setting up costs can be high, and there can be significant problems with blocking of the emitters from sand and silt, chemical precipitation from the water and algae.

Subsurface irrigation

With subsurface irrigation, as the name implies, irrigation water is applied below the ground surface. The water reaches the plant either through buried pipes or drains, or through seepage from irrigation or drainage canals. Subsurface irrigation is successfully practised in some humid areas, for example in the Netherlands. The use of this method in arid regions can cause serious salinity problems.

Drainage Systems

Drainage is often treated as the poor relation to irrigation. However, if it is not adequately catered for in the development of an irrigation system its absence can result in loss of agricultural production and potential failure of the scheme.

Drainage is required for a number of reasons:

- to make new lands available for agriculture;
- to remove excess surface water following irrigation or rainfall;
- to prevent or reduce waterlogging;
- to control salinity levels in the root zone.

Due to environmental considerations drainage of wetlands and marshes is less prevalent nowadays than it once was. In some countries, such as Guyana and the Netherlands, drainage is a prerequisite to agriculture, either irrigated or rainfed. In such locations the operation and maintenance of the drainage systems takes priority over that for the irrigation network. Basic drainage is required on irrigation systems to remove surface runoff and rainfall, to avoid the water ponding and causing damage to the crops. More elaborate drainage systems are required where over-irrigation and seepage losses from canals lead to a raising of the groundwater table and the occurrence of waterlogging in the irrigated area. Further elaboration is required where there is a need to control salinity levels in the root zone, caused either by a build-up of salts brought in by the irrigation water or through capillary rise from saline groundwater.

Waterlogging of soils restricts the rooting depth of the crops to the aerated zone above the water table. Other consequences of waterlogged soils are that they have a poorer soil structure, lower soil strength, increased susceptibility to damage by compaction, and a reduced availability of nitrogen. An additional problem is that toxic waste products are produced under the anaerobic conditions that exist in waterlogged soils, such that some compounds (e.g. ferric or manganese) that are not harmful under aerated conditions are converted to soluble toxic compounds (e.g. ferrous or manganous).

The benefits of drainage can therefore be summarized as:

- improved aeration of the soil, permitting optimum agricultural production;
- improved soil structure resulting from drier soils;
- leaching of unwanted salts from the root zone;
- leaching of certain soils (acid sulfate soils) to control soil acidity.

There are three main types of drainage system:

- surface drainage, comprising open drains to remove excess irrigation or rainfall;
- subsurface drainage, comprising a matrix of horizontal buried pipes set at 1–2 m below the surface and connected to deep open drains;
- pumped drainage, in which deep tubewells are used to draw down the groundwater, and saline water from tubewells is discharged into open surface drains.

The open drain network mirrors the irrigation canal network, with field drains picking up the irrigation and rainfall runoff and discharging into secondary and primary drains, before discharging either back into the river or to a sink. The subsurface drains are designed to prevent the groundwater table rising into the crop's root zone, and are typically buried at 1–2 m depth at a spacing of between 20 and 50 m depending on the hydraulic conductivity of the soil. Tubewells for pumped drainage are located at spacings that are also determined based on the hydraulic conductivity of the soil, and operate by drawing water out of the soil to create a cone of depression extending around the tubewell. Depending on the quality of the pumped water, it is reused directly for irrigation, or it is pumped into the irrigation canals and mixed with irrigation water to reduce the salinity levels, or it is pumped into the drainage network and disposed of away from the irrigated area.

Endnotes

[1] The term 'irrigation and drainage system' (I&D system) refers to the network of irrigation and drainage channels, including structures. The term 'irrigation and drainage scheme' (I&D scheme) refers to the total irrigation and drainage complex, the irrigation and drainage system, the irrigated land, villages, roads, etc.

[2] There are different terminologies used worldwide for the tertiary unit. Other terminologies include on-farm and watercourse unit.

[3] Again different terminologies are used worldwide for the main system. The term 'off-farm' is commonly used in Eastern Europe and Central Asia.

[4] Whether it is a full ministry or a department within another ministry depends on the country.

[5] There are many different terms for this person/position, including ditch rider, water bailiff, *mirab* (Central Asia), *ulu-ulu* (Indonesia, within the tertiary unit), *Juru Pengairan* (Indonesia, at the main system level).

[6] Note that where the operating authority relies on income from service fees, it will be interested in the financial well-being of the farmers and the farmers' ability to pay the service fees.
[7] The Warabandi system used in northern India and Pakistan is based primarily on equitable distribution of available water supplies, rather than the quantity of the supply (see Chapter 4 for more details).
[8] A project is a time-bounded activity for the development or rehabilitation of an I&D system.
[9] Puddling is the mechanical breaking down of the soil structure into small particles by ploughing and working the land with water ponded on the land. The working of the land creates a hard pan at 30–40 cm depth, and the fine soil particles settle and form a barrier to downward water flow.
[10] Also termed border checks or strip checks.

3

Management

Management is an essential component in any enterprise, but is not always given the consideration it deserves in the irrigation and drainage sector. Manuals are often written for operation and maintenance but tend to cover the technical aspects of operating and maintaining the irrigation and drainage system (I&D system), and do not address other management issues such as accounting and finance, administration procedures, financing, staff recruitment, human resource development and training.

This chapter sets out the elements of management in the irrigation and drainage sector, starting with a definition of management and the identification of the various management functions. This is followed by a discussion on the different management frameworks that exist and how these frameworks affect the approach to managing I&D systems. The different management functions including operation and maintenance, accounting and finance, staffing, administration, legal issues and public relations are then individually discussed in more detailed, together with a discussion on management information systems. The increasingly important role of the irrigation and drainage service fee is also discussed in some detail.

Management Functions

Management can be described as (Jurriens, 1991):

> The organised use of resources, in a given environment, for the planning, operation and monitoring of certain tasks to convert inputs into outputs according to set objectives.

Figure 3.1 shows the relationship between the key management processes involved in converting inputs into outputs. These processes can apply at different levels within the irrigation and drainage sector; at the national level the objectives may be to increase agricultural production and farmer livelihoods, at the system level the objectives will be to supply irrigation water in a reliable, adequate and timely manner to suit farmers' needs, and similarly to remove drainage water in an adequate and timely manner. To achieve the stated objectives plans have to be prepared; these may be longer-term at the national level, and seasonal or annual at the system level. The timespan for implementation will similarly vary, with the focus at system level being on the seasonal or annual cycle. Measurements need to be made of key performance indicators to see if the implementation is proceeding according to plan, and adjustments made as necessary. At the end of the cycle an evaluation should be carried out to ascertain if the objectives

©Martin Burton 2010. *Irrigation Management: Principles and Practices* (Martin Burton)

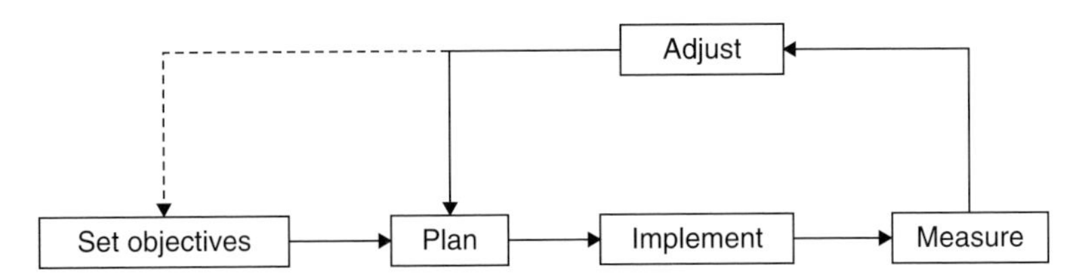

Fig. 3.1. Essential management processes.

have been satisfied, and adjustments made, if required, to either the objectives or the plan.

In the irrigation and drainage sector the key management functions include:

- identifying, setting and monitoring of objectives;
- operating and maintaining the I&D system;
- accounting and finance;
- employing, managing and motivating staff;
- administration;
- managing human resources, including training;
- legal issues;
- public relations.

These management functions will be discussed in more detail in later sections in this chapter.

Management Framework

At the scheme level three levels of management can be identified: the main system level, the tertiary unit level and the field level. In the simplest format the main system is managed by a government agency, the tertiary unit is managed by a water users association and the farmer manages at the field level. There are variations to this structure, for example a state farm or commercial estate, where the entire scheme is managed by one management entity.

In many cases the scheme-level management is part of a wider management structure. In the public irrigation and drainage sector there may be management units at district, regional/provincial and state or national level, while in the private sector there may be regional offices and a head office. At the other end of the scale, such as for traditional irrigation and drainage schemes (I&D schemes), there may be no higher management structure above the scheme level.

Higher-order management structures

The management framework, both at the scheme and higher levels, has a significant impact on the way in which individual I&D systems are managed. In countries or regions where irrigation is relatively extensive, an area-based (rather than scheme-based) organizational structure is generally adopted, with a national-level headquarters responsible for overall management and administration and regional and district offices responsible for management at their respective levels (Figs 3.2 and 3.3). In this case the main operation and maintenance (O&M) unit is the District Office, which may manage several systems within the District's administrative boundaries. It is common for the district (and regional) boundaries to follow local administration, rather than hydraulic, boundaries; thus an I&D system which cuts across these boundaries may be managed by two or more District Offices, or by the Regional Office. In addition to the Irrigation and Drainage Agency (I&D agency) there may be other agencies with separate organizational structures responsible for associated activities such as agricultural extension, veterinary services, credit provision and the like. Countries where such management structures exist include Egypt, Indonesia, the Kyrgyz Republic, Turkey and India.

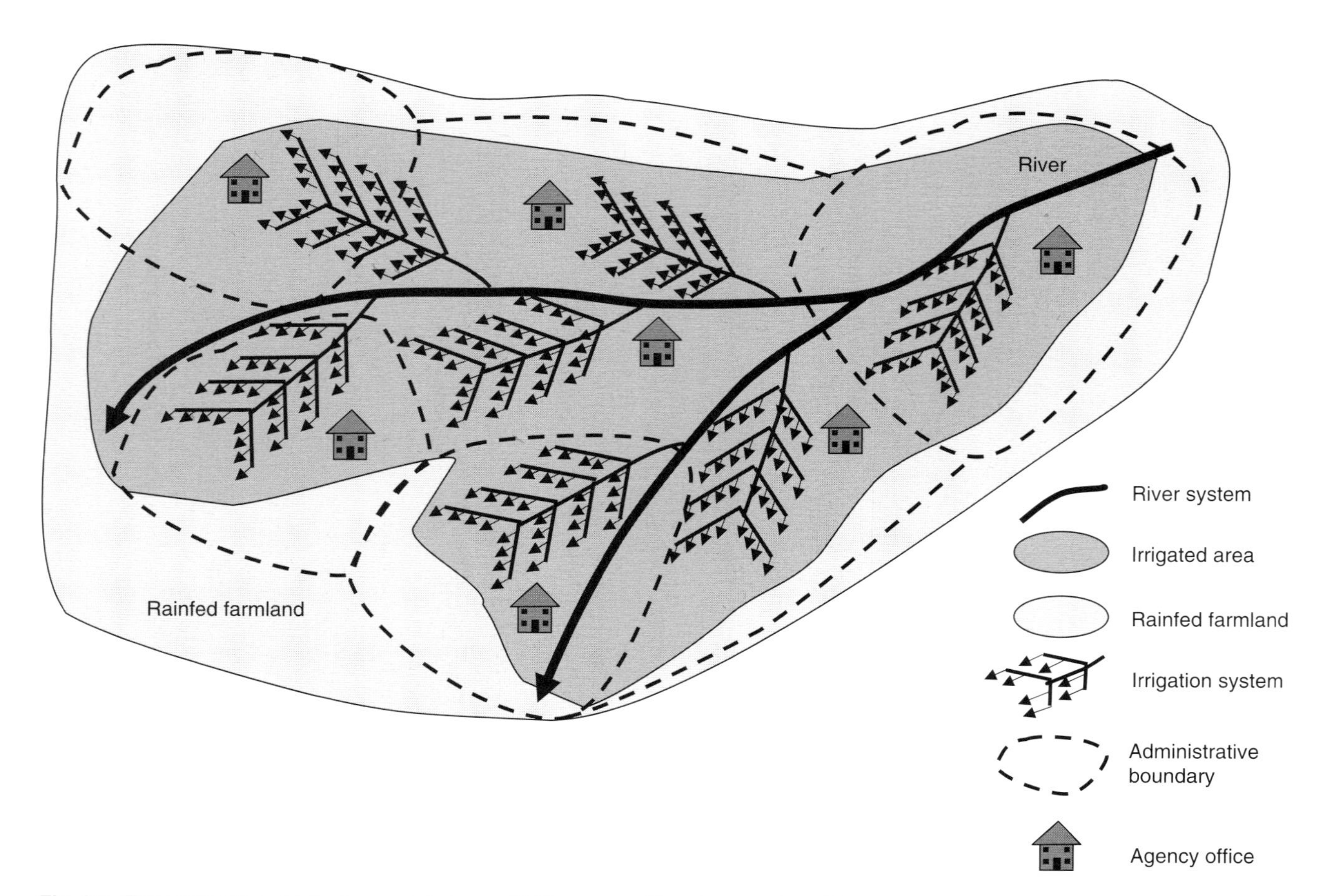

Fig. 3.2. Typical location of irrigation and drainage (I&D) agency offices in countries with extensive I&D systems.

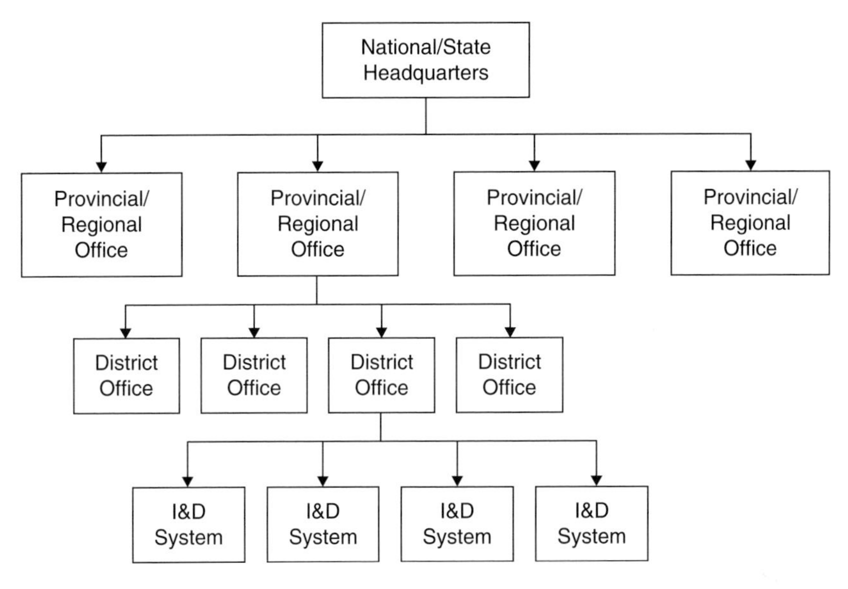

Fig. 3.3. Typical higher-level irrigation and drainage (I&D) agency management structure in countries with extensive I&D systems.

In countries where irrigation is not extensive and I&D systems are spread out (Fig. 3.4), an alternative organizational structure is often adopted. In these cases there is an office on each scheme, with a management team dedicated solely to that scheme. In addition to being responsible for operating and maintaining the I&D system, management may also be responsible for other functions, including provision of agricultural inputs and machinery, crop storage and marketing (Fig. 3.5). Until recently the Mwea Irrigation Scheme in Kenya was managed in this manner by the National Irrigation Board, which organized many of the farming operations and scheduling of irrigation water for the smallholder tenant farmers. Moves are now being made to encourage greater involvement and participation by farmers in scheme management through the formation of water users associations (WUAs). Other examples include sugar estates in a number of countries including Guyana, Ethiopia, Swaziland, Zambia and Kenya. These estates may be publically or privately owned, the key feature is that all farming operations from planting to harvesting and marketing are managed by the estate management. In many of these locations there are smallholder outgrowers who supply cane to the estates. The agricultural practices of these outgrowers are overseen by the estate in order that the cane is supplied at the scheduled time and to an agreed quality.

At state or national level the key management roles include policy formulation, budget allocation, planning and sourcing finance for further development and rehabilitation, specification of work functions and staffing for lower-order management units, and general overall management control and performance monitoring and evaluation. A key role at the national or state level is to liaise and work with other Ministries to coordinate programmes to support irrigated agriculture. A key task at the national/state level is to obtain an adequate budget for the management, operation and maintenance of I&D systems; this generally requires presenting a convincing argument to the Ministry of Finance and Government.

At the regional level the management functions relate to oversight and coordination of the lower-order (district) management units. Funds are generally dispersed from the national/state level to the regional offices, and then on to the district level. The Regional

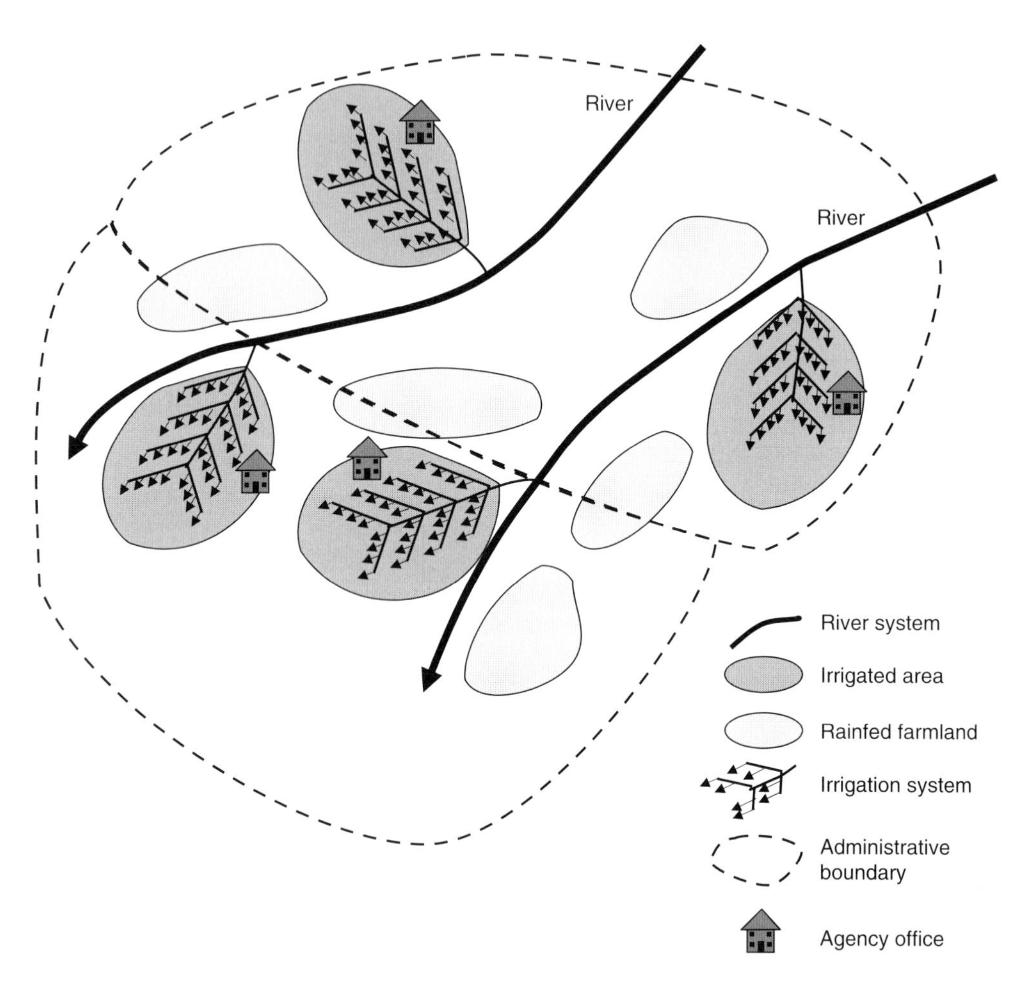

Fig. 3.4. Dispersed irrigated agriculture with irrigation and drainage agency offices located on individual schemes.

Office will be responsible for monitoring the performance of the District Offices, in terms of their technical, financial and administrative functions, and reporting back up to the national/state level. Annual reports for each region and district are a useful means of institutionalizing performance assessment and monitoring at the different levels.

Depending on the country the core task of operating and maintaining the main system is the responsibility of the I&D agency District Office, the Scheme Office, or the WUA Federation or WUA Office. In many countries such as India, China, Egypt, Indonesia and Sudan the main system is managed by government agencies. In other countries such as Mexico and Turkey, the management of the main system has been handed over to water users associations or federations of water users associations (see Chapter 8 for more information on this process). In other locations with traditional irrigation systems the main system is managed by the water users.

The main management functions at this level relate to ensuring adequate operation and maintenance of the I&D system. Procedures need to be in place for making an assessment at the start of each season of the anticipated irrigation demand and checking that this can be matched by the anticipated supply. If anticipated demand exceeds anticipated supply then the demand must be reduced, generally

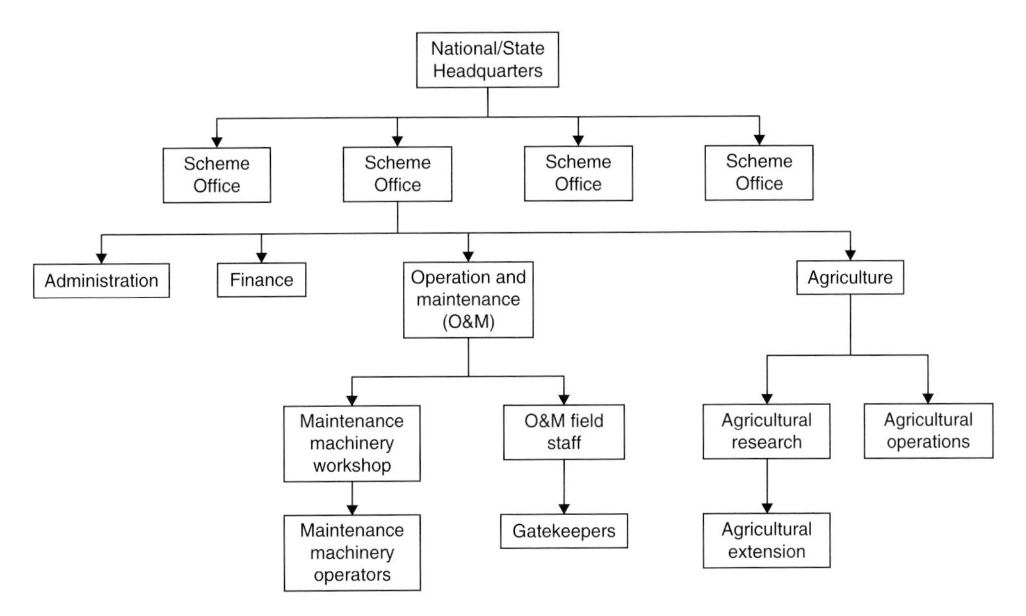

Fig. 3.5. Example of an irrigation and drainage agency management structure for dispersed irrigation and drainage schemes.

by reducing the area irrigated or the area of high water-demanding crops (such as rice). There also need to be procedures for matching supply to irrigation demand for each time period (this can be daily, weekly, 10-daily or bi-monthly) during the irrigation season. Allied to this is the need to manage the maintenance of I&D systems such that the operation of the systems is not compromised. Associated management functions will be financial management (salaries, expenses, O&M costs, etc.), general administration, staff management and motivation, liaison with other organizations and public relations.

A typical I&D agency organizational structure for management at this level is shown in Fig. 3.6. The organizational structure and command areas covered vary from country to country. In this example the main unit is a District Office responsible for approximately 30,000 ha, with administration, finance and O&M departments. Below the District Office there might be Section Offices responsible for irrigated command areas of approximately 5000 ha comprising three or four run-of-the-river irrigation systems. In each Section Office there are typically two or three office staff, six to eight Water Masters, and several gatekeepers and labourers. The gatekeepers are assigned to major headworks and to a water master's command area where they will adjust and monitor the tertiary unit intake gates following instruction from the water master. In addition one or more labourers may be assigned to work with the water master on regular maintenance of the system.

Tertiary unit (on-farm) level management structure

Increasingly, management at the tertiary unit/on-farm level is being carried out by water users associations. Some of these associations have been formed in recent years under project programmes, in other cases they are indigenous systems, which have existed for generations, having been formed by farmers themselves.

Cooperation and management at this level are crucial if the best use is to be made of the water delivered from the main system. Significant management effort thus needs to be put into communicating and liaising with water users such that they are fully involved

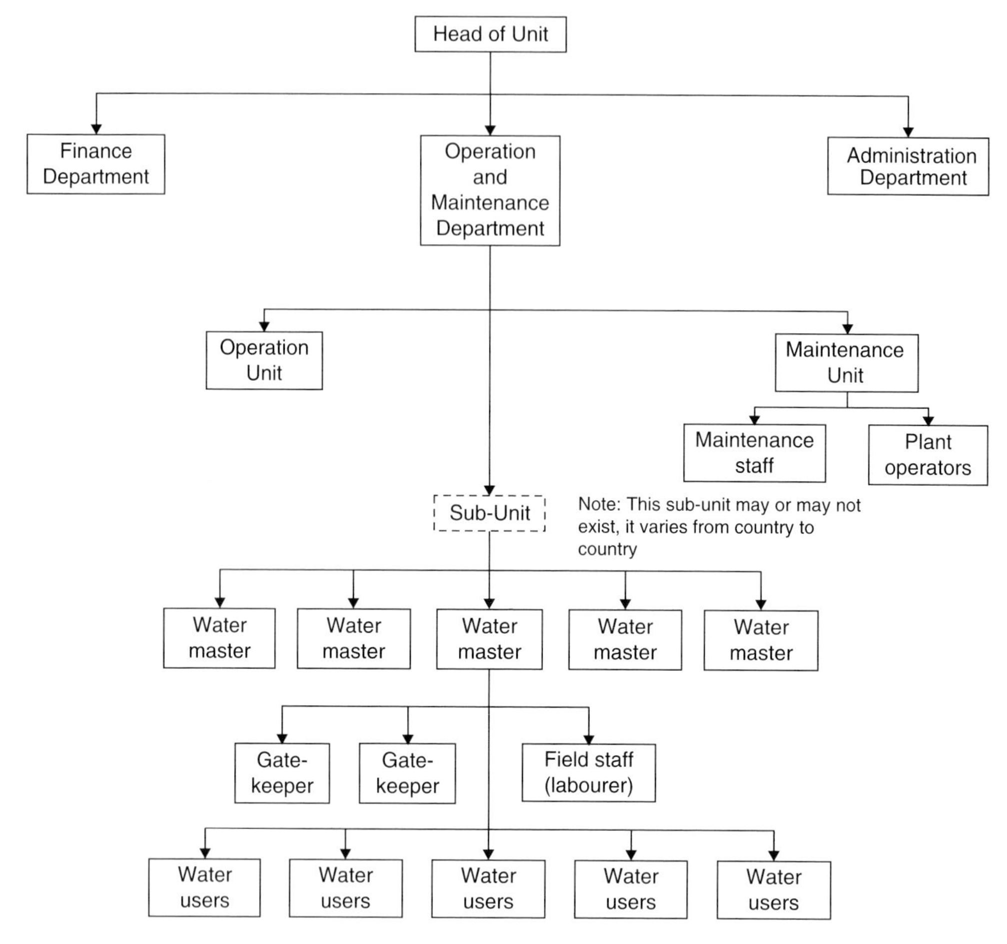

Fig. 3.6. Typical organizational structure for an irrigation and drainage agency management unit at the main system level.

and committed to the joint management of the I&D systems within the WUA service area. Cooperation is required in agreeing on irrigation turns and maintaining channels; management is required to organize meetings with water users, collect requests for irrigation water, organize schedules, and organize and supervise maintenance work. An additional function is to set and collect the irrigation and drainage service fees.

A typical structure for a Water Users Association is shown in Fig. 3.7. The core body is the General or Representative Council to which the WUA Management Board and committees report. The Management Board oversees the WUA Executive, which generally comprises an Executive Director, an Accountant, an O&M Engineer/Technician and field staff (Water Masters). In an Association with a General Assembly all members attend the Annual General Meeting and other general meetings; in an Association with a Representative Assembly, representatives will be elected by groups of farmers within the WUA command area and attend meetings on their behalf and report back to them. These representative groups are hydraulically based, usually on channel commands supplying 30–40 ha. Water users liaise directly with WUA field staff and the accountant on water delivery and fee payment, but will liaise with the Representative Council through their representative.

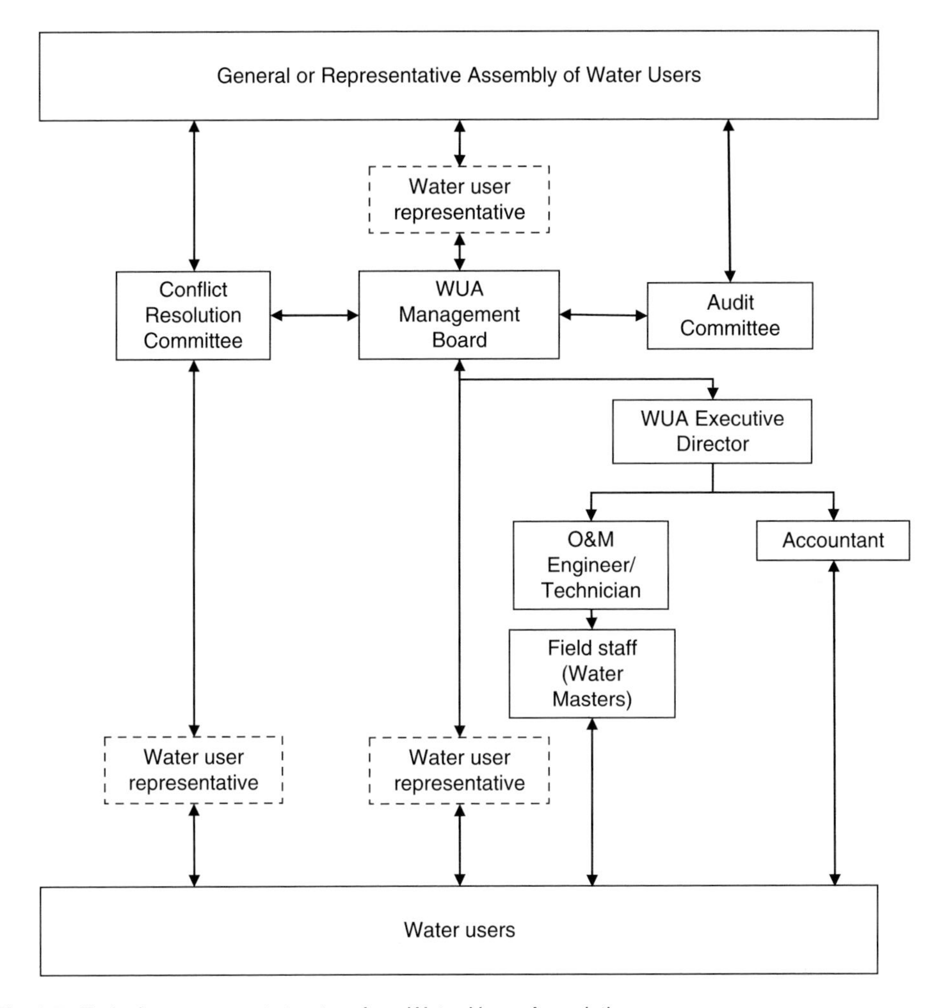

Fig. 3.7. Typical management structure for a Water Users Association.

More details on water users associations are provided in Chapter 8.

Field-level management structure

The management structure at the field level is relatively straightforward as it comprises the farmer, his or her immediate family, extended family, neighbours and employed labour. In smallholder irrigation schemes a large part of the farming activities are carried out by the farmer and his/her immediate family and extended family. Assistance may be sought from neighbours and other villagers at certain times, such as for land preparation and harvesting. This assistance may take the form of manual labour or the loan of agricultural machinery, and will be reciprocated in due course.

Setting Management Objectives

In any organization it is important to be clear about its purpose and objectives. In a commercial organization the purpose is to make money and remain in business; in a public sector organization the purpose will generally be more socially oriented. In a public sector I&D

agency for example, the purpose of the organization may be to stimulate social and economic development in rural areas through provision and support of irrigated agriculture.

Senior management needs to identify the *purpose, objectives* and *strategy* for the organization. The purpose sets out the primary goal(s), the objectives identify key achievements that are required to reach the goal and the strategy sets out the route to be taken. A simple set of questions can be used to prepare the purpose, objectives and strategies for an organization.

• Where are we now?	Evaluation and appraisal
• Where do we want to be?	Vision, purpose and objectives
• How are we going to get there?	Options and choices
• Which way is best?	Strategy
• How do we ensure arrival?	Implementation and control

Within the irrigation sector the purpose and objectives will vary at the different levels. Table 3.1 summarizes some of the possible objectives at these levels. At the national

Table 3.1. Possible objectives at different levels in the irrigation and drainage sector.

Level	Possible objectives
National	• To increase national agricultural production • To provide the population with agricultural products • To achieve self-sufficiency in food • To supply industry with raw materials • To generate foreign exchange earnings • To create employment • To limit rural migration to cities • To raise the income of the rural poor and to achieve a more equitable income distribution • To establish social stability or social control
I&D schemes	• To maximize agricultural output • To maximize the number of people settled on the irrigation/drainage scheme • To maximize the financial return on the capital investment in infrastructure • To maximize the financial return to farmers • To make efficient and productive use of land and water resources • To provide security against drought and famine • To minimize adverse environmental impacts • To cover the MOM costs through service fee recovery from water users
I&D system managers	• To provide an adequate level of service to water users • To optimize water distribution and minimize water losses • To avert or minimize waterlogging and salinization • To recover MOM costs • To maintain the irrigation and drainage infrastructure • To manage and motivate staff • To balance the accounts each year
Water users associations	• To liaise with water users to keep them involved, informed and committed to the Association • To liaise with the main system service provider and obtain reliable, timely and adequate water supplies • To provide an adequate level of service for water delivery • To maintain the on-farm system • To recover the MOM costs • To balance the books of the Association each year
Farmers	• To have a secure and stable life for themselves and their families • To be self-sufficient in food production • To earn a decent living (through the selling of agricultural products)

I&D, irrigation and drainage; MOM, management, operation and maintenance.

level government may be interested in using its (often scarce) land and water resources to satisfy a number of objectives. On the one hand it may support irrigation development to increase agricultural production and create employment and sustain the livelihoods of rural communities. On the other, it may wish to generate income and create employment by licensing commercial estates to produce agricultural products for industry and export.

Similar objectives will exist at the regional and district level. In countries where responsibility for rural development has been devolved to the regions or districts, the development of irrigation can be a major political objective at this level.

For individual I&D schemes the objectives may vary, depending on a variety of factors including its intended purpose (smallholder or estate), design and location. Some schemes are designed for 'protective irrigation' to sustain smallholder farmers in arid zones where droughts threaten livelihoods; other schemes are designed to provide commercial agricultural opportunities and to grow the local economy. Some schemes are concerned primarily with irrigation; in other schemes flood protection, drainage and protection from waterlogging and salinization are important.

For managers of I&D systems the primary objective is to provide an adequate level of service. This will require that irrigation water is provided in a reliable, adequate and timely manner, as efficiently and cost-effectively as possible, with equal opportunity for all users, irrespective of their status or location on the scheme. Associated with this service delivery is the recovery of service fees, where due, which will contribute to the costs for operating and maintaining the system. The system manager will have additional objectives associated with his/her managerial function, including managing and motivating staff, balancing the accounts each year and liaising with water users and other entities.

The objectives for the water users association are similar to those of the I&D system manager in terms of the operation and maintenance and cost recovery functions. However, an overriding objective must be to liaise and work with water users such that the association functions as a democratic, accountable and transparent organization, fully supported by all water users.

The objectives for farmers will vary depending on the circumstances. In some locations the main aim is survival; irrigation can provide reliable water supplies where rainfall is sparse, erratic or unreliable. In other locations income from irrigated agriculture may be competing with alternative forms of income and will thus need to be remunerative for the farmer for it to be an attractive option.

Figure 3.8 shows a linked hierarchy of objectives at different levels. Understanding such linkages, as with the nested systems outlined in Chapter 2, helps to understand where interventions can be made to improve the performance of I&D schemes.

Key Management Functions for System Operation and Maintenance

Detailed discussion on processes and procedures for system operation and maintenance is provided in Chapters 4, 5 and 6. The sections below summarize some key management principles and processes underlying the proper management, operation and maintenance of I&D systems.

Service delivery

At the centre of the management philosophy should be the principle of service delivery. This is particularly important in relation to the recovery of the irrigation service fee. Good service delivery is more likely to result in good levels of fee recovery; poor service delivery will almost certainly result in poor levels of fee recovery.

Figure 3.9 shows the core elements of service delivery, comprising the service delivery to the water user for which a service fee is paid to the service provider. The rights and responsibilities of each party are contained in the Service Agreement. The Service

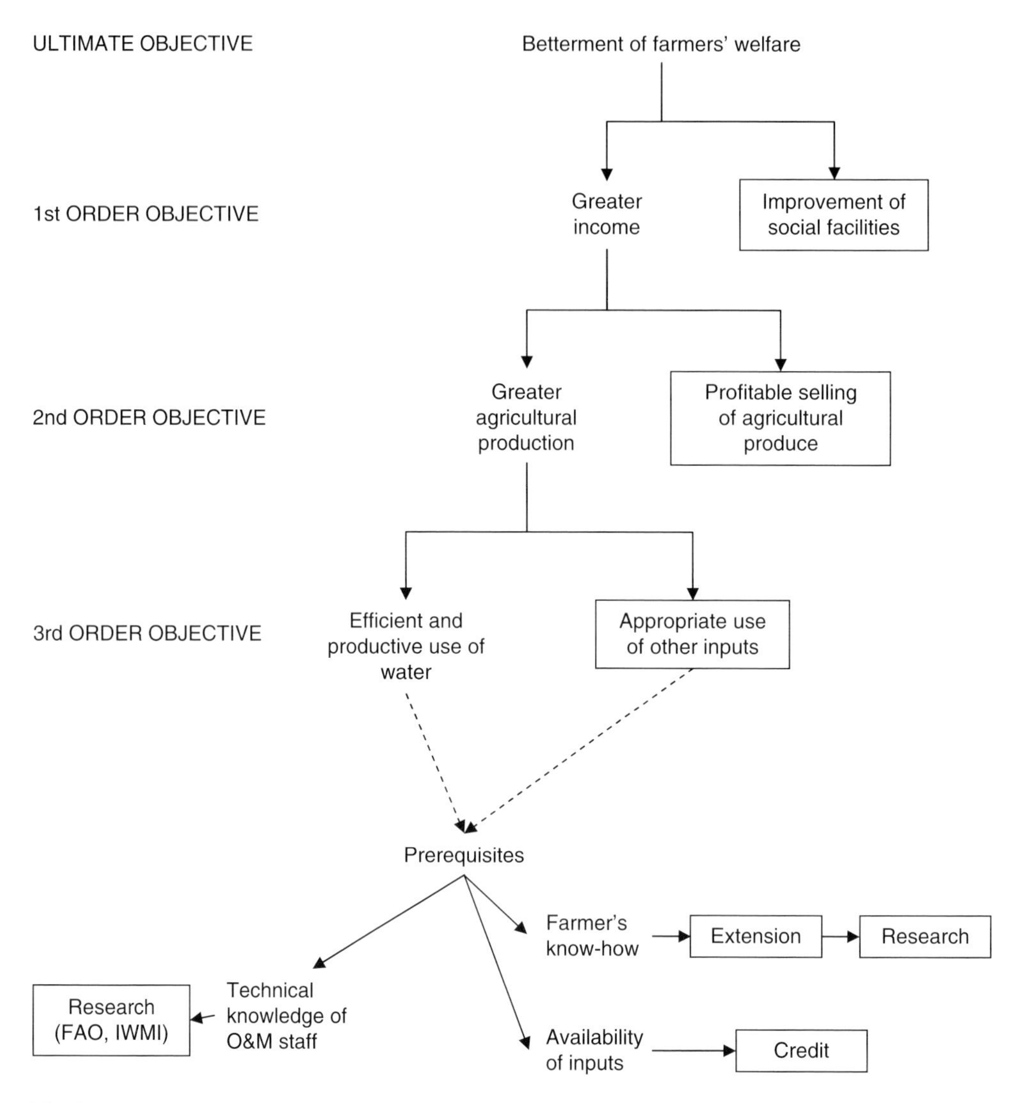

Fig. 3.8. Hierarchy of objectives. FAO, Food and Agriculture Organization of the United Nations; IWMI, International Water Management Institute; O&M, operation and management. (After FAO, 1982, with permission from the Food and Agriculture Organization of the United Nations.)

Agreement has two parts: (i) the Specification of the services to be provided; and (ii) the Conditions under which these services are provided. The Specification will typically detail the rate, duration and frequency of water supply, the method of verification of delivery, and the certainty or security of supply. The Conditions will stipulate the fee to be paid, the location of supply, procedures for ordering and notification of need for water, procedures in case of low or restricted flows, allocation priorities, and times and procedures for closure of canals for maintenance or in case of emergencies.

It is important that the Service Agreement is drawn up by an independent legal advisor and that it is seen to be reasonable and fair by both parties. In some instances the Service Agreement is written by the service provider with insufficient regard to the rights

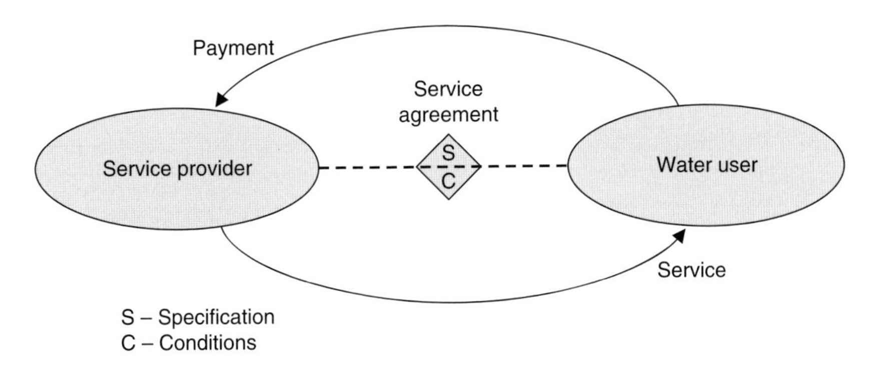

Fig. 3.9. Core elements of service delivery. (Modified from Huppert and Urban, 1998.)

of the water users. This form of agreement is counterproductive as it causes distrust and resentment among the water users, who may then not cooperate fully with the service provider.

Planning and management of water delivery

The management process for planning and organizing water delivery is shown in Fig. 3.10. There are six stages in the annual or seasonal management cycle, with a three-stage sub-cycle during the implementation phase. The components of each stage are outlined in the sections below.

Planning

Pre-season planning is required in order to match irrigation water demand with the anticipated supplies. This may require limiting crop areas in order to match supply and demand, or restricting the area grown to high water-demanding crops, such as rice. In schemes where water resources are limited pre-season planning is important. In schemes where water resources are abundant pre-season planning becomes less important and is sometimes difficult to impose upon farmers. Maintenance planning will also be required before the season to ensure that the system can deliver the required water supplies.

Budgeting

Budgeting is required at the start of the year for financial and other resources, including staff time and labour. The timing of financial payments can sometimes be a problem, with delays in payment from central government or head office causing delays in implementation of maintenance work.

Programming

A programme needs to be drawn up for the execution of operation and maintenance activities during the year or season. The timing of pre-season and in-season maintenance is particularly important, especially if canals have to be closed to carry out the work. Details of the programme will need to be discussed and agreed with water users.

Implementation

Once the irrigation season commences the system should be ready for farmers to plant their crops and receive their irrigation supplies. It is important to realize that within the irrigation season there is a further sub-cycle involving scheduling, allocation and monitoring of water supplies. This sub-cycle involves looking at the specific demands in the coming time period and then scheduling, allocating and monitoring the available water supplies.

The irrigation plan made at the pre-season stage will give the broad irrigation demands and locations of demand; the scheduling carried

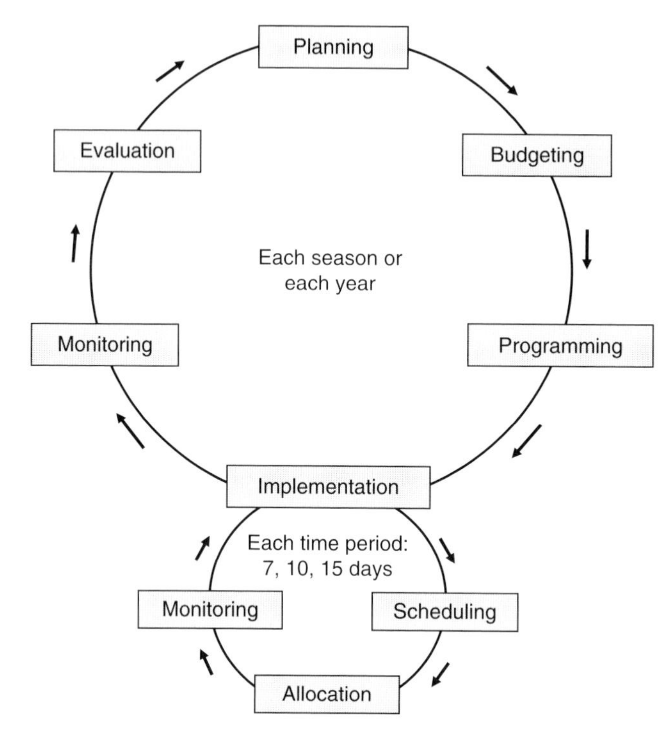

Fig. 3.10. Irrigation management cycle.

out within season gives specific discharges and volumes to be supplied to specific locations in the system for specific dates and times.

Monitoring of water deliveries is an essential part of the in-season management cycle, and is especially important at the main system/tertiary unit interface where the irrigation service fee is charged based on actual water delivered. In this situation it is usual for the daily record sheets to be signed by a representative of the main system service provider and the water users association.

Monitoring

During the season the implementation of the pre-season irrigation plan and work programme should be monitored. This might be especially important in the case of a system supplied from a reservoir, where it will be essential to keep careful track of the abstractions made, and to compare them with the planned abstractions and remaining supplies in the reservoir.

Evaluation

Evaluation is carried out at the end of the season to make several assessments.

1. *To compare the actual implementation against the plan.* This assessment looks at how closely the actual implementation complied with the plan, and how either the planning or the implementation needs to be improved in future to get a better match.
2. *To assess the viability of the plan.* This assessment looks at whether the plan was the right plan, or whether changes could be made to improve it. For example, in systems where authorization is required for growing high water-demanding crops, the authorized area should be reviewed at the end of the season or year to see if the authorized area was correct, too much or too little.

3. *To assess how implementation was carried out.* This assessment seeks to identify areas where the implementation can be improved, such as in matching supply and demand.
4. *To assess if the implementation met the needs of the water users.* This assessment is part of customer service and seeks to check if the service provided matched water users' expectations and needs.

The evaluation process does not have to be too laborious, it simply seeks to assess whether planning decisions made at the beginning of the season were correct, were implemented adequately, and whether any improvements can be made to improve agricultural and water productivity.

Further information on monitoring and evaluation is provided in Chapter 9.

Planning and management of maintenance

The management process for planning and organizing maintenance of the I&D system is shown in Fig. 3.11. The process commences with inspection and reporting, either on a regular basis from field staff, or as a result of seasonal or annual maintenance inspections. Preliminary costings are made and compared with the budget available, following which priority work is taken forward. It is important at this stage to consider preventative maintenance work that will avoid costly maintenance work in the future. Maintenance work is then planned and scheduled to fit with the irrigation season(s), and where necessary contracted out for implementation. Responsibility for supervision of the work should be clearly defined, and adequate time and resources committed to ensure adequate levels of supervision. Where contractors are involved, a final inspection for certification is carried out prior to payment. A final task, which is not always done well, is to record the work done. Good records of completed maintenance work can be invaluable in asset management planning and costing of future maintenance work. More detail on these processes is provided in Chapter 6.

Computer databases represent a powerful tool for maintenance management, enabling systematic recording of the work required and the work carried out.

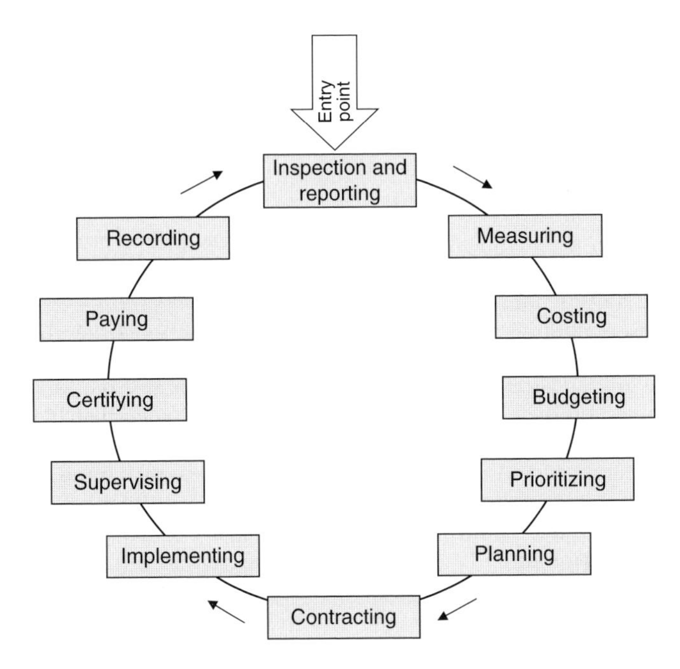

Fig. 3.11. Annual maintenance cycle.

Management Records and Information Systems

Management information systems and records are an essential feature of the management of I&D schemes. For an irrigation service provider the management records and information system might include:

- topographic maps of the command area;
- maps of the I&D system(s);
- schematic maps of the I&D system(s) showing key features (canals, control structures, command areas, etc.);
- schematic operational diagrams showing planned and supplied discharges, crop areas, etc. (see Chapter 4 for examples);
- forms for recording crop areas, discharges, climatic data, etc.;
- a maintenance register and forms for recording maintenance work required and implemented (see Chapter 6 for examples);
- an asset register for all infrastructure;
- engineering drawings of all the assets;
- administrative records;
- staff records (personal details, salary, annual reports, etc.);
- financial accounts and records.

For a water users association the management records are less extensive and might include:

- minutes of meetings;
- copies of the WUA registration documents;
- a map of the irrigation scheme, showing the boundaries, canals, drains, structure locations and, if applicable, the representative zones;
- if available, a full cadastral map showing all landholdings and their sizes;
- a register of members, with the names and landholding areas and locations for each member – details will also need to be kept for non-members if they require irrigation water supplies;
- accounts records including a cash book, a register showing the irrigation fees paid and an accounts book showing the income and expenditure;
- an asset register, detailing the lengths of canals and drains and the type and characteristics of all infrastructure;
- a maintenance register showing the maintenance work required and completed.

The design of an efficient management information system can add greatly to management productivity and efficiency. Figure 3.12 provides an example of the type of data that might be collected and processed to pass information up the management hierarchy. It is important that the amount of data passed on up the hierarchy is appropriate to that level. It is not sensible, for example, to send information on irrigation supplies to individual tertiary units up to the provincial level, these data should be kept at the section and district level. However, summarized information on abstractions from rivers for irrigation systems will be of interest and use at the provincial level. The frequency of the reporting is also important, and, as shown in Fig. 3.12, will vary according to the type of data collected and their use. In this example rainfall, river and canals flows are collected daily by the Water Master, summarized each 10 days and passed on to the Section Office, who in turn pass it on to the District Office each month. In contrast, data during a severe flood might be sent daily (or even hourly) to all levels.

Computers and appropriate computer software play an important role in management information systems. At the simplest level spreadsheets provide a simple and effective means of storing, summarizing and presenting data. Proprietary databases are now readily available and can be relatively easily programmed to enter, process, store and present data for individual situations. Alternatively, specialist database software can be purchased for specific applications such as processing, analysing and presenting data on rainfall, river and canal discharges, cropping, etc.

Geographic Information Systems (GIS) and remote sensing are increasingly important tools for data collection, processing, analysis and presentation in the irrigation and drainage sector. In some instances where

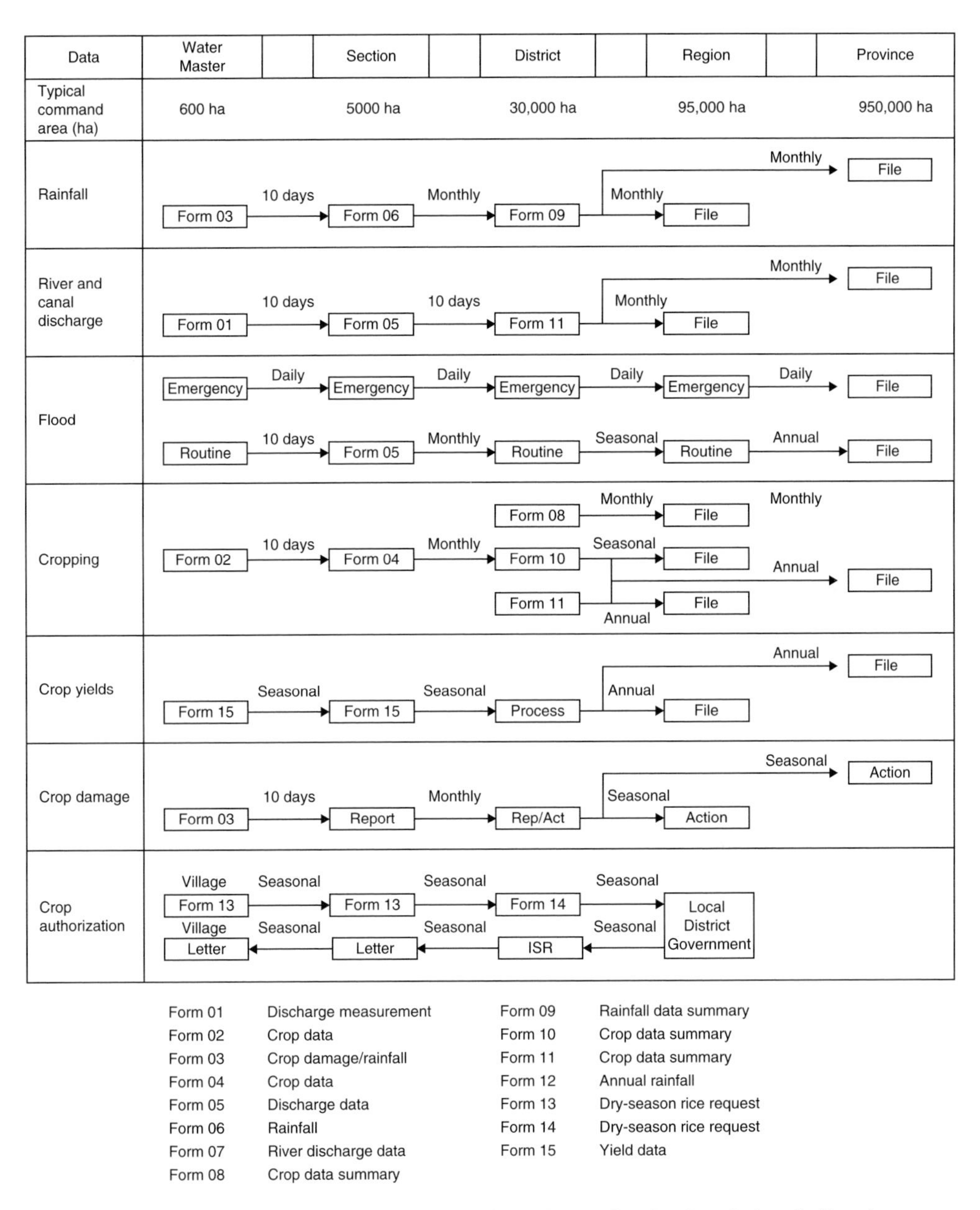

Fig. 3.12. Example of a data processing flowchart. ISR, Irrigation Service Regulations limiting dry-season rice area.

there are large irrigation systems, such as in India, remote sensing and GIS are increasingly being used to collect and process data on crop areas, crop types and even crop water demands and crop yields.

Accounting and Finance

There are two levels at which to discuss management accounting and finance: for the main system service provider and for water users associations. Processes and procedures for the main system service provider will be more extensive and complex than those for water users associations.

Main system service provider

The accounting and finance processes and procedures will vary depending on the management framework, and whether the main system service provider is a government or private entity. For a government agency the budget will be requested from government by the Head Office and will be part of the annual budget set by the national exchequer. The allocation of the budget is often established based on previous years' allocations and finances available, rather than on the actual needs, resulting in many cases in under-financing of the maintenance component. Head Office then disperses the finances to lower management levels according to their budget requests, which may be based on standard norms relating to staffing levels, size of systems, type and quantity of infrastructure (lengths of canals/drains, headworks, etc.). In some cases the funds are released by the Ministry of Finance annually or quarterly; in others they are released monthly. Delays in release of funds can sometimes be a problem, especially where funds are required at the start of the irrigation season for maintenance work.

A coding system (termed a chart of accounts) lies at the heart of any accounting and finance system. An example of a coding system is presented in Table 3.2 and shows how each item of expenditure can be coded and allocated to a budget line. Table 3.3 gives a breakdown of the associated accounts, showing that 30% is spent on management costs (items 1–6, staff salaries, office costs, etc.), 39% on operation costs (items 7–10, principally electricity costs for pump stations, with some transport and field trip costs) and 31% (items 11 and 12) on maintenance. The breakdown into management, operation and maintenance categories is useful as it helps identify where the costs lie.

Depending on the accounting and finance system used there are standard procedures for establishing the coding system. In establishing this coding system it is important for staff within the organization to easily and accurately account for expenditure such that the costs can be attributed and traced. As outlined above it is important to be able to attribute management, operation and maintenance expenditure, and to attribute these to different levels within the organization, and in particular to individual I&D systems. This then enables the

Table 3.2. Example of line budget categories and coding system.

Category	Code	Category	Code
Salary	1.1.0.0	Gas	1.3.3.7
Pension contributions	1.2.1.1	Communications	1.3.3.8
Travel expenses – head office	1.3.1.1	Other utilities and rent of buildings	1.3.3.9
Travel expenses – regional	1.3.1.2		
Procurement of technical equipment/materials	1.3.2.1	Procurement and services	1.3.4.1
Procurement of office supplies and equipment	1.3.2.2	Subsidies	3.1.1.1
		Capital repair	4.0.0.4
Meal allowances	1.3.2.3	Procurement of major equipment and goods	4.0.0.1
Clothing/uniforms	1.3.2.4		
Rent and maintenance of vehicles/transport	1.3.3.3	Building and structures maintenance/repair	4.0.0.2
Water	1.3.3.4	Civil construction schemes/projects	4.0.0.3
Electricity	1.3.3.5		
Heating	1.3.3.6	Design services	4.0.0.5

Table 3.3. Example of budget allocation at national level.

Item no.	Expenditure item	Total budget allocation (US$)	Budget allocation (%)
1	Salaries (incl. pensions)	580,563	28.6
2	Potable water supply	1,767	0.1
3	Electricity (offices)	2,843	0.1
4	Heating	1,202	0.1
5	Communications	14,046	0.7
6	Other expenses	4,104	0.2
7	Transport costs	162,418	8.0
8	Field trip expenses	13,086	0.6
9	Other services	211,513	10.4
10	Electricity (pump stations)	401,322	19.7
11	Equipment/materials	203,234	10.0
12	Capital repair and maintenance works	437,056	21.5
	Total	**2,033,155**	**100**

service provider to identify income from any given system, match it with expenditure and give an account to water users on each system of how the income from their service fees and other sources has been utilized.

Water users associations

Accounting and finance procedures for water users associations has to be far simpler than those developed for the main system service provider. The basic components will be:

- a register of members' fees due and paid;
- annual contract;
- crop area record book;
- irrigation invoices;
- register of irrigation invoices;
- cash book;
- bank documents (cheque book, paying-in book, monthly bank statements, etc.);
- payroll register;
- fines register;
- procurement register and procurement forms;
- inventory of assets;
- expenses register;
- general ledger;
- annual cash flow and balance;
- budget;
- annual financial report.

The register of members records the names of WUA members and their annual contribution. The annual contract is signed by the WUA management with each water user and sets out crop types, areas and irrigation to be provided. The crop area register records the crops actually planted and their area for all water users, and forms the basis for charging the irrigation service fee. Booklets of irrigation invoices are printed and an invoice issued for each irrigation event, either by the WUA Accountant or by the Water Masters. These booklets have duplicate copies for each invoice issued, which are returned to the WUA Treasurer and entered into the register of invoices. Figure 3.13 provides an example of a format for combining the crop and fee payment registers, which can be printed A3 size and distributed to WUAs. The register records the water user's name, total landholding area, crops grown, fee due and payments made, making it easy to see what is due and what has been paid. In a refinement of the register the first two columns could be printed on a sheet at the back of five or six sheets which have the remaining columns, with one sheet then being used for each year without the need to rewrite each farmer's name and landholding data each year.

Other documentation (cash book, payroll register, procurement register and forms, expenses register, etc.) is relatively standard. The inventory of assets refers to equipment

Crop and fee payment register

Year:

Name of water user	Total landholding area (ha)	Crop type and area planted								Total service fee due (MU)	Service fee payments															Total fee paid (MU)	Remaining to pay (MU)
											1			2			3			4			5				
		Barley	Wheat	Cotton	Vegetables	Alfalfa	Backyards		Total area cropped		Amount (MU)	Receipt no.	Date	Amount (MU)	Receipt no.	Date	Amount (MU)	Receipt no.	Date	Amount (MU)	Receipt no.	Date	Amount (MU)	Receipt no.	Date		
Farid Agayev	2.0	0.5		0.5	0.2	0.5			1.7	34	5	1345	1 May	10	2453	5 June	10	3453	15 July							25	9
Totals																											

Fig. 3.13. Example of a crop and fee payment register. MU, Monetary Units (dollars, shillings, rupees etc.).

and other assets purchased by the WUA, such as office furniture, office equipment, vehicles, motorbikes, maintenance equipment, etc. It does not refer to the infrastructure assets of the I&D system, which is a separate document.

The budget and annual financial report are obviously key parts of the WUA accounting process. An example of a WUA budget is presented in Fig. 3.14. The budget is fairly straightforward; the annual financial report (Fig. 3.15) is slightly more complex as it summarizes all the financial transactions during the year and shows the amount remaining in the bank and in cash.

Financing irrigation management, operation and maintenance, and cost recovery

Finding adequate funds to operate and sustain the system is the next most important management task after operation and maintenance. Under-investment in I&D system maintenance over the last 20–30 years has resulted in I&D systems falling into disrepair and requiring rehabilitation. Unfortunately, despite assurances by governments to funding agencies that adequate finance for management, operation and maintenance (MOM) would be provided following system rehabilitation, this has often not happened and rehabilitated systems have deteriorated, sometimes back to their pre-rehabilitation state.

Figure 3.16 presents a diagrammatic representation of the flow of finances for system MOM. In some countries funding and donor agencies have been providing finance to governments to support the MOM of their irrigation/drainage systems. If this finance is available it is added to the budget allocated by government to the line minister responsible for irrigation and drainage (this varies from country to country) who then assigns the finances to the I&D agency. Within the I&D agency the money is apportioned to the regional and district (or system) offices. Where irrigation and/or drainage fees are paid by water users this money is generally collected by the local office of the I&D agency. In some countries these service fees

WUA Annual Budget

WUA name: Zemokartli Year: 2008

Income	Amount ($)	Expenditure	Amount ($)
Membership fees	3,625	Salary for Manager	810
Irrigation service fees	6,562	Salary for Accountant	1,425
Grants	721	Salaries for Water Masters (4 no.)	1,125
Previous year's payments (2007)	150	Office costs	75
		Stationery, etc.	31
		Fuel	250
		Electricity for office	13
		Maintenance costs (contractor)	2,212
		Payment of ISF to Irrigation Agency	3,820
		Watchman for headworks	600
		Reserve funds	510
		Contingency	188
Total	11,058	Total	11,058

Fig. 3.14. Example of a water users association budget (ISF, irrigation service fee).

WUA Annual Financial Report

WUA name: Zemokartli Year: 2008

I. WUA command area and membership summary

a)	Total WUA command area (ha)	692 ha
b)	Total irrigated area during the year (ha)	692 ha
c)	Total number of farmers in the WUA command area (no.)	394
d)	Total number of WUA members (no.)	278

II. Financial summary

Cat. no.	Item no. Description	Value ($)	Total value ($)
1	Opening cash balance		
	1.1 Cash in bank	500	
	1.2 Cash held	100	
	Sub-total (1)		600
2	Income		
	2.1 Membership fees	3,450	
	2.2 Irrigation service fees	6,342	
	2.3 Fines	522	
	2.4 Grants	721	
	2.5 Other income	100	
	Sub-total (2)		11,135
3	Bank interst and donations		
	3.1 Bank interest	25	
	3.2 Donations	50	
	Sub-total (3)		75
	TOTAL INCOME (including opening balance)		11,810
4	Operating expenditure		
	4.1 ISF paid to Irrigation Agency	3,820	
	4.2 Salaries	3,960	
	4.3 Office costs	126	
	4.4 Transport costs	272	
	4.5 General expenditure (meetings, etc.)	40	
	4.6 Maintenance expenditure	2,325	
	4.7 Other	124	
	Sub-total (4)		10,667
5	Investments and loan repayments		
	5.1 Payment into Reserve Fund	450	
	5.2 Equipment and materials	346	
	5.3 Loan repayment	0	
	Sub-total (5)		796
	TOTAL OUTGOINGS		11,463
6	Closing balance		
	6.1 Cash in bank	242	
	6.2 Cash held	105	
	Sub-total (6)		347

WUA Accountant
Name: ____________________
Signature: ____________________
Date: __________

WUA Executive Director
Name: ____________________
Signature: ____________________
Date: __________

WUA Chairman
Name: ____________________
Signature: ____________________
Date: __________

WUA Seal:

Fig. 3.15. Example of water users association annual financial report. ISF, irrigation service fee.

are collected by the Ministry of Finance and paid into the general exchequer. This process is not recommended; it is far preferable that there is a direct link between the money paid by the water users to their local office, and the service that they receive, and can demand.

As can be seen from the plot in Fig. 3.16, the hope is that over time the fee payment levels will increase to cover a greater proportion

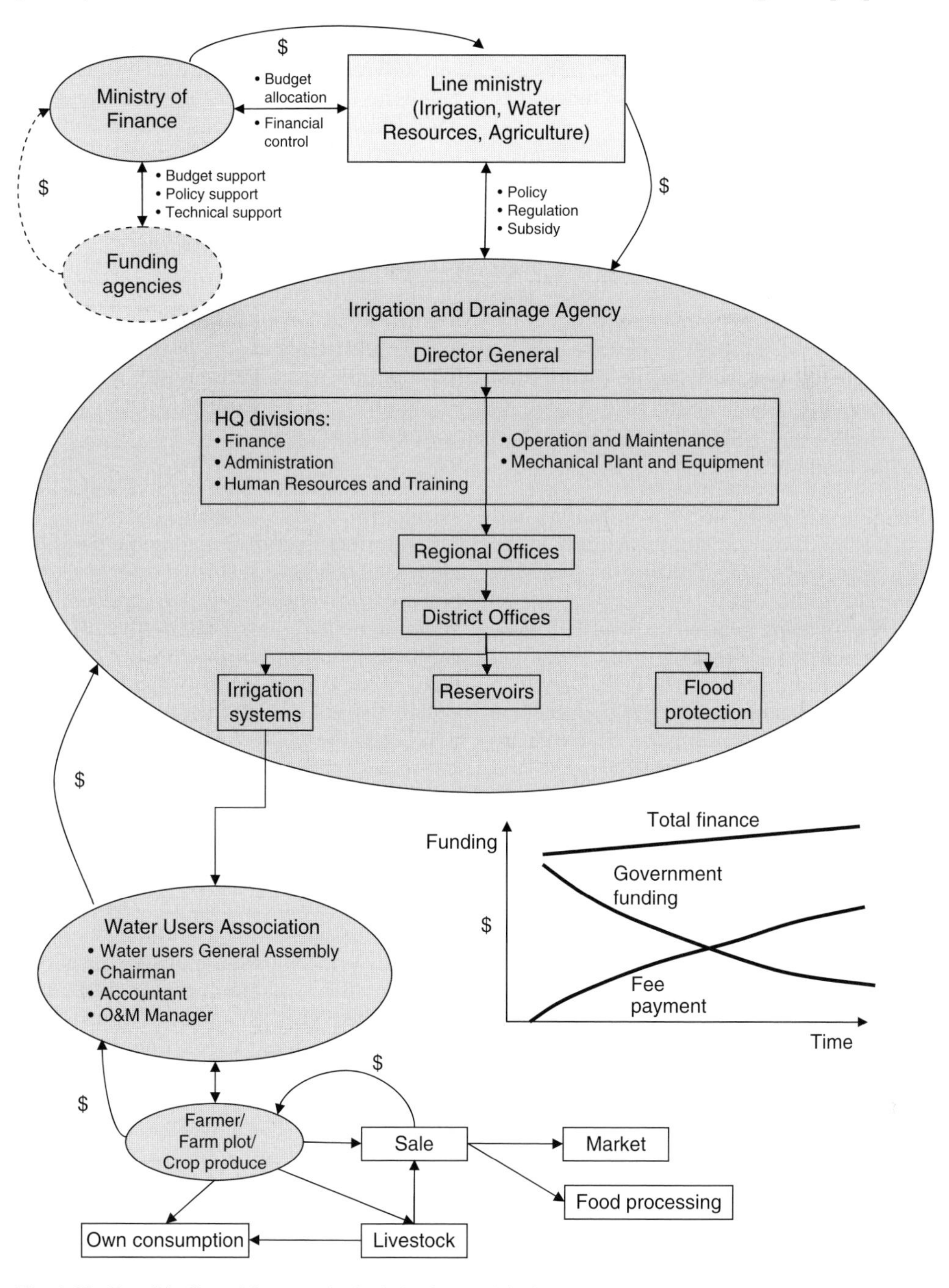

Fig. 3.16. Possible flow of finances in the irrigation and drainage sector.

of the total MOM finance required. In many countries this is still work in progress, as farmers have not been used to paying for irrigation and drainage services and a period of readjustment is required. In other countries, such as the USA and Australia, farmers are paying levels of service fees that are adequate to cover the day-to-day costs and to put aside money for asset replacement and renewal in the future.

In some countries, however, farmers find it difficult to pay high enough levels of service fees due to the small size of farmers' landholdings and the sometimes poor state of the market for agricultural goods. With landholding sizes of less than 1 ha and subsistence cropping it is sometimes difficult for farmers to find the cash to pay the service fee. In some cases, such as the Philippines, the irrigation agency has allowed farmers to pay in kind with agricultural produce. This approach has not generally been successful as the irrigation agency has to build and staff storage warehouses, and has to market and sell the produce, sometimes at rates lower than they traded the produce in from the water users.

The possible returns to irrigated agriculture and the ability to pay the irrigation service fee (ISF) are shown in Table 3.4, which provides an example of a crop budget for a 1 ha maize crop. Excluding the ISF the financial returns range between US$178/ha and US$432/ha for low to high yields if labour is costed, and between US$284/ha and US$586/ha if family labour is used and not costed. Including an adequate ISF of US$22.5/ha, the net returns fall to US$155/ha to US$409/ha with labour costed and US$262/ha to US$564/ha if labour is not costed. This is still a reasonable return for this crop; the ISF is only 9.5% of the total costs for the low-yield case and only 5% of the total costs for the high-yield case. In the low-yield case the ISF is equivalent to the expenditure on fertilizer; in the high-yield case the ISF is one-fifth of the expenditure on fertilizer. While a full analysis should be carried out on the basis of a farm, rather than a crop, this budget example shows that charging the full ISF is not unreasonable in terms of the returns obtained from supplying irrigation water.

Identifying the management and the operation cost components of the ISF is relatively straightforward at either the main system or tertiary unit/on-farm level. The main difficulty is in identifying and quantifying the maintenance costs as: (i) they are particular to individual systems; (ii) they vary from year to year depending on which infrastructure items need repair/maintenance; and (iii) it is difficult to know what should be the optimum level of maintenance. Under-spending on maintenance will result in deterioration of the physical system, and therefore a higher maintenance cost in the future. Getting the balance right is not easy.

In gravity-fed I&D systems a rule-of-thumb is that the maintenance expenditure should be about 70% of the total MOM expenditure. In organizations where the expenditure on management costs (mostly salaries) is more than that on maintenance there is more often than not a maintenance problem, leading to deterioration of the I&D system (Fig. 3.17).

A further issue with setting the ISF is the method of charging, that is whether it should be based on volume of water supplied, irrigable command area (irrespective of area cropped), cropped area, or crop area and crop type. In some locations the water users association charges on the time taken to irrigate, which has the positive effect of water users completing irrigation of their field as quickly as possible. A problem that has been encountered in several systems is that service fees are not paid where water is not supplied, either because it is not available (due to a drought) or due to adequate rainfall. Thus in very dry or wet years the service provider may not get an adequate income from providing irrigation water, yet they will still incur costs (staffing, maintenance, etc.). In order to cover these costs there is a reasonable argument that all landowners within an irrigation/drainage command area should pay a fixed annual area-based fee, irrespective of whether they irrigate or not. This area-based fee would be set to cover the fixed costs, and additional charges would then made to those who do irrigate for the variable costs of service provision.

The jury is out on whether irrigation management transfer will mean that I&D

Table 3.4. Example of a crop budget for a maize crop in US$/ha.

Item	Unit	Quantities			Unit prices (US$)		Financial costs and returns (US$/ha)			Economic costs and returns (US$/ha)		
		Low yield	Medium yield	High yield	Financial prices	Economic prices	Low yield	Medium yield	High yield	Low yield	Medium yield	High yield
Gross returns												
Grain output	kg	3,150	4,900	6,600	0.1	0.1	354.4	551.3	742.5	354.4	551.3	742.5
Maize stover	kg	4,725	7,350	9,900	0.0	0.0	59.1	91.9	123.8	59.1	91.9	123.8
Total							**413.5**	**643.1**	**866.3**	**413.5**	**643.1**	**866.3**
Costs of crop production												
Ploughing	times/ha	1.0	1.0	1.0	30.0	30.0	30.0	30.0	30.0	30.0	30.0	30.0
Discing/harrowing	times/ha	1.0	1.0	1.0	12.5	12.5	12.5	12.5	12.5	12.5	12.5	12.5
Furrowing	times/ha	1.0	1.0	1.0	12.5	12.5	12.5	12.5	12.5	12.5	12.5	12.5
Inter-row cultivating	times/ha	1.0	2.0	2.0	12.5	12.5	12.5	25.0	25.0	12.5	25.0	25.0
Seed	kg	30.0	30.0	30.0	0.3	0.3	7.5	7.5	7.5	7.5	7.5	7.5
Farmyard manure application	US$/ha	200.0	400.0	600.0			5.0	10.0	15.0	5.0	10.0	15.0
Ammonium nitrate fertilizer	kg	100.0	300.0	500.0	0.2	0.2	20.0	60.0	100.0	20.0	60.0	100.0
Fertilizer application	times/ha	1.0	1.0	1.0	10.0	10.0	10.0	10.0	10.0	10.0	10.0	10.0
Herbicide	times/ha	0.0	0.0	1.0	5.0	5.0	0.0	0.0	5.0	0.0	0.0	5.0
Herbicide application	times/ha	0.0	0.0	1.0	11.3	11.3	0.0	0.0	11.3	0.0	0.0	11.3
Pesticide	US$/ha	–	300.0	600.0				7.5	15.0		7.5	15.0
Labour (family & hired)												
Harvesting & shelling	labour-day	31.0	45.0	58.0	1.5	0.9	46.5	67.5	87.0	27.9	40.5	52.2
Other labour inputs	labour-day	40.0	45.0	45.0	1.5	0.9	60.0	67.5	67.5	36.0	40.5	40.5

Continued

Table 3.4. Continued

Item	Unit	Quantities			Unit prices (US$)		Financial costs and returns (US$/ha)			Economic costs and returns (US$/ha)		
		Low yield	Medium yield	High yield	Financial prices	Economic prices	Low yield	Medium yield	High yield	Low yield	Medium yield	High yield
Transport	US$/ha	300.0	450.0	600.0	0.0	0.0	7.5	11.3	15.0	7.5	11.3	15.0
Miscellaneous costs (5% of the costs above)							11.2	16.1	20.7	9.1	13.4	17.6
Total							**235.2**	**337.3**	**433.9**	**190.5**	**280.6**	**369.0**
Financial net returns before ISFs					With all labour costed		**178.3**	**305.8**	**432.4**	–	–	–
					With labour not costed[a]		**284.8**	**440.8**	**586.9**	–	–	–
Economic net returns excluding irrigation supply and distribution costs							–	–	–	**222.98**	**362.53**	**497.23**
Financial net returns after ISFs					With all labour costed		**155.8**	**283.3**	**409.9**			
					With labour not costed[a]		**262.3**	**418.3**	**564.4**			

ISF, irrigation service fee.
[a]Assuming that all labour is provided by unpaid household members, and no hired labour is used.

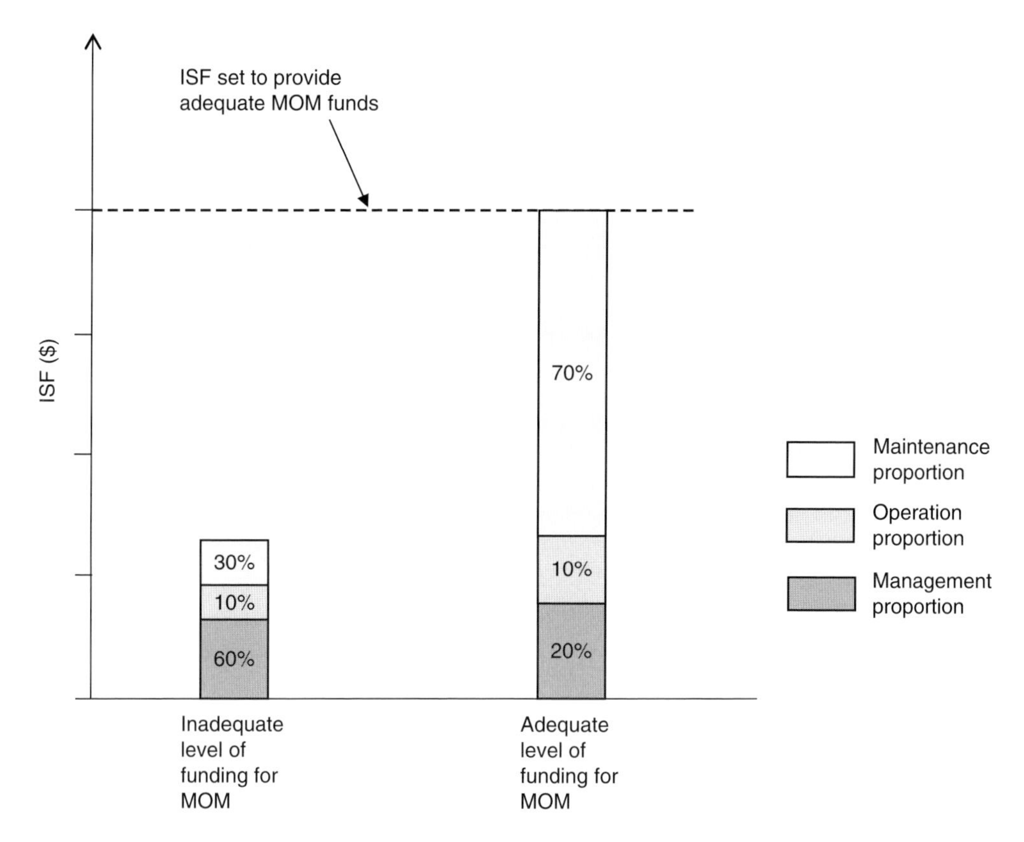

Fig. 3.17. Setting the irrigation service fee (ISF) to provide adequate funds for maintenance. MOM, management, operation and maintenance.

systems are adequately financed and the cycle of deterioration followed by rehabilitation halted. In some countries it may be that government will need to continue to support and subsidize the management, operation and maintenance of I&D systems, and recover the costs from other sources of taxation.

Staffing and Human Resource Development

An organization's human resource is an important asset. This is particularly the case with irrigation management where specialist skills are required for the central functions of planning, regulating, measuring and recording water distribution, and identifying, costing, implementing and recording maintenance activities.

In most I&D agencies there are norms governing the number and category of staff at different levels within the organization, together with job descriptions for each position. It is important that these staffing levels, and associated job functions, are periodically reviewed and updated. This will particularly relate to countries where standards of pay are rising and staffing costs increasing, and where new technology is bringing about changes in the way systems are managed, operated and maintained. Examples of where changes are being made include: (i) irrigation systems that are being modernized with automatic water level control devices, or remote gate operating systems; (ii) where staff are being provided with more efficient transport (motorbikes, instead of bicycles or travel by foot); and (iii) where computers are being introduced for data collection, processing and analysis.

In some countries the attitude towards human resource development (HRD) within the irrigation agency is still quite poor, with a reliance on top-down management and little encouragement, motivation or training for staff. The work is often seen as repetitious and therefore not requiring any significant inputs into staff motivation and training. Sadly, the significant HRD lessons learned over the last 20–30 years in the business and industrial sectors do not appear to be recognized or applied in the irrigation and drainage sector. This will need to change if irrigation agencies are to be made leaner and fitter for purpose, especially in relation to service delivery and customer satisfaction. If water users are expected to pay more for their water delivery and removal services they will expect far better levels of service and accountability than is the case in some situations at present.

Further discussion on training is provided in Chapter 7.

Administration

Efficient administration processes and procedures are the oil in any organization's machinery. Typical administrative responsibilities for an I&D agency include:

- procedures for recording, handling, storing and retrieving correspondence – this will include procedures for referencing all incoming correspondence, together with procedures for directing it to the responsible person for action and a tracking system to ensure that it is acted upon in good time;
- procedures for organizing staff travel and payment of relevant per diems and allowances;
- procedures for office communication, and communication between other offices within the organization;
- provision of support services, including secretarial support, IT support, draughtsmen, drivers and the like;
- provision of meeting and conference facilities, and procedures for organizing and running meetings and conferences;
- procedures for procuring office supplies and for office maintenance;
- printing and reproduction facilities;
- procedures for procuring equipment, spare parts, materials, supplies and support services;
- procedures for storing and inventory control of equipment, supplies and materials.

For water users associations the procedures are much simpler, and relate mainly to ensuring that the association's books are kept safe, the office is adequately maintained and meetings properly organized, with adequate notice being given, minutes kept and information disseminated to water users.

Legal Issues

There are a number of areas where legal issues occur, for an irrigation and drainage agency, for a water users association or for individual water users. These include:

- drafting of new, or redrafting of existing, legislation to establish water users associations and transfer management, operation and maintenance to water users;
- drafting of new, or redrafting of existing, legislation on the water law – this may include establishment of water rights for individuals and groups of water users, establishment of river basin councils, establishment of new agencies for water resources management;
- drafting of service agreements between service providers and water users;
- enforcement of service agreements in the civil courts, either by water users in relation to lack of service delivery, or by service providers in relation to failure by water users to pay service fees;
- enforcement of penalties for unauthorized abstraction or use of irrigation water, or damage to irrigation and drainage infrastructure, applied for either by main system service providers on water users associations or individual users, or

by water users associations on individual water users;
- action to obtain usufruct rights or full legal title to physical irrigation and drainage infrastructure;
- advice and lobbying to protect water users associations from some elements of taxation, including property taxes for physical infrastructure and VAT on membership and service fees.

In a large I&D agency there may be a small legal team, or a legal specialist, who will be engaged to advise on legal matters. For water users associations, legal advice is often provided as part of a WUA establishment project. In Kyrgyzstan the recently formed National Union of Water Users Associations has engaged a legal specialist to advise individual associations on procedures for taking over responsibility for the management of main system canals from the I&D agency.

Public Relations

Good public relations (often shortened to 'PR') are a useful management tool for any organization. For water users associations, good public relations are particularly useful in the early days of forming and establishing the WUA, and are often a key component of any WUA-related project. Promotion of the WUA concept on television and radio, and through newspapers, improves the understanding of WUAs and helps in gaining support for these new management entities. Promotion to gain acceptance and support of the WUA concept by politicians has to be a major public relations exercise of any WUA project, and should continue to be part of WUA activities after the project has finished. When established, WUAs need good public relations in order to ensure their access and rights to water, and to ensure that they are taken seriously as a voice for the irrigation community. Prior to the formation of WUAs the government line agency responsible for irrigation and drainage will have protected the water rights for irrigation water users. With management transfer these line agencies are less closely involved, and WUAs need to be aware that they must now protect their own interests.

Good public relations are essential for WUA management. They must communicate, liaise and work closely with water users if they are to retain the support of the water users. Good communication and liaison is the glue that binds these associations; if it is weak or non-existent then the association will fail.

For the I&D agency good public relations with water users associations and water users makes life easier and irrigated agriculture more productive. Irrigation and drainage service delivery differs from domestic water supply and provision of electricity in several important ways. First, irrigation is an open-access resource, which is very difficult to police and protect full-time (especially at night). Second, irrigation and drainage is often fundamental to people's livelihoods, it is not an option as may be the case with electricity. Third, whereas domestic water supply and electricity can be provided on demand, this is rarely the case with irrigation; close cooperation and communication is required between the user and the supplier if supplies are to be reliable, adequate and timely. Good irrigation and drainage service delivery is about working in partnership with water users, not in conflict with them.

Good public relations are useful for the I&D agency in liaising and working with other government agencies and organizations, such as local and regional governments, and national government. Good public relations can strengthen the position and standing of the I&D agency; similarly, poor public relations can weaken its standing in the community, and within government.

4

Operation of the Main System

This chapter describes the fundamentals of operation at the main system level, and the processes and procedures followed where the design of the system has dictated how the system is operated. The chapter begins with a discussion on different forms of irrigation scheduling, and provides a framework for categorizing different forms of scheduling. This is followed by discussion of different control systems, as it is these systems that dictate how the system can be operated. Discharge measurement is then discussed, followed by discussion of how the flow in canals should be regulated to avoid fluctuations in canal levels and water delivery. Next an outline of the procedures for planning and monitoring of water distribution at the main system level is given, followed by examples of different processes and procedures for operation of the main system.

Overview

Operation of irrigation systems can usefully be divided into three levels: (i) the main system, comprising primary and secondary canals; (ii) the tertiary unit or on-farm system, comprising the tertiary and quaternary canals; and (iii) the field level, comprising the field channels and ditches. The drainage system mirrors these divisions, though there is generally not much operation involved unless there is pumped drainage.

The reason for dividing the operation into these three levels is that the management is different, both in terms of the organizations and people involved and in terms of the processes and procedures. Generally the main system is managed by a government agency, though this is changing as systems are being transferred to management by water users associations (WUAs) or federations of WUAs. The next management unit is the tertiary unit or on-farm level, which is generally managed by water users, either directly if they own or farm all the land at this level, or by groups of water users through WUAs or similar farmer groups. The lowest management level is the field, where the farmer manages the application to the land of the water provided by the other two management levels.

There are a number of approaches used worldwide for operation at the main system level. The operational processes and procedures used depend on decisions made at the design stage and will include consideration of:

- the number, capability and cost of staffing available;
- the finances available for construction of the irrigation and drainage system;
- the anticipated finances available for management, operation and maintenance;

©Martin Burton 2010. *Irrigation Management: Principles and Practices* (Martin Burton)

- the nature and availability of the water resource;
- the level of technology employed at field level and the capability of farmers;
- the benefits and returns to irrigated agriculture.

A further factor is also the 'school' of irrigation engineering with which the designer is familiar, be it based on experience in the USA, Europe, India, Russia, China, Egypt or elsewhere.

Main System Operation Processes

The aim of the main system operation is to match the supply of water at the hand-over point to the water user to the demand at that point at a given time and date. The water demands made by the water users may be determined by using sophisticated techniques including soil moisture probes, or they may be based simply on demands by farmers for water to irrigate specific plots of land, without any detailed calculations of the actual crop water needs.

The three variables governing the supply of irrigation water are the flow rate, the duration of flow and the frequency of supply (interval between deliveries). At the main system level the ease of varying the flow rate, duration and frequency of supply is governed by the type, number and location of control and measurement structures and the skill of the staff responsible for operation of the system. In order to specify the values of these key variables set processes and procedures are required, as will be discussed in the sections below.

As outlined in Chapter 3 the main system operation processes comprise pre-season planning, in-season operation and post-season evaluation (Fig. 4.1). Data are collected prior

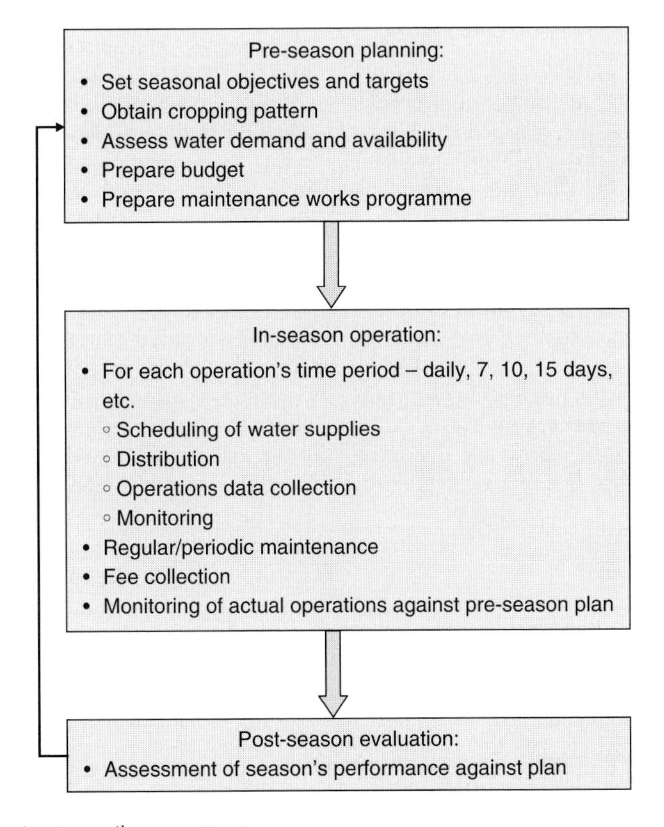

Fig. 4.1. Main system operation processes.

to the irrigation season on planned crop areas, available water supply, budget allocation and maintenance needs.[1] During the season the irrigation water schedules have to be planned, the water distributed and data collected to monitor the implementation of the plan. Regular/periodic maintenance is carried out during the season to maintain the flow regime, and fees are collected, where due. At the end of the season the plan and the implementation are evaluated against data collected during and at the end of the season. The purpose of this evaluation is to identify shortcomings in either the planning or the implementation in order to make improvements for the future.

Planning takes place at two levels:

- before the irrigation season, to obtain information from water users on their planned cropping patterns and irrigation water requirements;
- during the irrigation season, to plan the water allocation and discharges at control points for each irrigation time period.

Pre-season planning often takes place in systems where the main system service provider needs to ascertain if the required discharges can be met from the predicted available water supplies. Such procedures are common in former Soviet Union countries such as Azerbaijan and the Kyrgyz Republic. In these countries water users associations have been formed, and water users submit an application to the WUA detailing their planned cropping pattern. This application is checked by the WUA and a contract agreed between the WUA and the water user to provide irrigation water according to the norms for the specified cropping pattern. The WUA compiles the irrigation requests and submits the cropping pattern and monthly irrigation demands to the main system service provider before the irrigation season commences, and signs a contract with them for provision of this water supply.

In any irrigation system in-season planning will be required, irrespective of whether a seasonal plan has been prepared or not. While the seasonal plan is useful in setting the boundaries and pattern of the flow profile required during the irrigation season, it is not sufficiently detailed to be used without adjustment. Factors that will cause the seasonal plan to be different from the in-season plan may include changes in cropping by some farmers, changes in climatic conditions (hotter/colder, more/less rainfall than planned), changes in water supply availability, etc.

The in-season planning will take place at the start of each irrigation time period (often each 7, 10 or 15 days) and will use information collected from the previous time period, including requests from water users and information on actual discharges supplied (Fig. 4.2).

In a system with arranged-demand scheduling, this in-season planning is essential as the requests made by the water users have to be collated and the required discharges planned and allocated. In a system with a fixed rotational pattern the in-season planning is much simpler, as the supply is fixed and the water users have to adjust their irrigation to suit the supply available. In a system with demand irrigation the planning and in-season operation procedures are simpler still as the system will automatically respond to the irrigation demands by the water users. Examples of the planning, implementation

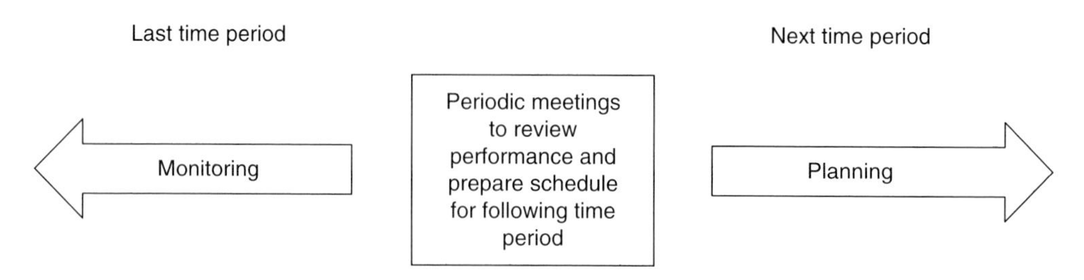

Fig. 4.2. In-season planning meeting(s) to review performance and plan ahead.

and monitoring procedures for these different approaches are given at the end of this chapter.

During the irrigation season irrigation supplies will need to change in order to match the changing crop water demands. In non-automated systems this requires that the main system service provider collects data on the irrigation demands on a regular basis and then prepares a schedule to match supply and demand. In some schemes the irrigation demands are determined by the main system service provider based on the farmers' cropping patterns; in other schemes the irrigation demands are prepared by the water users and given to the main system service provider at intervals during the crop season. The frequency of the changes to the schedule varies from scheme to scheme; in some cases schedules are prepared daily, in others they are prepared each 7, 10 or 15 days. In the Goulburn–Murray scheme in Australia for example, water users submit requests when they need water, with a lead in time of 2–3 days required for the main system service provider to schedule and deliver the requested water. In East Java, Indonesia, cropping pattern data are collected in one 10-day time period based on which a schedule is prepared for the next 10-day time period. In the Goulburn–Murray scheme irrigation water supplies are varying throughout the scheme on a daily basis, in the East Java case the discharges are changed once each 10 days and then held at those values over the 10-day period. There is a significant difference in the amount of management effort that goes into these two different, but similar processes.

It is necessary to monitor and record the water allocation during the irrigation season in order to:

- know what water has been allocated where, and if the planned allocations have been made;
- know what fee to charge the water user or group of water users for irrigation water delivered;
- feed back into the planning process for the next time period;
- monitor and evaluate operational performance.

Monitoring and recording of discharges as part of the fee collection process is one of the main tasks of water masters in many irrigation systems. Monitoring and recording information for operation performance assessment is increasingly important. Table 4.1 outlines the range of operating and monitoring procedures for different types of system, and the relationship between the design of the system and the level of operations management required.

It is important that proper systems are established for recording discharges. These will include standardized forms, printing of stage–discharge charts or discharge tables for measuring structures, and procedures for joint recording of measurements taken between the water user(s) and the service provider's field staff to avoid disputes over the readings and quantities taken. An example of a standardized recording form is provided in Fig. 4.3, while Fig. 4.4 is a schematic map showing the performance of the system in relation to the delivery performance ratio (DPR).

Figure 4.3 is used to record the base data required to schedule irrigation water for the time period under consideration (10 days, 11–20 July), for 15 tertiary units on one secondary canal (B3) with four water users associations. The base data include the area irrigated, the requested discharge and duration. The planned allocation is then prepared based on the available water supply; in this example there is sufficient water and the planned allocation is able to match the requests. When the time period is over, the actual average discharge and flow durations are recorded for each offtake and the DPR (actual/planned) calculated. The planned discharge, actual discharge and DPR can then be entered in the relevant boxes on the schematic diagram of the system (Fig. 4.4), and colour coding used to highlight the pattern of the DPR (blue – excessive supply, green – adequate supply, red – inadequate supply).

Further examples of such recording and processing forms are provided at the end of this chapter on the procedures used in East Java, Indonesia, using the relative area method of main system operation.

Table 4.1. Linkage between type of irrigation system and operational planning and monitoring.

			System components for operation							
System type	Description	Example location	Control structures	Measuring structures	Cropping	Technology level	Staffing level	Operational planning	Operations data collection	Operational monitoring
Proportional distribution (constant-amount, constant-frequency)	Water distributed in proportion to opening – used in hill irrigation systems in Nepal	Hill irrigation, Nepal	Simple ungated proportional division structures	None	Arrange cropping pattern to match supply pattern	Low	Low	None	None	Monitor structures and ensure no blockages. Volume delivered controlled at design stage by proportional size of opening. Equitable distribution of available supplies the primary objective
	Water distribution on main system in proportion to CCA. Water allocation within watercourse allocated on a time-share basis in proportion to the area of each farmer's plot	Warabandi system, northern India and Pakistan	APM at watercourse intake. Simple on/off division boxes in field	Slotted flume on tail of secondary canal (distributary)	Arrange cropping pattern to match average annual water supply pattern	Medium	Low	Medium (to prepare seasonal Warabandi schedule)	Limited (plot and watercourse command areas)	The design requires that the secondary canal flows at design discharge (FSL) in order to maintain command over the APM. Canal water levels monitored at the head of the secondary canal. Frequency and duration of supply to each farmer monitored within the tertiary unit. Rate not monitored

Relative crop area method (restricted-arranged)	Water allocated based on factoring the crop area in relation to the crop's water requirement relative to the base crop. Used in Indonesia, referred to as the Pasten or relative area method	East Java, Indonesia	Gated control structures	Required	Varied	High	High, but rela-tively low skill levels needed for O&M	High	High	Weekly or 10-daily planning of water allocation based on calculated demand. If water short reduce supply equally to all users. Monitor discharges at primary, secondary and tertiary intakes, compare actual water delivered with plan each week/10 days. Equitable distribution of available water the primary objective, followed by secondary objective of delivering adequate supplies (when water available)
Limited-rate, arranged	Water allocated based on calculations of irrigation water demand by farmers using standard calculation procedures such as water balance sheets and climatic data	Goulburn–Murray, Australia	Gated control structures	Required	Varied	High	Medium	High	High	Regular (daily, weekly, 10-daily) updating of irrigation water demand by farmers and planning of water allocation. Distribute water to match demand. Primary objective to match supply with demand

Continued

Table 4.1. Continued

			System components for operation							
System type	Description	Example location	Control structures	Measuring structures	Cropping	Technology level	Staffing level	Operational planning	Operations data collection	Operational monitoring
Demand	Water distributed in response to opening of the outlet gates to farms	Aix-en-Provence, France	Automated control structures	Required	Varied	Very high	Low number, but high skill levels	Low	High, but auto-mated	Continuous monitor-ing of water levels and discharges through automated control systems. Immediate response to irrigation demand. Monitor system to ensure control systems are functioning, and monitor to ensure that total demand can be matched by available supply at water source. If system is computer-control-led, monitor discharges and water levels

CCA, cultivable command area; APM, adjustable proportional module; FSL, full supply level; O&M, operation and maintenance.

FORM 04

WATER REQUEST, ALLOCATION AND ACTUAL SUPPLY SUMMARY

Division: Region 3

Canal name: B3 Branch Canal

Period: From 11 July to 20 July

Note: These last columns are completed at the end of the time period

Water user association	Primary/ secondary canal	Command area (ha)	Design canal capacity (l/s)	REQUEST			PLANNED ALLOCATION			ACTUAL		MONITORING
				Area irrigated (ha)	Discharge (l/s)	Duration (h)	Discharge (l/s)	Duration (h)	Handover discharge (l/s)	Discharge (l/s)	Duration (days or h)	Delivery performance ratio (actual/ planned)
Col. (1)	Col. (2)	Col. (3)	Col. (4)	Col. (5)	Col. (6)	Col. (7)	Col. (8)	Col. (9)	Col. (10)	Col. (11)	Col. (12)	Col. (11) / Col. (8)
	B3	1668	2852	236	1282	24	1282	24	1282	1273	24	0.99
Cane Grove	B3-1	110	132	20	66	24	66	24		64	24	0.97
	B3-2	90	108	18	60	24	60	24		70	24	1.17
	B3-3	80	96	15	50	24	50	24		60	24	1.21
	Sub-total	280	-	53	175	24	175	24	1031	194	24	1.11
Crabwood Creek	B3-4	140	168	17	56	24	56	24		60	24	1.08
	B3-5	167	200	20	66	24	66	24		61	24	0.92
	B3-6	125	150	15	50	24	50	24		62	24	1.25
	B3-7	170	204	20	68	24	68	24		70	24	1.04
	Sub-total	602	-	72	239	24	239	24	689	253	24	1.06
Fellowship	B3-8	102	122	18	60	24	60	24		48	24	0.81
	B3-9	50	60	15	50	24	50	24		53	24	1.07
	B3-10	240	288	29	95	24	95	24		97	24	1.02
	B3-11	65	78	14	46	24	46	24		52	24	1.12
	Sub-total	457	-	76	251	24	251	24	331	250	24	1.00
Golden Grove	B3-12	95	114	18	60	24	60	24		54	24	0.91
	B3-13	54	65	12	40	24	40	24		35	24	0.88
	B3-14	95	114	21	70	24	70	24		55	24	0.79
	B3-15	85	102	19	63	24	63	24		50	24	0.80
	Sub-total	329	-	70	232	24	232	24	0	194	24	0.84
	Total	1668	-	271	897		897			891	24	0.99

Fig. 4.3. Example of a data processing and analysis form for 10-daily water allocations.

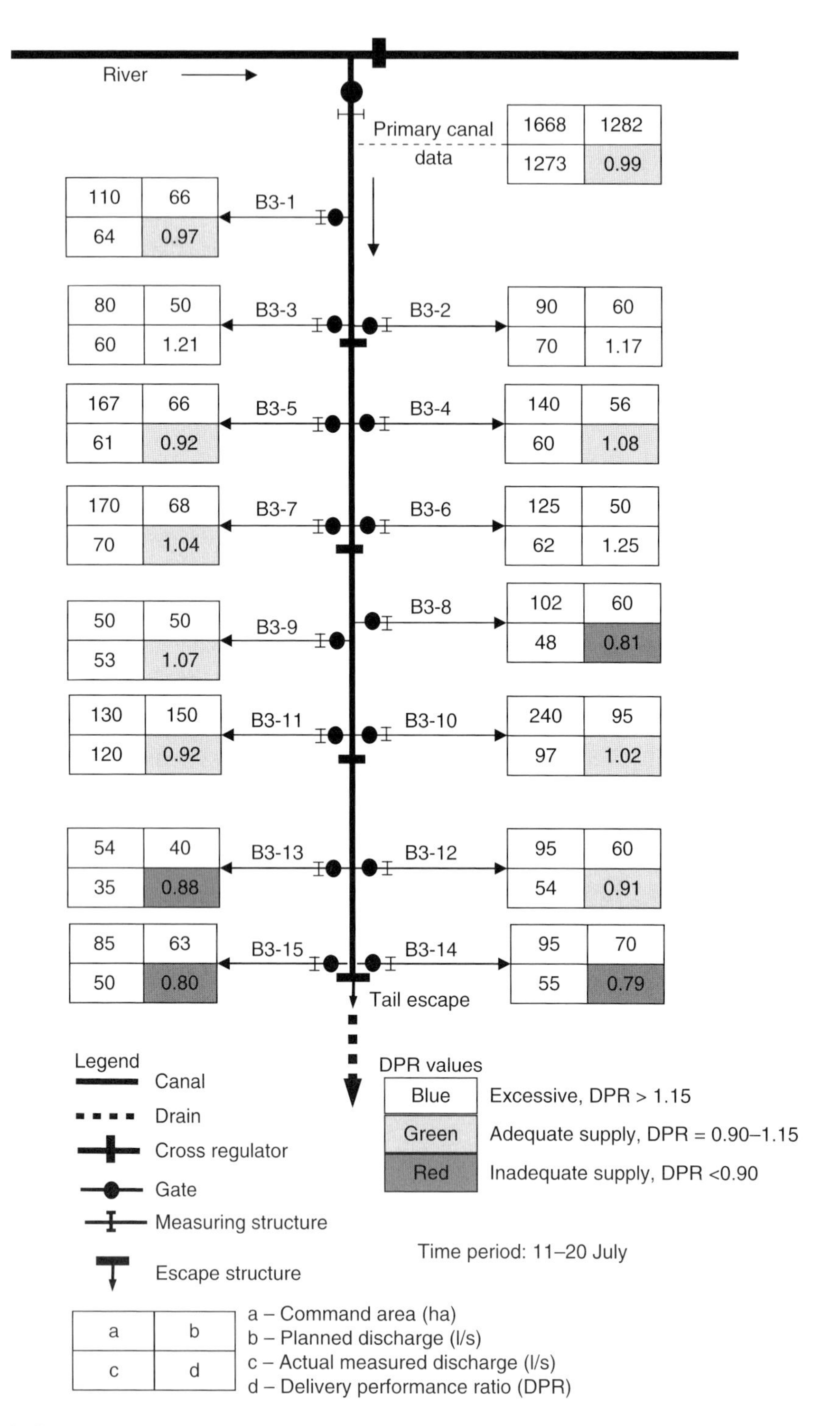

Fig. 4.4. Example of schematic diagram for operational monitoring of the delivery performance ratio (DPR) at each time period.

Irrigation Scheduling

Scheduling of irrigation is the core function of operation of an irrigation system, at any level. The three main variables involved in scheduling of irrigation applications are:

- frequency;
- flow rate;
- duration.

Frequency (or interval) is how often the water is supplied, for example every day, every week, every 2 weeks; *rate* is the quantity of flow; and *duration* is the period (in seconds, minutes, hours, days) for which the water is available. Multiplying the rate and duration gives the volume of water supplied during an irrigation event.

Different combinations of these variables give three commonly used forms of irrigation water supply:

- continuous flow;
- rotational flow;
- on-demand flow.

In *continuous flow*, as the name implies, the flow is continuous, 24 hours per day, 7 days per week. The main variable considered is the flow rate, the other two are already determined. With *rotational flow* irrigation, supplies are rotated between delivery points, with canals running at full or partial discharge, or closed. The frequency and duration of flow become the key variables, together with the rate. With *on-demand flow* the supply can be continuous or intermittent; it is entirely up to the demands made at the point of delivery.

Decisions at the design stage on the form of the rate, frequency and duration of supply at different locations in the irrigation system govern the selection of control and measurement structures, the capacity of canals and the operational procedures to be followed. For example, a decision to rotate irrigation supplies at any location in the system will mean an increase in the capacity of the canals below that location and the provision of a control (and possibly measurement) structure to allow the flow to be regulated.[2] Provision of this control structure will require someone to operate it and management procedures to determine how the structure should be operated.

It should also be noted that storage has an important part to play in relation to the rate, frequency and duration of irrigation water supply in an irrigation system. The storage can be on the main system, within the tertiary unit, on the field or in the root zone, and may be for storage of water overnight or for several days.

Using these three variables all water delivery schedules can be categorized, and can be broadly divided into two types (after Replogle and Merriam, 1980):

1. Rigid schedules;
2. Flexible schedules.

Table 4.2 summarizes the various schedules and their composition in terms of the three variables. The table goes from the most flexible (demand) at the top, to the most rigid (constant-amount, constant-frequency) at the bottom. Examples of these schedules can be identified in different countries and different schemes within countries (see Table 4.1).

Rigid schedules

Rigid, predetermined, supplier-controlled schedules are:

- constant-amount, constant-frequency;
- constant-amount, variable-frequency;
- varied-amount, constant-frequency.

These three schedules are best explained in diagrammatic form (Fig. 4.5). The diagrams show the volume of water delivered in comparison to the crop's irrigation demand. As shown, the more rigid schedule (constant-amount, constant-frequency; Fig. 4.5a) is less able to match the pattern of irrigation water demand, with either over-supply or under-supply at some growth stages depending on the actual volume delivered during each irrigation event. The last case (varied-amount, constant-frequency; Fig. 4.5c) is better able to match irrigation water requirements but requires more management input.

Table 4.2. Types of irrigation schedule.[a] (After Replogle and Merriam, 1980, with permission.)

Schedule name	Frequency	Rate	Duration	Example
Demand	Unlimited	Unlimited	Unlimited	Aix-en-Provence, France. Downstream level control systems designed to supply maximum demand
Limited-rate, demand	Unlimited	Limited	Unlimited	
Arranged	Arranged	Unlimited	Unlimited	
Limited-rate, arranged	Arranged	Limited	Unlimited	Golbourn–Murray, Australia
Restricted-arranged	Arranged	Constant	Constant	Relative area method, East Java, Indonesia
Fixed-duration, restricted-arranged	Arranged	Constant	Fixed by policy	Fairly commonly used in the USA
Varied-amount, constant-frequency (modified-amount rotation)	Fixed	Varied as fixed	Fixed	
Constant-amount, varied-frequency (modified-frequency rotation)	Varied as fixed	Fixed	Fixed	
Constant-amount, constant-frequency	Fixed	Fixed	Fixed	Warabandi system in NE India and Pakistan; hill irrigation systems, Nepal

[a]Terminology: *unlimited*, unlimited and controlled by the user; *limited*, maximum flow rate limited by the physical size of the system turnout capacity but causing only moderate to negligible constraints in farm operations, the applied rate is controlled by the user and may be varied as desired; *arranged*, day or days of water availability are arranged between the service provider and the user; *constant*, the condition of rate or duration remains constant as arranged during the specific irrigation turn; *fixed*, the condition is determined by the service provider.

Flexible schedules

Flexible (on-demand) schedules are user-controlled, though there is often a need for compromise between the irrigation service provider's ability to supply water and the farmer's demand.

There are various flexible schedules:

- demand;
- limited-rate, demand;
- arranged (as to date);
- limited-rate, arranged;
- restricted-arranged (in which both the rate and duration are fixed and remain constant as arranged);
- fixed-duration, restricted-arranged schedule (in which the fixed duration is set by policy, usually 24 hours, and the date and constant rate are arranged).

Demand

With this schedule there are no restrictions on the frequency, rate or duration. Automation of the control systems is essential to implement this schedule, and storage often has an important role to play.

Limited-rate, demand

The flow rate may be restricted by supply capacity, but there is no restriction on the frequency or the duration. Again, automation is essential to implement this schedule.

Arranged

There are no restrictions on the frequency, rate or duration, only that these have to be agreed prior to delivery with the water service provider. This process requires an adequate communication, data collection and data processing system.

Limited-rate, arranged

The flow rate is restricted, otherwise as for the arranged schedule above.

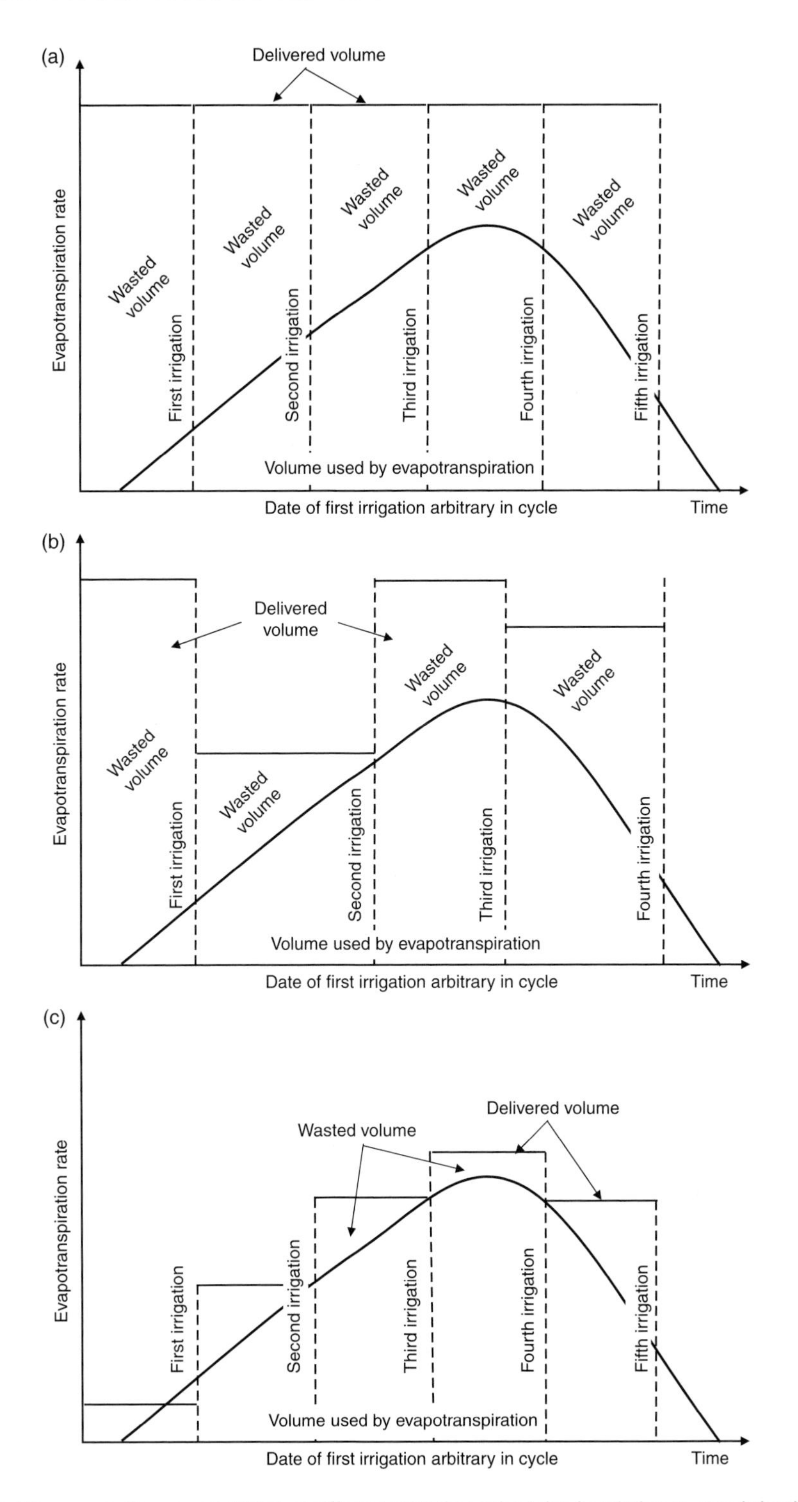

Fig. 4.5. Diagrammatic representation of different irrigation schedules in relation to crop irrigation demand; (a) constant amount, constant frequency (rotation) schedule; (b) constant amount, variable-frequency (modified frequency rotation) schedule; (c) varied amount, constant frequency (varied amount rotation) schedule (after Replogle and Merriam, 1980, with permission).

Restricted-arranged

Further restrictions are made on the arranged schedule. The date, rate and duration have to be discussed and agreed beforehand, once agreed they cannot be changed by either party. This schedule requires the highest level of management by the farmer, who has to plan well ahead (and have the data to do so). The system does not have to be automated, but it is more efficient if it is.

Fixed-duration, restricted-arranged

The duration is fixed by policy (usually 24 hours), the rate and date are arranged. This schedule allows the water masters to plan their work and reduces the number of manual changes in flow rate. This form of schedule does not require automation of the system, and is fairly commonly used in the USA.

Implementation of schedules

The above schedules can be implemented in different ways and are determined by the design of the system, as shown in the example below where the tertiary unit is supplied on a continuous flow basis. Within the tertiary unit irrigation water can be supplied either continuously, day and night (Case 1), or the flow can be rotated on (day) and off (night) with the use of an on-farm storage reservoir (sometimes termed a night-storage reservoir), as in Case 2. This arrangement is quite common, with water being supplied to a (night-storage reservoir on a continuous 24-hour basis, and then withdrawn during the daytime by the users within the tertiary unit.

Case 1
Main system
Continuous flow IN
Continuous flow OUT
Tertiary unit
Continuous flow OUT

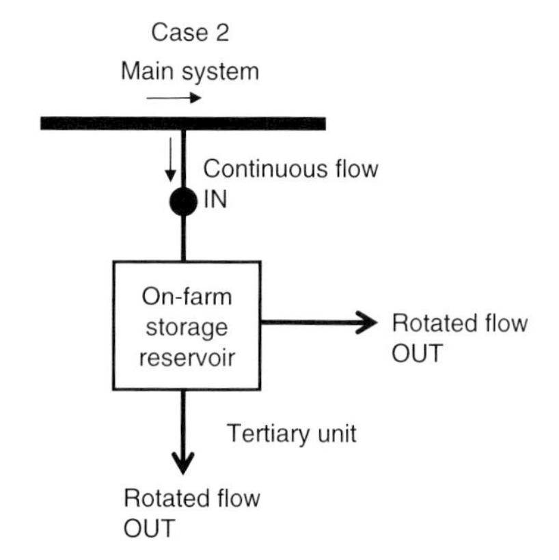

With a similar physical layout, a rotated flow to the tertiary system can provide a rotated flow within the tertiary unit (Case 3) or a continuous flow if an on-farm storage reservoir is provided (Case 4).

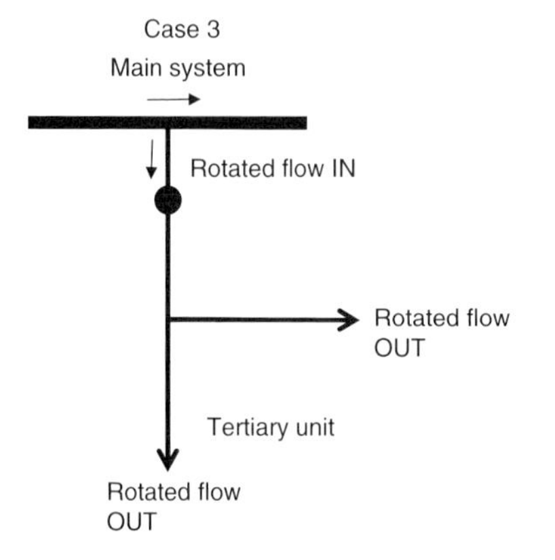

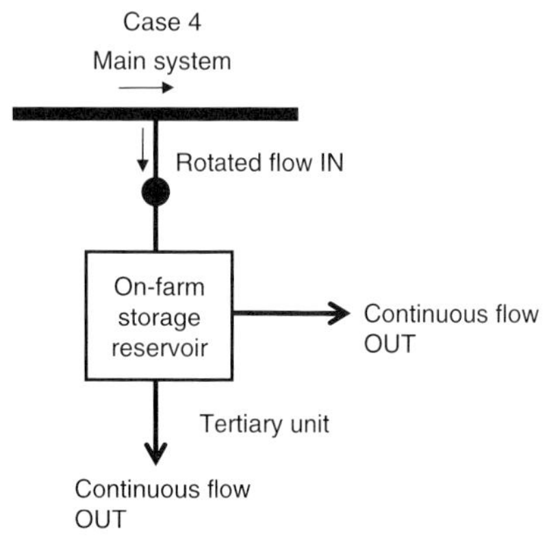

It is important to note that the arrangements shown in this example, where a reservoir is used to change the water supply characteristics, can also occur higher up the system between the primary canal and offtaking secondary canals or at a lower level between the quaternary canal and the field.

The consequences of rotating flow at the main system level are illustrated in Fig. 4.6, where the maximum discharges required in each canal reach are calculated for continuous flow and two different rotation flow options. As can be seen, the continuous flow option gives lower maximum flow levels in the middle and lower reaches of the primary and secondary canals. The option of rotating flows to all tertiary units at the same time on one secondary gives the highest maximum flow rates in the primary and the secondary canals. This illustration clearly demonstrates the importance of considering at the design stage how the system will be operated.

The decision on which type of scheduling system to adopt will depend on the design of the irrigation system (the types of control and measuring structures), the staffing levels and capabilities, and the operation procedures. Typically the demand patterns within the irrigation system will vary depending on the type, area and planting date of the crops in the fields. This variation in demand needs to be matched with the supply available; the accuracy with which the demand is matched will depend on the system design, the staffing levels and their capabilities, and the operational procedures.

Figure 4.7 illustrates this situation. In time period 8 (measured from the start of the irrigation season) the irrigation demand in each of the tertiary units is different, due to the different command areas, cropping patterns, soil types, etc. The main system service provider thus has to determine the demand at each tertiary unit intake during the time period and seek to match this with the flow at the system intake from the river. The control gates on the primary canal need to be operated to pass the required secondary canal discharge. Likewise the tertiary unit gates need to be operated to pass the required tertiary unit discharge. These discharges need to be maintained nearly constant during the time period, requiring regular adjustment by the gate operators. The process requires ascertaining the demands, planning the supply, regulating and measuring the flow, and reporting back. A failure to properly operate the gate, particularly on the primary canal, will result in a shortage of water at some locations, and an excess at others.

Control Systems

Control systems and structures are required to enable the system managers to divert, distribute and measure irrigation water supplies to water users. Poor control of the irrigation water may result in over-supply to some parts of the system and a water deficit at other parts of the system.

Control systems and structures enable the management of the frequency, rate and duration of the water supply. A further variable is the water level; control structures manage the water level in order to maintain sufficient command at key locations in the irrigation network.

The type and distribution of control and measurement structures are determined at the design stage, and will determine the operation and maintenance procedures for the lifetime of the project. Existing control and measurement systems may be upgraded at some point in time, for instance by converting manually operated gates to remotely controlled operation. The sections below outline the range of possible control methods and the benefits and limitations of each in order that suitable choices can be made either at the design stage for new schemes, or when considering upgrading or changing control systems for existing schemes.

An example of where the originally designed control systems were changed is the Ganges–Kobadak Irrigation Scheme in Bangladesh. This is a pumped irrigation scheme with a command area of some 116,000 ha. The original design was based on demand irrigation with downstream level control, but had to be changed to upstream level supply control when it was found that tail-end farmers were not closing the tertiary unit intake gates when they did not require water, with the excess water flowing into the

Data: Continuous flow water duty 1 l/s/ha
Options: 1. Continuous flow to all tertiary units
2. 1 in 3 day rotation to 1 tertiary in each secondary
3. 1 in 3 day rotation to all tertiary units on 1 secondary

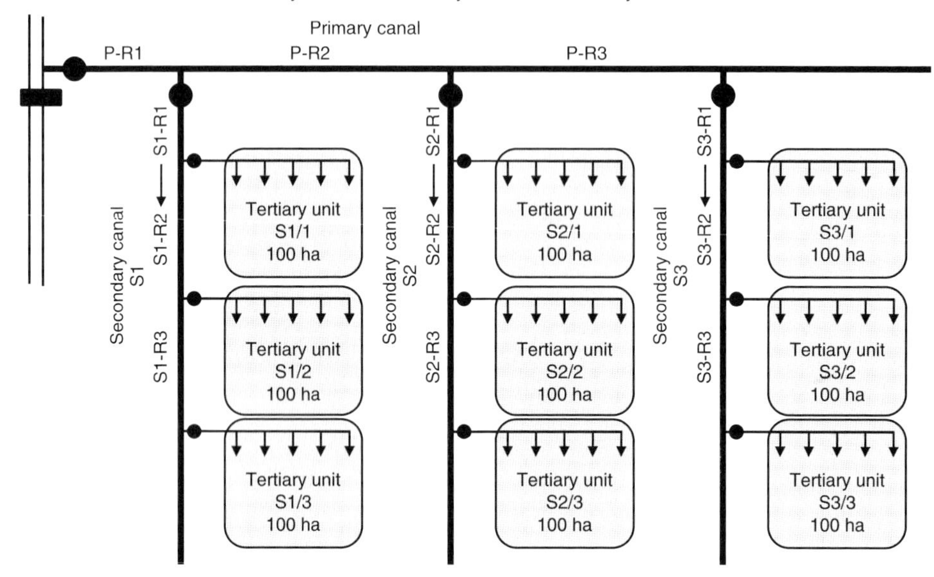

Water allocation and flow rate calculations

Canal/Canal reach	Option 1 – Continuous flow to all tertiary units			Option 2 – 1 day in 3 rotation to 1 tertiary in each secondary			Option 3 – 1 day in 3 rotation to all tertiary units on 1 secondary		
	Day 1	Day 2	Day 3	Day 1	Day 2	Day 3	Day 1	Day 2	Day 3
	Discharge (l/s)								
Primary canal Reach P-R1	900	900	900	900	900	900	900	900	900
Primary canal Reach P-R2	600	600	600	600	600	600	900	900	0
Primary canal Reach P-R3	300	300	300	300	300	300	900	0	0
Secondary canal Reach S1-R1	300	300	300	300	300	300	0	0	900
Tertiary S1/1	100	100	100	0	0	300	0	0	300
Secondary canal Reach S1-R2	200	200	200	300	300	0	0	0	600
Tertiary S1/2	100	100	100	0	300	0	0	0	300
Secondary canal Reach S1-R3	100	100	100	300	0	0	0	0	300
Tertiary S1/3	100	100	100	300	0	0	0	0	300
Secondary canal Reach S2-R1	300	300	300	300	300	300	0	900	0
Tertiary S2/1	100	100	100	0	0	300	0	300	0
Secondary canal Reach S2-R2	200	200	200	300	300	0	0	600	0
Tertiary S2/2	100	100	100	0	300	0	0	300	0
Secondary canal Reach S2-R3	100	100	100	300	0	0	0	300	0
Tertiary S2/3	100	100	100	300	0	0	0	300	0
Secondary canal Reach S3-R1	300	300	300	300	300	300	900	0	0
Tertiary S3/1	100	100	100	0	0	300	300	0	0
Secondary canal Reach S3-R2	200	200	200	300	300	0	600	0	0
Tertiary S3/2	100	100	100	0	300	0	300	0	0
Secondary canal Reach S3-R3	100	100	100	300	0	0	300	0	0
TertiaryS3/3	100	100	100	300	0	0	300	0	0

Fig. 4.6. Variation of canal discharges with rotation of irrigation supplies.

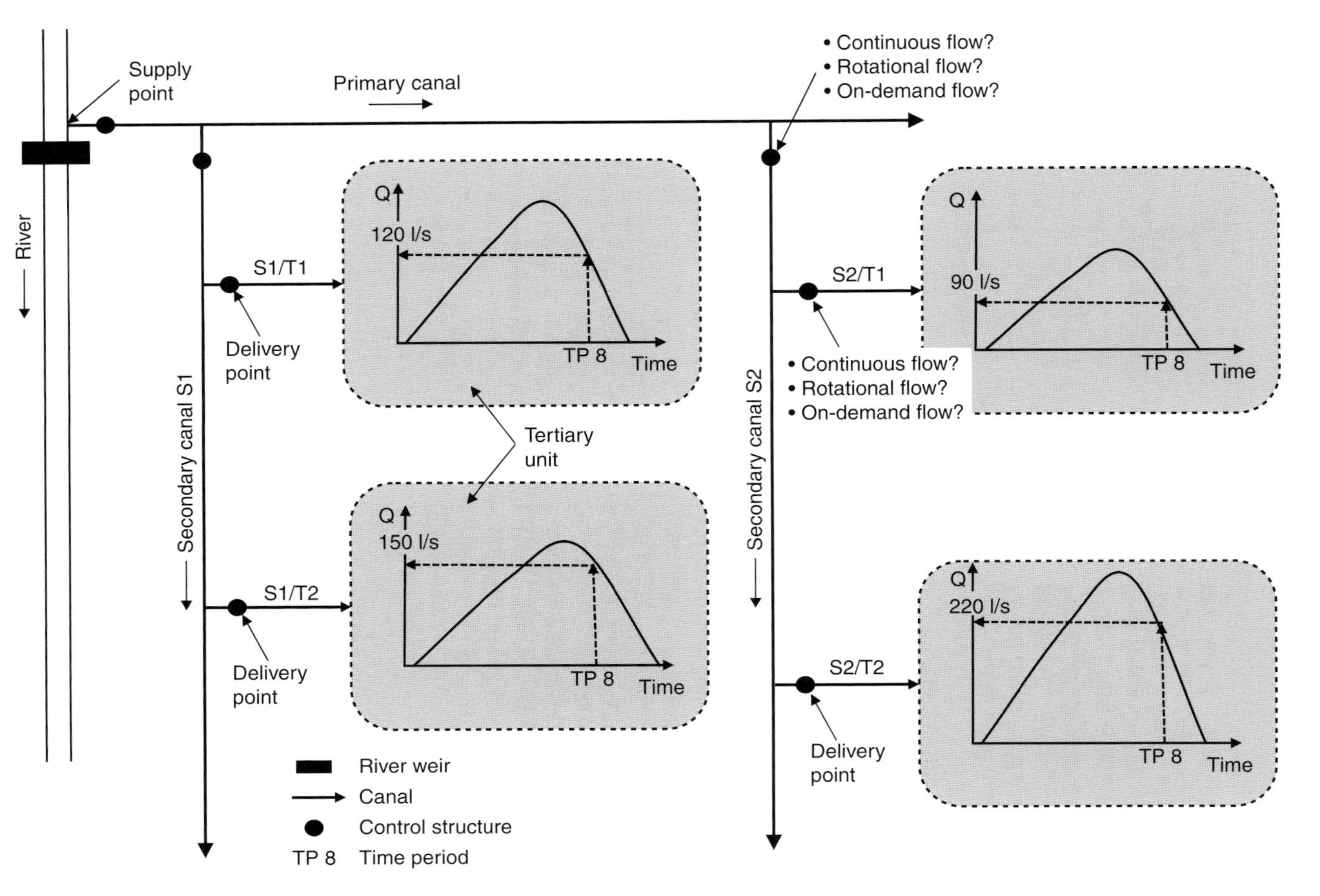

Fig. 4.7. Typical patterns of irrigation demand within an irrigation system.

drainage system. This resulted in excessive pumping of irrigation water in some periods (at a cost of some US$10,000/day). The re-engineering of the system to upstream level supply control was relatively straightforward, but did mean that more management effort was required to schedule and regulate the water supplies.

Control loops

Figure 4.8 shows generalized control loops in an irrigation system for system management and control structures. The management control loop starts with calculating the demand, adjustment to match the supply available, setting of control structures, monitoring the distribution, evaluation of planned with actual distribution and collection of data for the next time period. For the control structures the control loop moves from the present setting to the required setting, either by means of manual adjustment by an operator or automated adjustment based on downstream or upstream water level changes.

Systems, structures and equipment

Control systems are separate from *control structures*. Various types of structure may be used for the same system and different systems may use similar structures. When considering changes to a control system account must be taken of the structures currently used in the canal system and whether these may be improved or need to be replaced. The *equipment* necessary for water control in open channel systems includes the following.

- Hand- or motor-operated gates:
  - overflow gates as head or cross regulators;
  - undershot (orifice type) gates as head cross regulators;
  - power supply, electric controllers, motors.
- Automatic water level control:
  - self-regulating gates;
  - automatic motor-operated gates.
- Flow rate control:
  - proportional – flow dividers, fixed or adjustable;
  - manual;
  - automatic.
- Measuring equipment:
  - gate setting sensors;
  - water level sensors;
  - flow measuring systems (weirs, flumes, etc.);
  - interfaces, for display, processing or transmission.

Canal control systems

Canal control systems vary from those with *upstream control*, which are supply-oriented, to those with *downstream control* that are demand-oriented. The different canal control systems available for the operation of main and secondary irrigation systems are listed below and briefly described in the following sections and Table 4.3.

- Fixed upstream control.
- Gated upstream control:
  - with flow rate control;
  - with water level control;
  - with structures for manual operation;
  - with structures for automatic operation.
- Downstream control with level-top canals.
- Upstream and downstream combined control.
- Centralized control:
  - with non-responsive scheduling;
  - with responsive, arranged delivery.
- Responsive systems for sloping canals:
  - general;
  - with local independent controllers;
  - with dynamic regulation.
- Pressurized system.

The control system does not necessarily define the *water distribution method* (scheduling), although downstream control tends to be flexible and demand-oriented and upstream control is usually associated with more rigid supply-oriented water delivery.

The *level of technology* required for each control system varies. Fixed upstream control

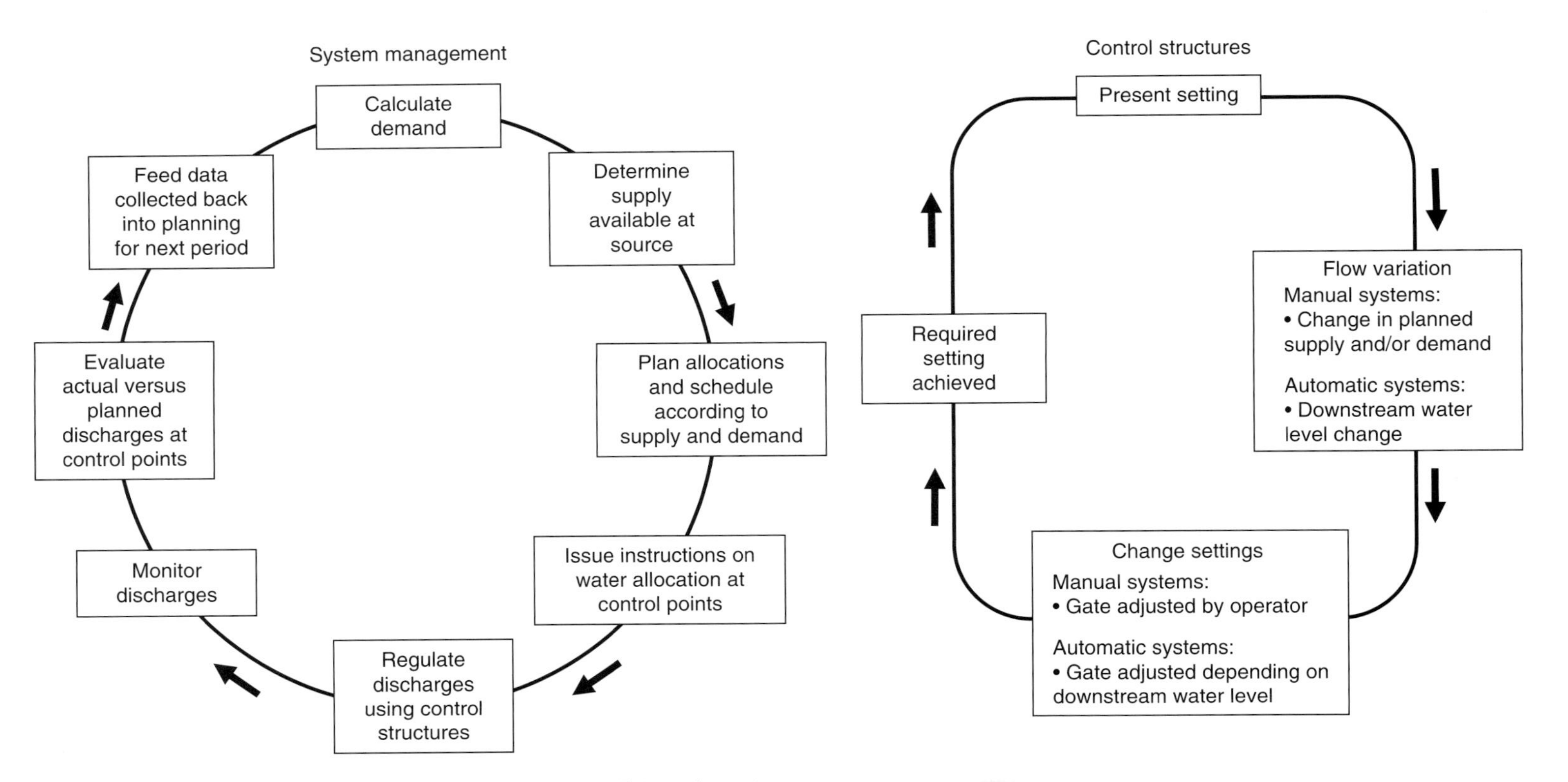

Fig. 4.8. Generalized control loops for system management and control structures.

Table 4.3. Key characteristics of different canal control methods.

Canal control method	Water control	Water delivery[a]	Automation	Control location	Control equipment
Fixed upstream control	Upstream water level	C	–	–	Proportional dividers (weirs)
Manual upstream control	Upstream water level	C, R, A	Manual	Local	Manual or motorized sluices or radial gates
Auto-electric upstream control	Upstream water level	C, R, A	Auto-electrical	Local	Undershot or overshot gates with electrical controllers
Auto-hydraulic upstream control	Upstream water level	C, R, A	Auto-hydraulic	Local	Automatic gates for constant upstream water level
Auto-hydraulic level-top canals	Downstream water level	D	Auto-hydraulic	Local	Level-top canals with automatic gates for constant downstream water level
Auto-electrical level-top canals	Downstream water level	D	Auto-electrical	Local	Level-top canals with electrical controllers
Combined upstream and downstream control	Upstream and downstream water level	A	Automatic	Local	Any combination of the above arrangements for automatic control (usually hydraulic)
Centralized arranged control	Upstream and downstream water level	A	Auto-electrical	Central	Electrically operated gates operated by central computer program
Centralized responsive independent control	Flow or volume in downstream pool	D	Auto-electrical	Central	Sloping canals with locally independent electrical controls and sensors at each gate with microprocessors
Centralized dynamic regulation	Hydraulic simulation	D	Auto-electrical	Central	Almost all systems are electrically controlled by a central computer
Pressurized systems	Flow	A, D	Automatic	Central	Pipelines

[a]C=continuous; R=rotation; A=arranged; D=demand.

(proportional distribution) is technologically very simple in terms of construction, operation and maintenance, whereas responsive centralized control requires sophisticated computer equipment, regular maintenance and skilled operators.

Fixed upstream control

Technical features

With these systems water distribution is controlled by dividing incoming flow into predetermined and generally fixed proportions (usually based on the area served) by means of proportional dividers at each bifurcation point. Control structures are designed to divide flow proportionally whatever the flow rate arriving at the structure. Variations include the Warabandi system of north-west India and Pakistan, and traditional farmer-managed systems found in Nepal, Bali and northern Tanzania. In the Warabandi system flow is proportional down to tertiary level (proportionally fixed by size of outlet, based on command area) and is then rotated between farmers within a block (proportionally fixed by time share based on landholding size). In the farmer-managed systems in Nepal, Bali and northern Tanzania

Fig. 4.9. Proportional flow control on farmer-managed irrigation systems in Nepal.

Fig. 4.10. Proportional flow control on farmer-managed irrigation systems on the slopes of Mount Kilimanjaro, northern Tanzania.

the flow is divided in approximate proportion to the area supplied, using simple proportional division structures (Figs 4.9 and 4.10).

Consequences and impacts

All structures throughout the system (or the part of the system with fixed upstream control) are non-adjustable and therefore operational requirements are minimized. The service provider needs only control the flow into the system and fulfil the maintenance requirements of the system. This makes the system relatively inexpensive to run.

The system is theoretically entirely equitable, although in practice equity is hard to achieve because structures rarely divide flow in the correct proportions over a wide range of flows. Also the range of flow conditions results in varying flow levels, leading to damage to canal sides. Siltation, although limited by this type of control, will cause variations in the behaviour of control structures. Of the two examples shown, above the weir in Fig. 4.9 may work reasonably well if the sediment is removed upstream of the weir to allow an even and slow velocity of approach. The flow splitter in Fig. 4.10 will not work that well as there is a poorly defined upstream pond and little apparent head loss across the structure, which will result in the flow in each channel being influenced by the downstream conditions. Properly designed flow splitters can be very effective; poorly designed ones can be relatively ineffective (though this observation can be applied to all control structures).

Because there is no control in the canal system it is difficult to respond to sudden events (such as a canal breach) along the distribution system. Water cannot be used efficiently in terms of crop production per unit water as the fixed control is inflexible and unable to respond to the varying demands of farmers with differing water needs.

Gated upstream control

Technical features

Water distribution is controlled by adjusting gates within the system to provide the required flow at each offtake. At the inlet to the canal system gates are adjusted to allow the required flow into the system. All cross regulators downstream of the inlet should then be adjusted to maintain a specified water level in the main canal immediately upstream of the structure with offtake gates then adjusted to pass the required discharges. Depending on how the flows are regulated there may be problems with fluctuations at these division points, which can cause variations in the flows entering the offtaking canals. This issue is discussed in more detail later (see 'Regulation of Canals' and 'Flexibility' sections).

Some systems are designed to minimize the adjustment required at each control point.

These include using Neyrpic gates on the offtakes and long weirs in the parent canal, which are designed to minimize the impact on offtake flow of the variation of upstream head over the gate and to make flow adjustments in steps (e.g. fully open or closed gates of 10, 20 and 30 l/s capacity, which can be opened/closed to provide offtake flows in the range of 0–60 l/s or more, depending on the size and configuration of the individual gates).

A range of different gates (sluice gates, radial gates, underflow, overflow, etc.) can be used, which are either manual (possibly motorized) or automated. These are adjusted according to schedules determined either by the irrigation agency or by the agency in conjunction with the water users.

Consequences and impacts

This system may be used for a range of delivery schedules except demand schedules. It is best suited to arranged delivery, as adjustments can be made according to farmers' predetermined requirements and gate settings coordinated throughout the system. However, this requires good communication between the farmers and the irrigation agency. If manually operated, gated control also requires a large number of dedicated staff to operate gates throughout the system.

Although this type of control is relatively cheap to install, the high staffing levels required make it expensive to operate (Fig. 4.11). As labour rates increase, this level of staffing may become financially unsustainable. If there is good communication between the control centre and the water master then this system of control is able to respond quickly to sudden changes in circumstances. As the gates are operated independently, one part of the irrigation system may be shut down without affecting other parts of the system.

Upstream control requires a known flow rate delivered to specific offtakes (see previous discussion associated with Fig. 4.7). However, when extra flow is added to the system it takes hours or days to arrive at the desired location. The supply and demand cannot be exactly matched. At the tail end of the canal any errors in gate adjustment will be magnified, leaving either a deficiency of water or wasting water into the drainage system. Corrections are difficult to make accurately.

For most structures used in upstream gated control, a large number of small adjustments are necessary in order to achieve the desired water level. This makes automation a desired method of control. Automated gated control requires a higher degree of maintenance than manual gate control. Staff need to be well-trained in the operation of automatic gates and in preventative maintenance of control structures. An unreliable power supply or a poor control programme will also lead to poor operation of the system.

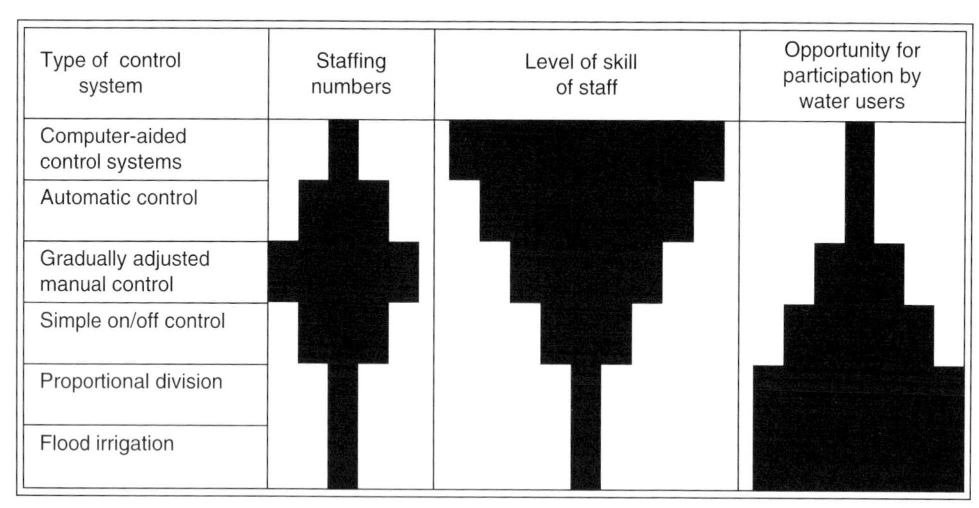

Fig. 4.11. Staff and farmer participation for canal control. (Adapted from Horst, 1990.)

Automated control is more expensive to install but the staffing costs are lower than for manual control. Adjustments to gates can be more precise using automated control rather than manual control.

Downstream control with level-top canals

Technical features

Downstream control is entirely demand-oriented. When a farmer opens an outlet the change in flow rate within the system causes upstream gates to make corresponding adjustments automatically until, eventually, gates at the source respond. Structures on the main canals must have some way of sensing the change, either hydraulically or electronically, in level or flow rate immediately downstream of the structure. Each structure has a set target level, which it automatically maintains.

As demand can vary at any time even with an arranged schedule (with the option for the farmer of turning off his supply when desired) all structures must be automatic. In order to be able to regulate for a flow rate of zero the canal banks must be level although the bottom of the canal may have a standard slope.

Consequences and impacts

Although control is by demand downstream of each control structure, this does not necessarily mean that demand schedules are being used. Canal capacities may not facilitate even a limited-rate demand schedule; however, water supplies may be turned off by the farmer at will, without risking damage to the canal system.

As canals must have level tops, the canal bed slope between structures should be kept to a minimum. On steeper gradients the additional earthworks necessary to maintain a level top become prohibitively expensive.

Because of the responsive nature of downstream control, exact flow rates and delivery times do not have to be calculated. This reduces the need for data collection and processing, and communication systems, thus lowering staffing costs. Level-top control does not require electronic communication systems to coordinate gate opening and closing because all structures are connected hydraulically through the canal system.

Upstream and downstream combined control

Technical features

Combined control uses upstream control for the headworks and along the major canals. A storage reservoir is then required where upstream control converts to downstream control. Below these reservoirs downstream control is exercised by water users taking water either on demand or by arrangement.

Control structures are as described for upstream and downstream control in the sections above. The reservoirs are generally located off the main line of the canal to avoid excessive siltation, though in the case of the Gezira Irrigation Scheme in the Sudan the minor (secondary) canals are over-sized to allow storage of water during the night. Off-stream reservoirs need only be able to store 1–2 days' supply of water provided communication is sufficient to alert the main system managers of fluctuations in water level in the reservoir.

Consequences and impacts

This system of combined control allows the flexibility of downstream control without the cost of providing for maximum capacities in the larger canals. Although there is an additional cost for building storage reservoirs, overall construction costs are lower than for a completely downstream-controlled system.

It is particularly useful in situations where there is an initial steep topography prohibiting the use of level-top canals but where downstream control is desired.

Accurate monitoring and good communication are necessary between water users and the irrigation agency in order that demand can be roughly predicted and upstream gates adjusted to maintain necessary flow, with the storage reservoir either supplying or absorbing the difference between expected and actual demand. Empirical data will assist in the prediction of required supply for reservoir recharge.

There can be more than one change from upstream to downstream control within an irrigation system provided that each change from upstream to downstream control includes a storage reservoir. This is not necessary for changes from downstream to upstream control as downstream systems respond to demand.

Centralized control

Technical features

All centralized control methods exercise control from a single centre where all data are collated and processed and all gate adjustments are made.

For non-responsive scheduling, operators are instructed, according to a pre-arranged plan, to change gate settings without any input from the water users. This system may be automated but is usually manual.

Normally centralized control is used with monitoring to provide an arranged system based on water users' needs. Gates are electrically operated and adjustments made from the control centre using water level or volume data from monitoring points along the canal and water orders from users. Computer models of the irrigation system may also be used for setting gates.

Consequences and impacts

Centralized control enables the irrigation agency to coordinate the operation of an irrigation system much more rapidly because gates are not independent and therefore gate settings can be predictive, reducing response times through the canal system. Because changes can be made simultaneously throughout the system water users at the tail end of the system do not have to wait for 2 or 3 days for a change in delivery to reach them (unless the changes in flow required are significantly greater than the storage available within the system, in which case the routing of the flow will take longer).

For systems which are not fully automated, data are generally processed using a simulation program and then instructions for gate settings are given to operators who manually adjust the gates. This requires well-trained, dedicated and motivated staff to ensure accurate operation of the system.

Centralized control may use a computerized automated system. This requires robust electronic equipment, reliable power supplies to each gate, and skilled operators and maintenance personnel. Maintenance also needs to be preventative rather than curative as manual override of malfunctioning gates is not always possible.

Responsive systems for sloping canals

Technical features

Responsive systems require centralized monitoring although gates may be either independently controlled or moved together. Measurements are taken every few seconds or minutes and water-use predictions updated. A computer program examines water levels in pools and actual flow rates are compared with a statistical prediction of demand, then gate movements are dictated from the central facility.

Consequences and impacts

The centralized systems described above require arranged delivery schedules whereas responsive systems allow much greater flexibility and are demand-oriented. The risk of failure is high if personnel, maintenance, power supply, initial equipment quality and communications do not perform very well, and so a skilled and efficient operational environment is needed to ensure rapid response to problems.

There is minimal human intervention in the operation of the canal system, which can operate fast and effectively in response to users' needs. It combines the advantages of downstream control with a coordinated centralized system. Canals do not have to be as large or as level as for level-top canals and therefore this control system may be used on steeper topography.

The equipment necessary is complex, sophisticated and expensive, although savings are made in canal design and reservoirs are not needed in the system.

Discharge Measurement

In many irrigation and drainage systems measurement of discharge is an essential component of the operation process.

Discharge measurements need to be made in rivers, canals, drains and pipelines and can be made in a variety of ways using:

- velocity–area methods;
- dilution techniques;
- hydraulic structures;
- slope–hydraulic radius–area method;
- flowmeters.

The most commonly used techniques are the velocity–area method, hydraulic structures and flowmeters.

Velocity–area methods

Velocity–area discharge measurement involves the measurement of the channel cross-sectional area and the average velocity of flow. The cross-sectional area is measured using a tape and level staff or depth gauge, the average flow velocity is determined with a current meter or a float. When measurements have been taken at a given location for a variety of flow conditions a stage–discharge curve can be formulated to enable discharge to be determined from the depth alone (Fig. 4.12). The stage–discharge curve must periodically be checked and if necessary recalibrated. When establishing the stage–discharge station it is important to ensure that the channel flows at normal depth, and that flow is not impeded by downstream obstructions, such as cross regulator structures, culverts or vegetation growth (Fig. 4.13).

Measurement technique – float method

The float method is a simple, yet relatively effective way to determine the discharge of a flow stream. It is not that accurate (possibly to within ±20–30%), but is a considerable improvement on a visual estimation of discharge. It is well suited to smaller channels, less so to larger channels where the variation in velocity across the channel will change significantly.

The procedures for simple float measurement[3] are outlined below.

1. Select a fairly straight, uniform and clear (of weeds) reach of channel 20–30 m in length, away from areas of turbulence (e.g. the measurement point should not be located immediately downstream of a gate or drop structure). The length of the measured section should be 10–20 times the water surface width.
2. Place pegs in the bank to mark the start, middle and end of the section.
3. Select a float. An orange or a small plastic container or bottle weighted with sand or stones is suitable as it floats just below the surface and is not influenced by wind.

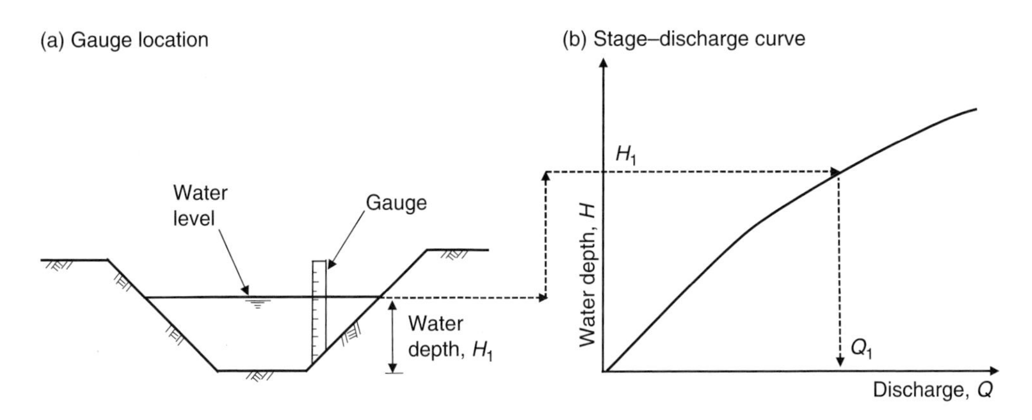

Fig. 4.12. Use of a calibrated gauging site for discharge measurement.

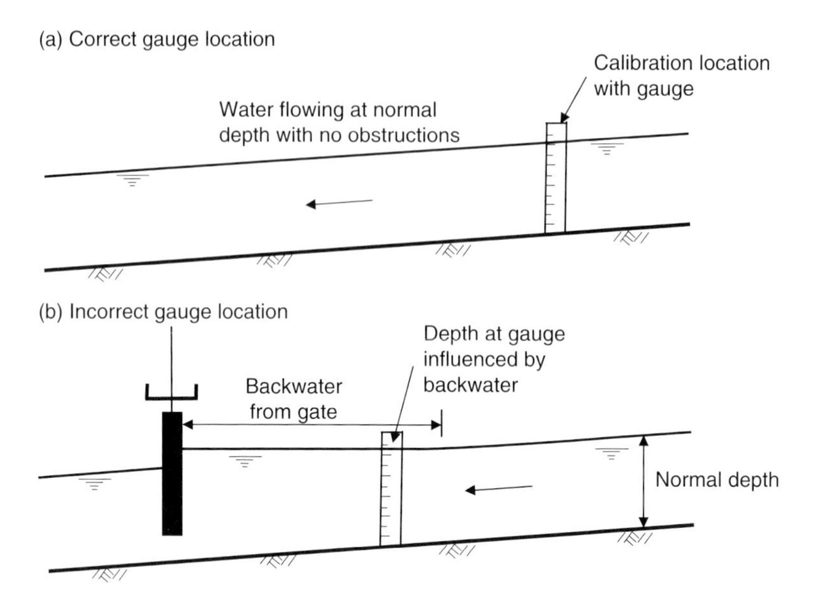

Fig. 4.13. Correct and incorrect positions to establish a stage–discharge curve.

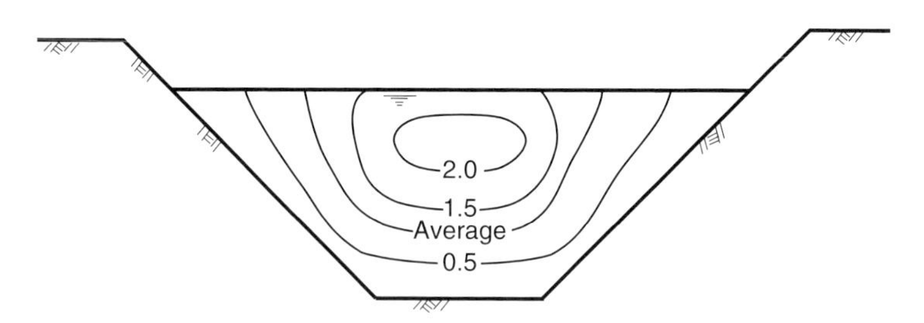

Fig. 4.14. Typical velocity distribution within a channel.

4. Release the float in the centre of the channel about 2–3 m upstream of the start peg to allow the float time to adjust to the flow rate. Measure the time it takes the float to travel over the measured section, and repeat the exercise at least three times to obtain an average surface flow velocity in metres per second.
5. Measure the cross-sectional areas at the start, middle and end of the section using a level staff or graduated rod and tape. Calculate the average cross-sectional area of the measured section.
6. The float measures the surface velocity, which is higher than the mean velocity of flow (Fig. 4.14). The mean velocity is given by multiplying the surface (float) velocity by a reduction factor; a value of 0.7 is typically used. The discharge is then obtained by multiplying this mean velocity by the average cross-sectional area (Fig. 4.15).

Though the theory is straightforward it is not always that easy in practice to implement. The float may tend to drift towards either bank or get snagged on vegetation in the canal. Several measurements will be needed to get a valid average, and care taken to discard measurements where the float movement is impeded.

Current metering

Current metering when carried out correctly can be an accurate method of determining discharge. Using the two-point method, measurements can be accurate to within ±5% of the true discharge; using the single-point method measurements can be within ±10% of

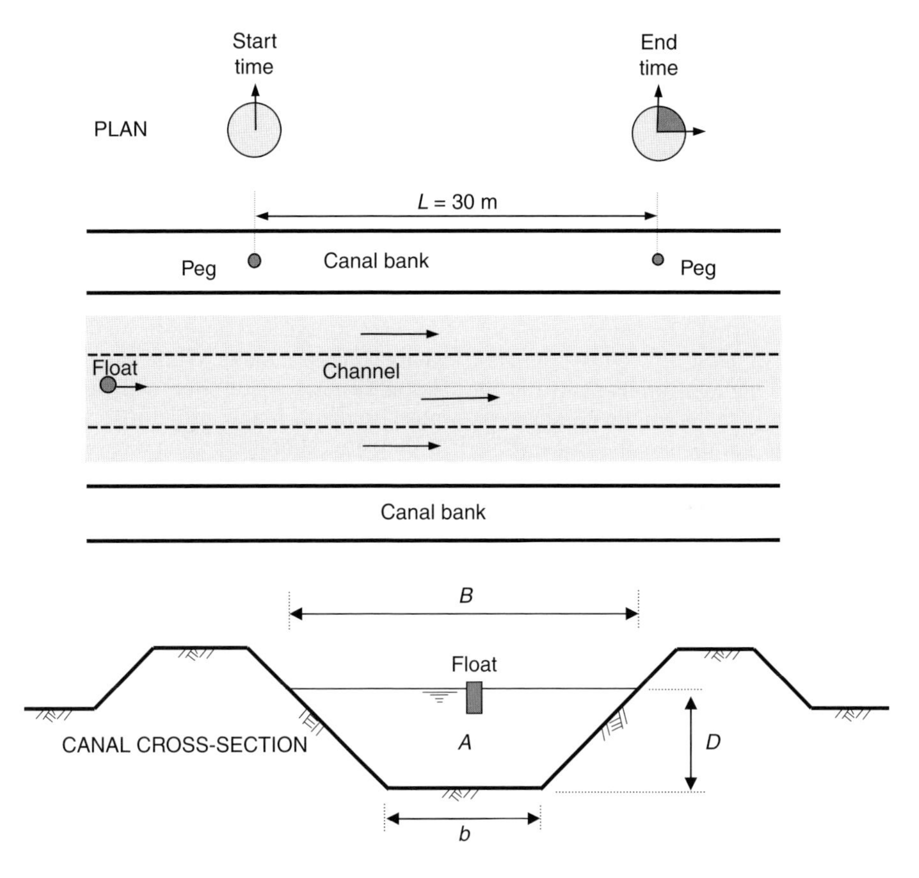

Description	Datum	Formula	Calculation	Result
Canal bed width, b(m)	0.60	b		
Flow top width, B(m)	2.00	B		
Flow depth, D(m)	0.70	D		
Area, A (m^2)		$D \times [(b+B)/2]$	$0.70 \times [(0.60+2.00)/2]$	0.91
Float time, T(s)	70, 74, 72	$(T_1+T_2+T_3)/3$	$(70+74+72)/3$	$T_{av} = 72$
Distance between pegs, L(m)	30	L		
Float velocity, V (m/s)		L/T_{av}	30/72	0.42
Float coefficient	0.7			
Average flow velocity (m/s)			0.7×0.42	0.29
Discharge (l/s)			$0.29 \times 0.91 \times 1000$	264

Fig. 4.15. Example of calculations for discharge measurement using a float.

the true discharge. The procedures for current metering are outlined below.

1. Select a fairly straight, uniform and clear (of weeds) reach of channel, away from areas of turbulence (such as immediately downstream of a gate or drop structure). The length of the measured section should be 10–20 times the water surface width.

2. Stretch a guide rope or tape across the water surface, perpendicular to the streamflow.

3. Measure the total surface width and divide it up into equally spaced sections such that no section occupies more than 10% of the flow area (Fig. 4.16). If using a rope, place tags on the rope to mark the boundaries or mid-point of each section.

4. For the two-point method measure the flow velocity using a current meter at 0.2 and 0.8 of the stream depth (measured from the surface). For the one-point method, measure the flow velocity at 0.6 of the stream depth (measured from the surface). Take measurements at the (horizontal) centre point of each section, taking at least two measurements at each point. If the measurements differ by more than ±10% take a third. Ensure that the current meter is parallel to the streamflow and is clear of any weeds.

5. Calculate the flow velocities at each point using the current meter calibration tables. For the two-point method calculate the mean velocity by taking the average of the 0.2 depth and 0.8 depth readings.

6. Measure the depth and horizontal position at each vertical division of the sections. Multiply each section's area by its average velocity to obtain the section discharge and summate all to obtain the total discharge.

7. It is a wise precaution to always carry out the calculations before leaving the site and to check the value obtained against a rough estimate made by a simple float measurement.

8. Monitor water levels at the start and end of the flow measurement period by taking a reading of a nearby gauge board, or by placing a peg at the water's edge at the start. Note any changes in level. Significant variation in water level during the flow measurement period will obviously adversely affect the accuracy of the discharge value obtained.

Hydraulic structures

Hydraulic structures are commonly used to measure discharge at control points. If constructed to the standard designs they provide an easy-to-use and accurate method of discharge measurement. While standard measuring structures are often installed for flow measurement, many structures can be used if they are calibrated (using a current meter). Such structures include gates, drop structures and division structures.

There are five main categories of hydraulic structures used for measurement:

- broad-crested weirs;
- short-crested weirs;
- flumes;
- orifices;
- sharp-crested weirs.

It is important to note that the discharge measuring structure does not reduce the flow entering the canal; this is often a cause of concern among farmers who may sometimes damage a measuring structure as they think it is impeding the flow. The structure raises the water level upstream by 5–10 cm,[4] and increases the velocity of flow in the canal section over the weir crest. The discharge is the same as in the canal without the measuring structure.

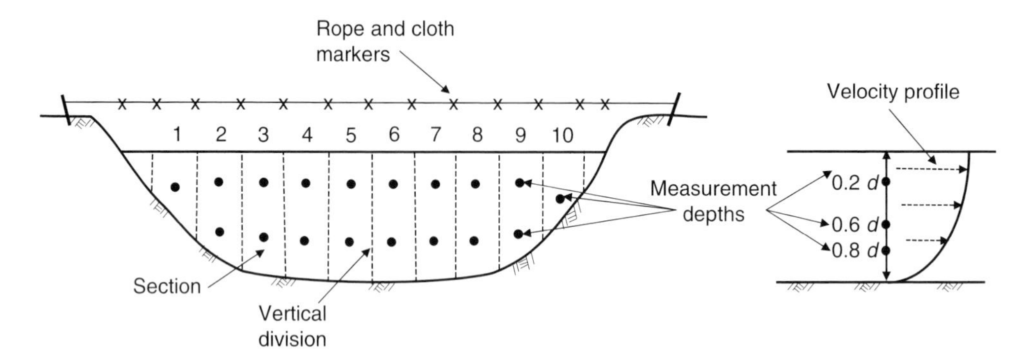

Fig. 4.16. Sectioning of a channel for flow measurement.

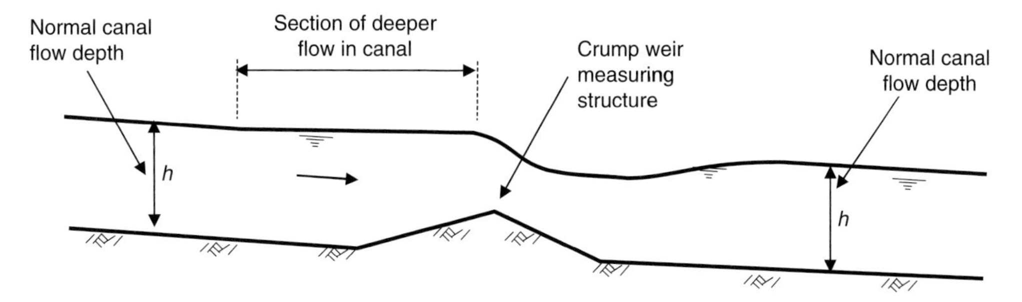

Fig. 4.17. Zone of interference from a discharge measuring structure.

Where the canal is on a slope the measuring structure will create a higher water level for a short distance upstream, after which the flow depth will return to the normal depth for the canal (Fig. 4.17). At some distance from the measuring structure the flow depths upstream and downstream of the measuring structure will be the same, with no interference from the measuring structure.

Theory for hydraulic structures

The theory behind hydraulic structures is complex and lengthy, and is well described elsewhere (Bos, 1989). A brief summary of some of the key points is, however, of value in understanding the practical functioning of such structures.

For practical purposes with discharge measuring structures we are interested in relating a single measurement of water depth to the discharge flowing over the structure. For some measuring structures the relationship between depth and discharge can be derived mathematically; for others it must be determined empirically through measurements in a laboratory where standard depth–discharge tables can be derived.

For broad-crested weirs, flumes, orifices and short-crested weirs the head–discharge relationship can be derived mathematically; for short-crested weirs hydraulic model tests are required. Analysis of broad-crested weirs and flumes is similar, while sharp-crested weirs can be considered to behave as orifices with a free water surface.

For broad-crested weirs and flumes flow is contracted such that the flow passes from subcritical through critical depth and back to subcritical. In the weir the base of the channel is constricted; in the flume the sides and possibly the channel base are constricted (Fig. 4.18). The key features are that the approach velocity approximates to zero and

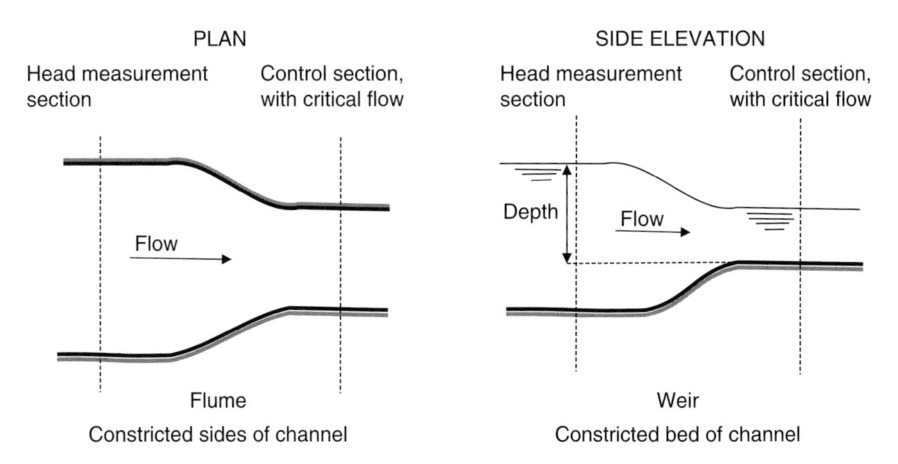

Fig. 4.18. Constriction of channel section to form a control section for discharge measurement.

the control section (the weir crest/flume throat) is sufficiently long to enable the critical depth to be achieved.

From an analysis of specific energy[5] in an open channel it is apparent that the specific energy for a given discharge is a function of water depth. A plot of water depth against specific energy gives a curve as shown in Fig. 4.19. This shows that for a given specific energy level, except the minimum, there are two alternate depths of flow. These correspond to 'subcritical' and 'supercritical' flow conditions. In the subcritical state the flow is slow and deep, in the supercritical state it is fast and shallow. It can be seen from Fig. 4.19 that at the minimum specific energy level there is only one value of depth, referred to as the *critical depth*. By the nature of the characteristics of critical flow a change in downstream water level cannot influence the upstream water level if critical flow conditions exist between the two sections considered. In a measuring structure the channel cross-section is constricted such that the specific energy level is reduced from subcritical through the minimum to supercritical. The transition from supercritical back to subcritical occurs downstream of the control section in the form of a hydraulic jump.

Thus for a *broad-crested weir* with a rectangular cross-section, from the relationship between the velocity at the critical depth, Bernoulli's equation and the continuity equation, the general head–discharge equation can be determined as:

$$Q = C_d C_v \frac{2}{3} \left(\frac{2g}{3}\right)^{0.5} b h^{1.5}$$

where

Q=flow rate
C_d=discharge coefficient
C_v=velocity head coefficient
b=breadth (width) of the weir
h=upstream head over the weir crest
g=acceleration due to gravity.

The discharge coefficient C_d depends on the shape and type of the measuring structure while C_v is a correction coefficient used to compensate for neglecting the velocity head in the approach channel. The value of C_v is dependent on the approach velocity, which in turn depends on the upstream channel cross-sectional area. The dimensions of the upstream approach section are decided and fixed at the design stage, and, for accurate measurement, must be maintained throughout the working life of the structure.[6] Under these conditions the above equation for a round-nosed broad-crested weir reduces to:

$$Q = 1.71 b h^{1.5}$$

For *short-crested weirs* the streamlines are not parallel, thus the mathematical derivation of the head–discharge relationship is more complex and cannot be resolved by current

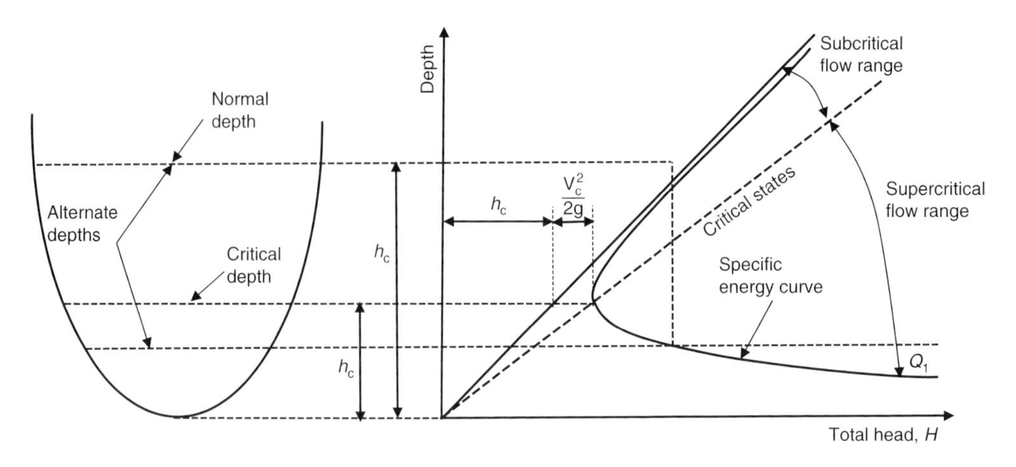

Fig. 4.19. Specific energy–depth relationships.

theory. In this case experimental data can be made to fit the head–discharge relationship for broad-crested weirs, with the discharge coefficient expressing the influence of streamline curvature in addition to the factors it accounts for with broad-crested weirs.

For *orifices* the velocity of flow through the orifice v is directly related to the head thus:

$$v = (2\mathbf{g}H_1)^{0.5}$$

where H_1=total head (static head plus velocity head).

Introducing C_v and C_d to correct for assumptions regarding the velocity, head and the location (relative to the channel sides) and condition of the orifice, the above equation becomes:

$$v = C_v C_d (2\mathbf{g}h_1)^{0.5}$$

where h_1=static head measured on the gauging post.

Allowing for the difference in size between the vena contracta and the orifice, the discharge through the orifice can be expressed as:

$$Q = C_v C_d A(2\mathbf{g}h_1)^{0.5}$$

or

$$Q = C_e A(2\mathbf{g}h_1)^{0.5}$$

where

A=cross-sectional area

C_e=effective discharge coefficient.

For the derivation of head–discharge relationships *sharp-crested weirs* can be likened to an orifice with a free water surface (Bos, 1989), and, with a number of assumptions and a rectangular control section, the discharge equation reduces to:

$$Q = C_e \frac{2}{3}(2\mathbf{g})^{0.5} bh_1^{1.5}$$

where the effective discharge coefficient C_e corrects for the assumptions made. For a Cipoletti weir the formula reduces further to:

$$Q = 1.86bh_1^{1.5}$$

For both the broad-crested and the sharp-crested weir the head–discharge relationships can be presented in standard tables (Fig. 4.20) for use by field staff.

The above relationships and equations apply provided that the control section of the measuring structure is not submerged or drowned out. For sharp-crested weirs the nappe must be aerated, for flumes and broad-crested weirs the hydraulic jump must be able to form. The upper limit of the submergence ratio (downstream flow depth to upstream flow depth) at which the hydraulic jump forms is known as the *modular limit*, and is often taken as 0.75. Some measuring structures, such as the Crump weir, have a high modular limit (0.80) and can accurately measure flow in the non-modular range if additional measurements are taken of downstream water levels. Such a facility is useful if the water level in the downstream section rises due to siltation or vegetation growth, and allows reasonably accurate measurement to take place until the silt or vegetation is removed and the weir can return to the designed (normal) operating mode well below the modular limit.

Design, siting and construction of measuring structures

It is important to note that there are detailed procedures to be followed in the design, siting and construction of measuring structures. Reference should be made to the relevant design manuals or reference works (e.g. Bos, 1989; Skogerboe and Merkley, 1996) for these details. Common problems with the design, siting, construction and use of measuring structures include the following.

- Setting the crest or invert of the structure either too high or too low relative to the downstream flow depth. If set too high, the discharge may be limited if the level in the parent canal is restricted; if set too low, then the measuring structure will be drowned out. If there is any danger of the measuring structure being drowned out

	Discharge (l/s)														
h	Cipoletti weir $Q = 1.86\, b\, h^{1.5}$							Broad-crested weir $Q = 1.71\, b\, h^{1.5}$							*h*
(cm)	0.50	0.60	0.80	1.00	1.25	1.50	2.00	0.50	0.60	0.80	1.00	1.25	1.50	2.00	(cm)
5	10	12	17	21	26	31	42	10	11	15	19	24	29	38	5
6	14	16	22	27	34	41	55	13	15	20	25	31	38	50	6
7	17	21	28	34	43	52	69	16	19	25	32	40	48	63	7
8	21	25	34	42	53	63	84	19	23	31	39	48	58	77	8
9	25	30	40	50	63	75	100	23	28	37	46	58	69	92	9
10	29	35	47	59	74	88	118	27	32	43	54	68	81	108	10
11	34	41	54	68	85	102	136	31	37	50	62	78	94	125	11
12	39	46	62	77	97	116	155	36	43	57	71	89	107	142	12
13	44	52	70	87	109	131	174	40	48	64	80	100	120	160	13
14	49	58	78	97	122	146	195	45	54	72	90	112	134	179	14
15	54	65	86	108	135	162	216	50	60	79	99	124	149	199	15
16	60	71	95	119	149	179	238	55	66	88	109	137	164	219	16
17	65	78	104	130	163	196	261	60	72	96	120	150	180	240	17
18	71	85	114	142	178	213	284	65	78	104	131	163	196	261	18
19	77	92	123	154	193	231	308	71	85	113	142	177	212	283	19
20	83	100	133	166	208	250	333	76	92	122	153	191	229	306	20
21	89	107	143	179	224	268	358	82	99	132	165	206	247	329	21
22	96	115	154	192	240	288	384	88	106	141	176	221	265	353	22
23	103	123	164	205	256	308	410	94	113	151	189	236	283	377	23
24	109	131	175	219	273	328	437	101	121	161	201	251	302	402	24
25	116	140	186	233	291	349	465	107	128	171	214	267	321	428	25
26	123	148	197	247	308	370	493	113	136	181	227	283	340	453	26
27	130	157	209	261	326	391	522	120	144	192	240	300	360	480	27
28	138	165	220	276	344	413	551	127	152	203	253	317	380	507	28
29	145	174	232	290	363	436	581	134	160	214	267	334	401	534	29
30	153	183	245	306	382	458	611	140	169	225	281	351	421	562	30
31	161	193	257	321	401	482	642	148	177	236	295	369	443	590	31
32	168	202	269	337	421	505	673	155	186	248	310	387	464	619	32
33	176	212	282	353	441	529	705	162	194	259	324	405	486	648	33
34	184	221	295	369	461	553	737	170	203	271	339	424	509	678	34
35	193	231	308	385	481	578	770	177	212	283	354	443	531	708	35
36	201	241	321	402	502	603	804	185	222	295	369	462	554	739	36
37	209	251	335	419	523	628	837	192	231	308	385	481	577	770	37
38	218	261	349	436	545	654	871	200	240	320	401	501	601	801	38
39	227	272	362	453	566	680	906	208	250	333	416	521	625	833	39
40	235	282	376	471	588	706	941	216	260	346	433	541	649	865	40
41			391	488	610	732	977			359	449	561	673	898	41
42			405	506	633	759	1013			372	469	582	698	931	42
43			420	524	656	787	1049			386	482	603	723	964	43
44			434	543	679	814	1086			399	499	624	749	998	44
45			449	561	402	842	1123			413	516	645	774	1032	45
46			464	580	725	870	1161			427	533	667	800	1067	46
47			479	599	749	899	1199			441	551	689	826	1102	47
48			495	619	773	928	1237			455	569	711	853	1137	48
49			510	638	797	957	1276			469	587	733	880	1173	49
50			526	658	822	986	1315			484	605	756	907	1209	50
(cm)	0.50	0.60	0.80	1.00	1.25	1.50	2.00	0.50	0.60	0.80	1.00	1.25	1.50	2.00	(cm)

Fig. 4.20. Discharge measurement tables for Cipoletti and broad-crested weirs (Q = discharge (l/s); b = weir crest width (m); h = upstream head over weir (cm)).

the water level in the downstream section must be checked over the full range of anticipated discharges. Care needs to be taken to check for any conditions downstream, which might adversely affect the canal geometry, such as weed growth, siltation, culverts, cross regulator structures, etc.

- Not allowing sufficient width and depth upstream of the measuring structure to slow the approach velocity.
- Siting the measuring structure too close to the head regulator gate. It is obviously useful for the gate operator to have the measuring structure located close to the head regulator gate. However, it

should not be so close that the accuracy of discharge measurement is adversely affected by the turbulent flow through the head regulator.

- Using the wrong type of structure, for example using a weir rather than a flume where there is a heavy sediment load in the canal, or using a sharp-crested weir where there is limited head available.
- Using incorrect dimensions for the structure, for example a weir width being constructed as 1.05 m width rather than the required 1.0 m width and the gate operator using standard discharge measurement tables for a 1.0 m weir. Crest widths of measuring structures should be checked during and following construction or repair to avoid this problem.
- Setting the gauge too close to the measuring structure, and/or not setting the gauge zero at the correct level relative to the measuring structure control section.

Types of measuring structure

Broad-crested weirs

Broad-crested or long-based weirs are structures that induce the streamlines to flow parallel to each through the control section. To achieve this, the length L of the weir must be sufficiently long in relation to the upstream head h_1 (Fig. 4.21).

Broad-crested weirs are more robust than sharp-crested weirs though they are not as accurate. They have a high modular limit and thus do not require such a high head loss across the structure. For example, in a channel with an operating discharge range of 30–120 l/s, a sharp-crested weir would require a minimum head loss of 0.15 m (0.10+0.05) and a maximum head loss of 0.31 m (0.26+0.05 – see Fig. 4.20, Cipoletti weir, crest width 0.50 m, allowing 5 cm aeration under nappe). On the other hand a round-nosed broad-crested weir with the same width would require a minimum head loss of 0.03 m and a maximum head loss of only 0.09 m (0.27/3 – see Figs 4.20 and 4.21). Though broad-crested weirs can be difficult to construct (ensuring parallel faces, a uniform and horizontal crest, and smooth, even upstream curves in the case of round-nosed weirs) they are very functional measuring structures.

An adjustable form of the broad-crested weir is the Romijn weir (Figs 4.22 and 4.23d), which combines a flow regulation and a measurement function. Developed and extensively used on irrigation systems in Indonesia, the Romijn gate is adjusted up or down to pass the required discharge over its crest. A brass gauge attached to the weir measures the head over the weir from which the discharge value is derived. Similar overshot gates have been developed and are in use worldwide though the head–discharge relationships will vary depending on the leading dimensions of the gate. Such moveable weirs are also used as cross regulators, facilitating control of water level and discharge together with discharge measurement.

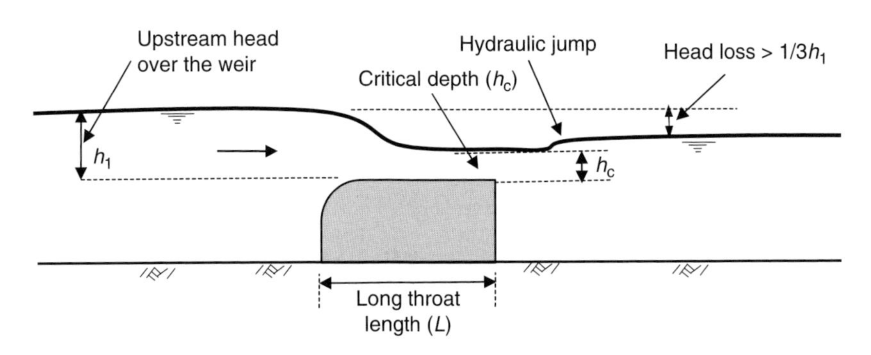

Fig. 4.21. Essential features of broad-crested weirs.

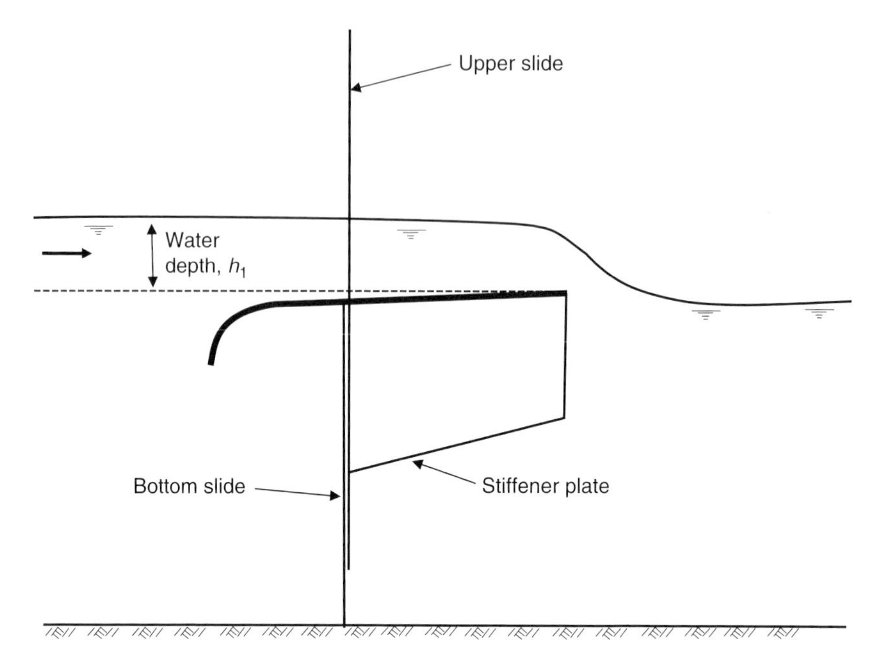

Fig. 4.22. Romijn weir.

Short-crested weirs

With short-crested weirs the streamlines are not parallel over the crest, as is the case with the broad-crested weirs. The streamline curvature has a significant influence on the head–discharge relationship. A typical short-crested weir is the Crump weir (Fig. 4.24). This weir is suitable for many sizes of canals and rivers, is accurate, relatively cheap and easy to construct, and has a high modular limit. With crest tappings to measure the pressure head over the weir crest, discharges can be determined beyond the modular limit. An additional important benefit is that the structure passes sediment freely.

Another popular form of the short-crested weir is the Replogle weir (Fig. 4.23g and h). The weir is similar to a Crump weir in having a sloping front face, but differs in having a short horizontal crest section with either a vertical or sloping back face. Like the Crump weir the Replogle weir is easy to construct and is particularly suited to trapezoidal or parabolic lined channels.

Flumes

Flumes are similar in principle to weirs except that the constriction of flow is obtained primarily by narrowing of the vertical walls of the structure rather than raising the bed level. Flumes can be divided into two categories:

- long-throated;
- short-throated.

A *long-throated flume* (Fig. 4.25) is a geometrically specified construction built in an open channel where sufficient fall is available for critical flow to occur in the throat of the flume. The theory for critical depth flumes is the same as for broad-crested weirs as they both constrict the streamlines to parallel flow in the control section. As a result the design of the structure can be treated analytically.

Short-throated flumes produce a large curvature in the water surface and the flow in the throat is not parallel to the flume invert. Their design cannot be treated analytically and it is not possible to predict the stage–discharge relationship, this has to be done through laboratory and field calibration. Examples of short-throated flumes are the Parshall flume, H-flume and the cut-throat flume (Fig. 4.26).

Flumes are ideal measuring devices where there is a high sediment load or a relatively low head loss is required. They

Fig. 4.23. Forms of flow measurement. (a) Measuring the discharge in a primary canal using a current meter (Indonesia). (b) Measuring the discharge in a canal using floats (Tanzania). (c) Cipoletti weir – note the tranquil flow upstream of the weir and the location of the gauge (Indonesia). (d) Adjusting a Romijn gate – the gauge is located on the metal frame to the right of the gate operator (Indonesia). (e) Neyrpic proportional distribution modules for flow control and measurement (Morocco). (f) Crump weir (UK). (g) Well-functioning Replogle weir on a main canal; note the even flow over the weir crest, the head loss and hydraulic jump on the downstream face (Albania). (h) Construction of a Replogle weir in an existing lined canal (Albania).

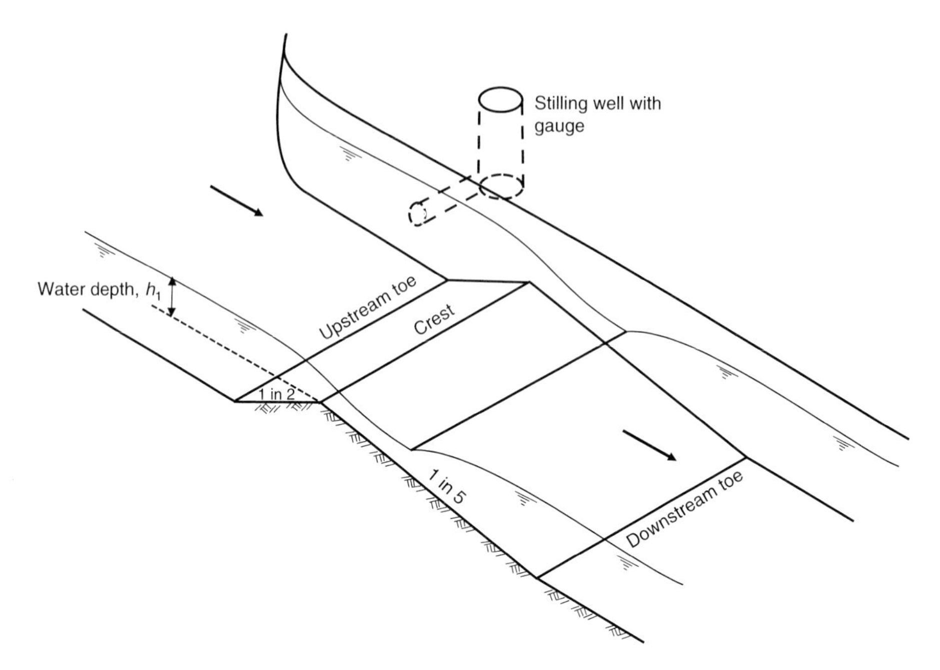

Fig. 4.24. Crump weir.

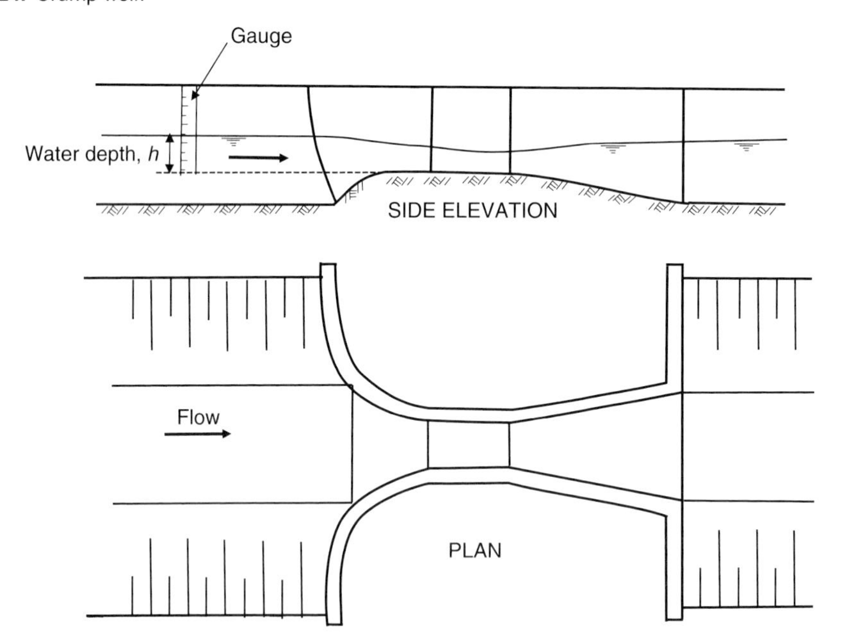

Fig. 4.25. Long-throated flume.

can be difficult and expensive to construct (especially the Parshall flume) and for short-throated flumes having to derive the head–discharge relationships empirically means that the range of sizes available is limited.

Orifices

Free-flow or submerged orifices can be used for measurement purposes. There is a wide range of orifice structures, some of which are designed specifically for measurement

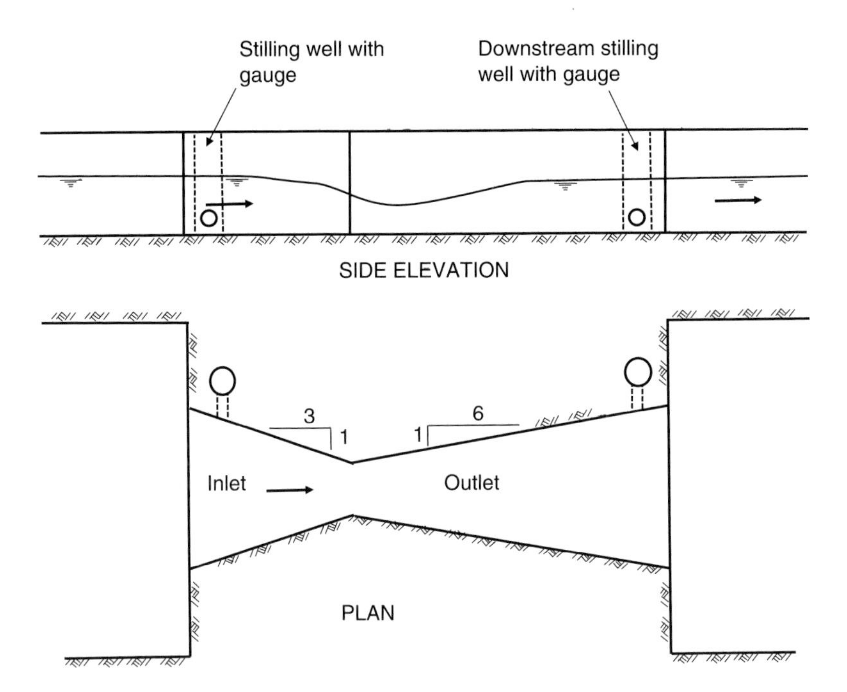

Fig. 4.26. Cut-throat flume.

purposes (such as the constant-head orifice and the Metergate) and some which, though not designed specifically for the purpose, can be calibrated and used for measurement (undershot gates are an example).

The constant-head orifice is a combined regulating and measuring structure, which originates from the USA. It uses an adjustable submerged orifice for measuring the flow and a (downstream) adjustable gate for flow regulation. Operation of the structure is based on setting and maintaining a constant head differential across the measuring orifice. Discharges are varied by changing the area of the orifice and then adjusting the downstream gate to produce a 6 cm differential across the orifice. They are relatively robust and are said to be easy to set and use, though in some locations it is found that only one of the gates is actually adjusted and then for flow regulation, not discharge measurement.

A simplified structure using one set of gates only is the Neyrpic module, which is designed to pass an almost constant discharge for a relatively wide range of variation in upstream head (Fig. 4.23e). Variations in upstream head of between 0.20 and 0.50 m result in variations of discharge through the module of only ±10%. A variety of discharges can be passed by opening a combination of gates. Discharge is proportional to gate width, and each module has a set of gates of different widths. A module with five gates – two of 30 l/s, one of 20 l/s and two of 10 l/s – will pass any discharge in units of 10 l/s from 10 to 100 l/s. The module is robust, has a relatively high modular limit (0.6) and is easy to use and to install. It has the added advantage that the discharge is 'visible' to water users in that it is proportional to total open gate width, a concept that traditional water users are familiar with. Their main disadvantages are that they are costly and are prone to clogging by debris and thus need fairly regular clearing.

Ordinary flow regulation gates can be used for discharge measurement (Fig. 4.27), though they generally have to be individually calibrated due to the variation in the flow conditions through the gate opening (dependent on gate thickness, side wall and bed shape and condition). The parameters required to determine the discharge are:

- upstream head;
- downstream head (if the gate is drowned);
- gate width;

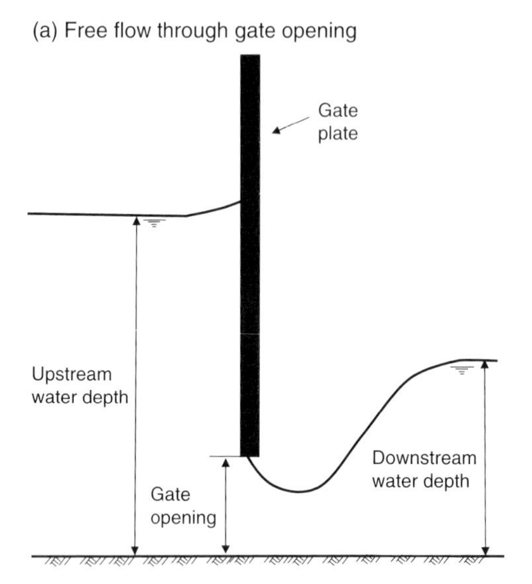

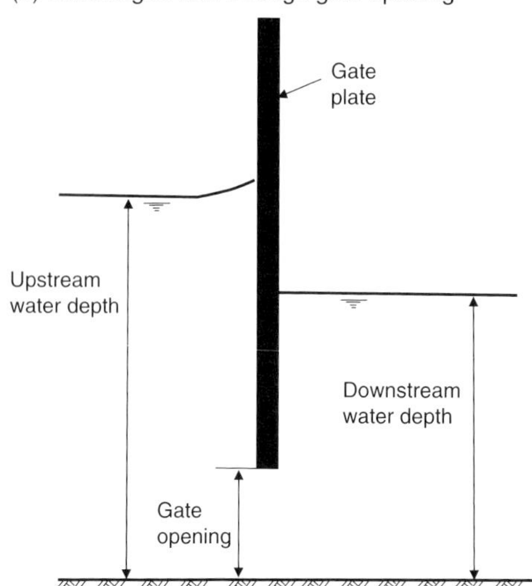

Fig. 4.27. Discharge measurement through a gate.

- gate opening;
- discharge coefficient.

By calibrating the gate for different upstream and downstream heads and gate openings a discharge coefficient graph can be obtained. This can then be incorporated into the general orifice discharge equation to enable the discharge to be determined for any setting and condition (free-flow or drowned). Calibrating the gates can be difficult. Using such procedures for discharge measurement is not generally recommended for regular operating purposes; a standard measuring structure is preferred.

Sharp-crested weirs

Sharp-crested or thin-plate measuring devices include Cipoletti weirs (Fig. 4.23c), V-notch weirs and rectangular weirs. The principal features of such devices are a sharp edge, uniform approach streamlines upstream and an aerated nappe.

These devices, if correctly installed and well maintained, are extremely accurate (±5%). Their disadvantages are that the sharp crest is prone to damage by floating debris, a relatively large head loss is required for correct operation and they are prone to sedimentation upstream, and thus inaccuracies in measurement.

Flowmeters

Flowmeters are propellers or vanes which, like a current meter, rotate as a result of the forces acting on them by flowing water. There are two common models:

- the propeller meter;
- the Dethridge meter.

The *propeller meter* is commonly used for flow measurement in pipes. It is a totalizing meter in that the number of revolutions is proportional to the total flow passing. The propeller should always be fully submerged. An alternative to the propeller meter following recent developments in electronic engineering is the use of non-intrusive ultrasonic and electromagnetic flow measurement devices. These operate in a variety of ways, either through Doppler shift or the accurate measurement of time of travel of ultrasonic signals located on opposite sides of the pipes.

The *Dethridge meter* is widely used for flow measurement in Australia at turnouts into farm units. It is of interest as it is one of the few totalizing (or volumetric) measuring devices available for open channel flow. It is an undershot water wheel with eight blades, which rotates as the water flows under the cylinder. The discharge is regulated by a small sluice gate located upstream of the water wheel. The discharge range is 40–140 l/s for the large meter and 15–70 l/s for the smaller meter. It is accurate (±5%), fairly simple and robust, and operates with a relatively low head loss (approximately 0.15 m). Its great advantage is that it measures the volume of water delivered to the farm unit and thus enables the water supply agency to charge the water user for the volume of water used. It does, however, require regular maintenance to function correctly.

Regulation of Canals

The purpose of canal regulation is to maintain a steady state within the irrigation network. The steady state comprises three main variables:

- canal discharges;
- canal water levels;
- volume of water stored in each reach of the canal network.

A canal network is in a steady state when the flow into the canal equals the flow out, and the volume stored in the canal reaches is stable. In a gradually varied flow state, the flow in equals the flow out plus the change in reach storage.[7]

To change from a steady state with a given inflow and outflow to a steady state with another value of inflow and outflow requires that the canal network passes through a state of gradually varied flow. In order that the canal network stabilizes as quickly as possible gates have to be adjusted in a set order; that is, the canal flows have to be regulated.

An example is shown in Fig. 4.28 of how the flow might be regulated at one cross regulator and associated offtakes in order to maintain a stable flow pattern in the canal following a need to alter the flows as a result of changed demands at the offtakes.

With this sequence the surge of water is 'routed' through the system, with changes being made as the surge of water reaches each control point. If the sequence is not followed then the change of storage in each reach and the change of discharges are not regulated and fluctuations will occur in the canal network. These fluctuations can take some while to stabilize, during which time the water delivery to the offtakes will also fluctuate.

With this system of regulation from top to tail end the key variable is the time of travel of the surge of water. The time of travel for each reach is often known by the canal operation staff; for example the Australian State Rivers and Water Supply Commission Water Bailiff's Manual (SRWSC, 1980) gives a figure of 13 km/h as the estimated travel time for routing of main canal discharges. The procedures recommended in this manual for regulating flows from top to bottom (the 'down method') and from bottom to top (the 'up method') are outlined below as further examples of the process.

Down method

In the down method the water is passed from one reach to another and one water master to another as the surge passes down the canal (Fig. 4.29). Before the intake gate to the canal is opened all channels must be at supply level. The intake gate is opened and the first offtake is adjusted to the correct setting immediately the surge reaches the first cross regulator, and the predetermined number of bars is taken out of the check structure[8] to pass the discharge downstream. This procedure is followed by the water master at each successive check structure until the end of their section, after which they transfer the regulation of the surge on to the next water master.

If this process is not carried out correctly then fluctuations will occur and the canal will not remain at full supply level. If the check structure is adjusted before the surge arrives then the water level will drop and it takes time to build up to supply level, and the surge is attenuated. If the check structure is regulated after the increase in the flow then when the bars are removed, an initial flush of water greater than the required discharge will be sent down the canal, and there could be a series of flushes. Early or late flushes will adversely affect downstream water masters, who will then have to cope with a fluctuating flow to make their adjustments.

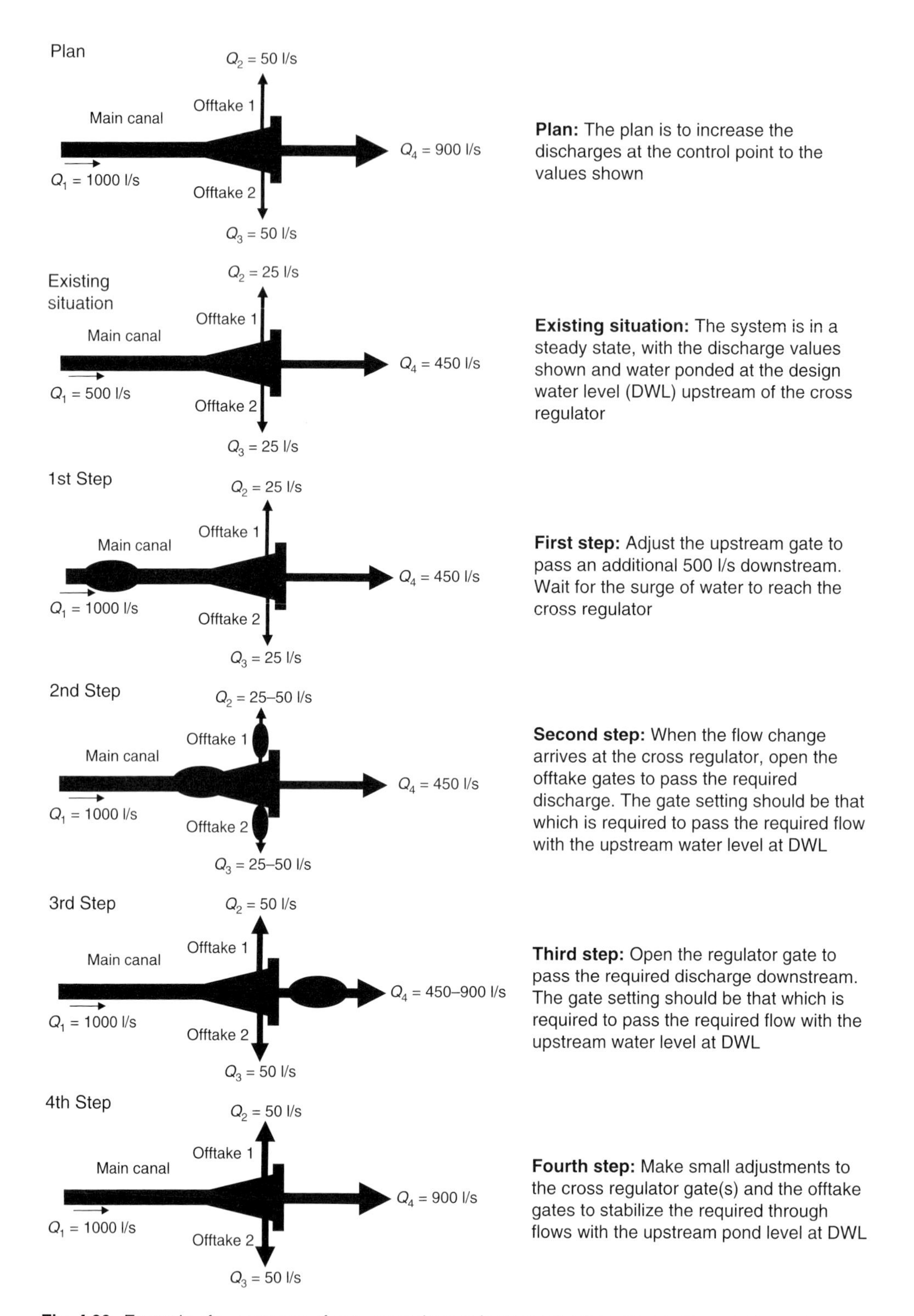

Fig. 4.28. Example of a sequence of gate operations to increase the flow passing through the system.

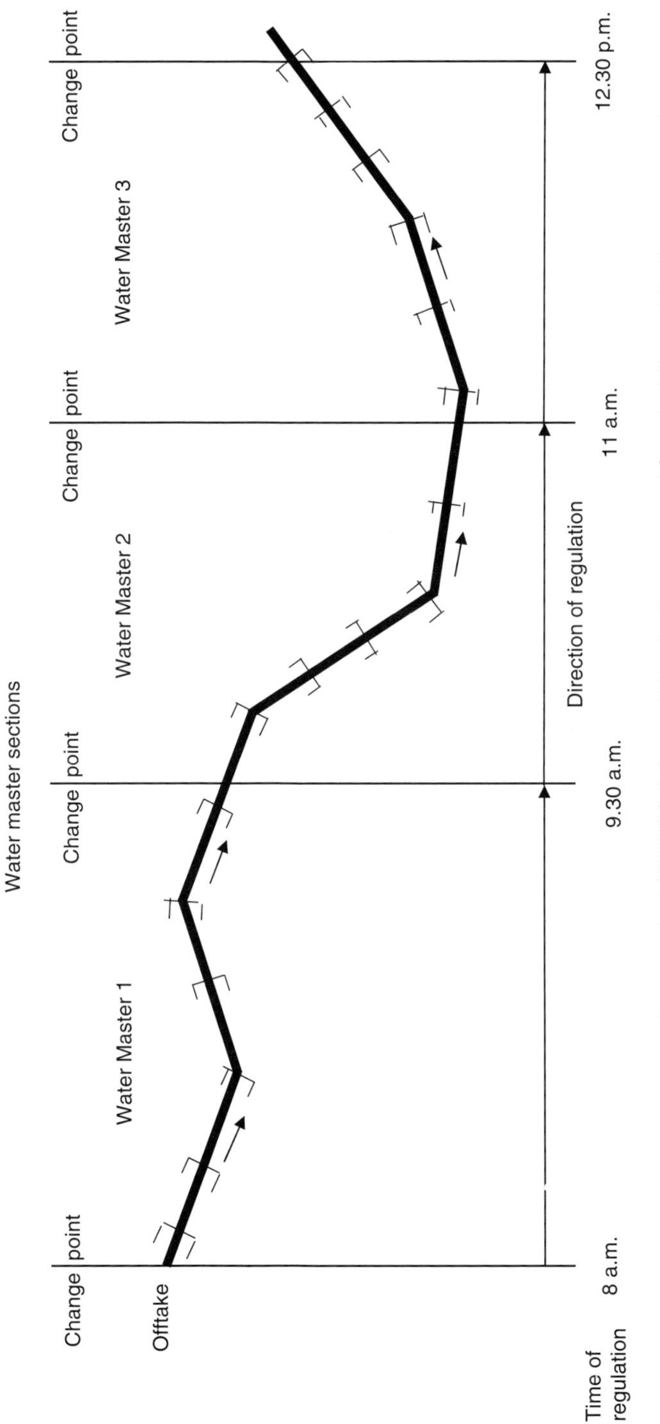

Fig. 4.29. Regulation of a canal using the Down method (SRWSC © State of Victoria, Department of Sustainability and Environment, 1980).

Up method

In the up method the required numbers of bars are removed from the downstream check structure to allow the required discharge to pass downstream. The water master must then move as quickly (and safely) as possible to the next upstream check to remove the required number of bars there, to replace the flow just released at the previous (downstream) check. This procedure is repeated up the canal to the intake, where the volume of water released at the lowest check structure is replaced by opening the intake gate to the required setting. Speed is of the essence in the up method if the size of the fluctuations in the canal reaches is to be kept to a minimum.

Down and up method

The down method requires a long time to pass the surge of water down the system. This time can be considerably reduced by combining the down and up methods of canal regulation. This method requires that the water masters coordinate their actions,[9] starting their adjustments such that the water master moving upstream meets with the water master moving downstream at the same time at the change point.

The process is shown in Fig. 4.30. The first water master regulates his check structures in succession moving downstream, and at the same time the next water master regulates his checks moving upstream to meet the first water master at the prescribed time. A similar process is followed by the third and fourth water masters, and the fifth and sixth water masters, with the start and meeting times being agreed by the water masters beforehand based on their experience of travel times within their systems.

This method is considerably quicker than the down method alone, and results in less fluctuations than the up method alone. Keeping the time to change the control structure settings to a minimum is essential if the canal flows are changed each day, as it then leaves time for the water masters to carry out their other work.

Flexibility

The flow in a canal is very rarely steady. Fluctuations occur for a variety of reasons, including changes in the flow at the intake, gate adjustments or inflow from drainage. If the configuration of the control structures at bifurcation points is not designed correctly these fluctuations can have an adverse impact on the flows throughout the system. To describe the impact of these fluctuations the term *flexibility* is used (which is defined in the Equation at the bottom of the page).

The two principal control configurations (Fig. 4.31) are orifices (e.g. undershot gates) and weirs (e.g. overshot gates, or fixed-crest cross regulators). For orifices the discharge is a function to the power 0.5 of the head over the structure; for a weir the discharge is a function to the power 1.5 of the head over the structure. Thus the flexibility can be written in terms of the head over the structure:

$$F = \frac{1.5h_p}{1.5h_o}$$

where

h_o = upstream head over the offtaking structure
h_p = upstream head over the parent canal structure.

In this configuration where the control structures on the offtaking and ongoing parent canal are the same and the crest or sills are at the same level then the flexibility is equal to 1, and the fluctuations in the offtaking canal will match those in the ongoing parent canal (Fig. 4.31a).

If, however, the control structures are different then the fluctuations in the offtak-

$$\text{Flexibility, } F = \frac{\text{Rate of change of discharge for the offtaking canal}}{\text{Rate of change of the ongoing parent canal}} = \frac{dQ_o / Q_o}{dQ_{ds} / Q_{ds}}$$

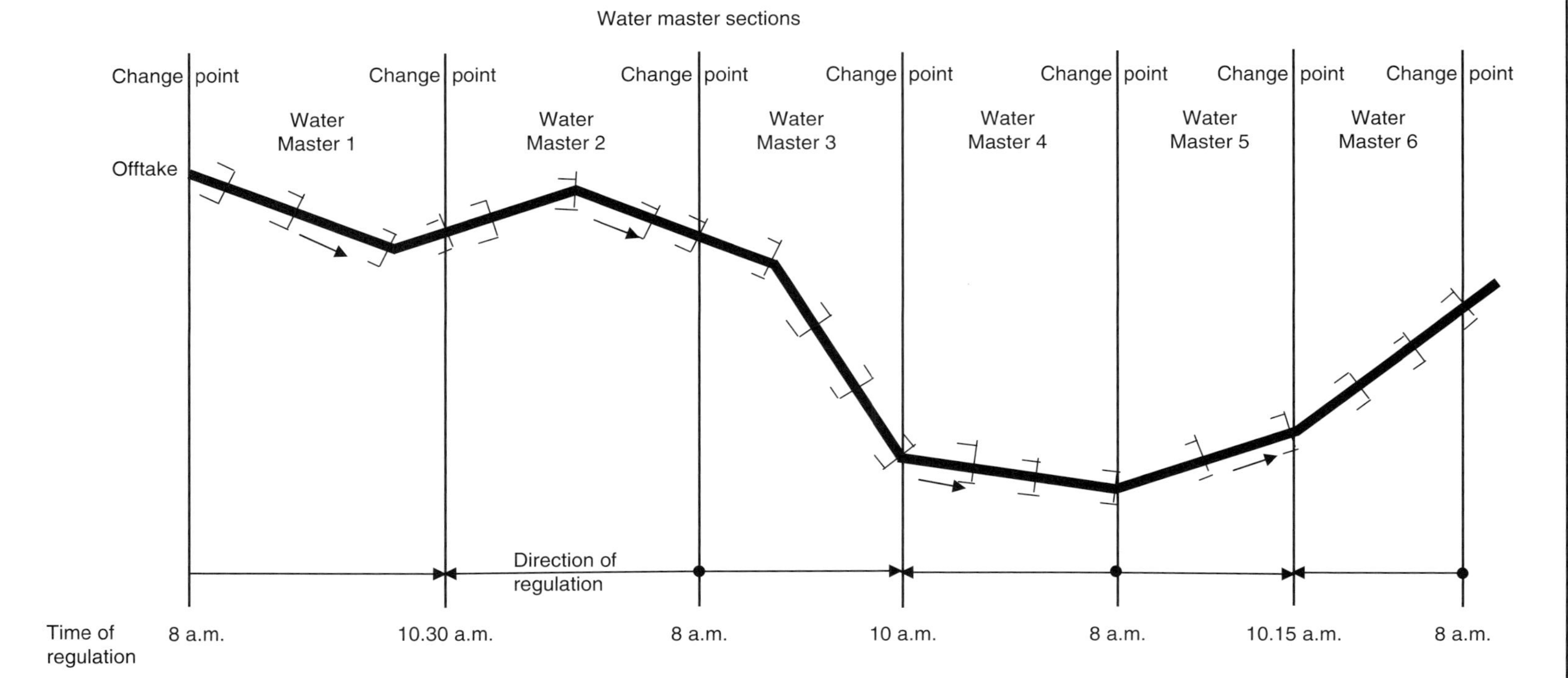

Fig. 4.30. Regulation of a canal using the Down and Up method (SRWSC © State of Victoria, Department of Sustainability and Environment, 1980).

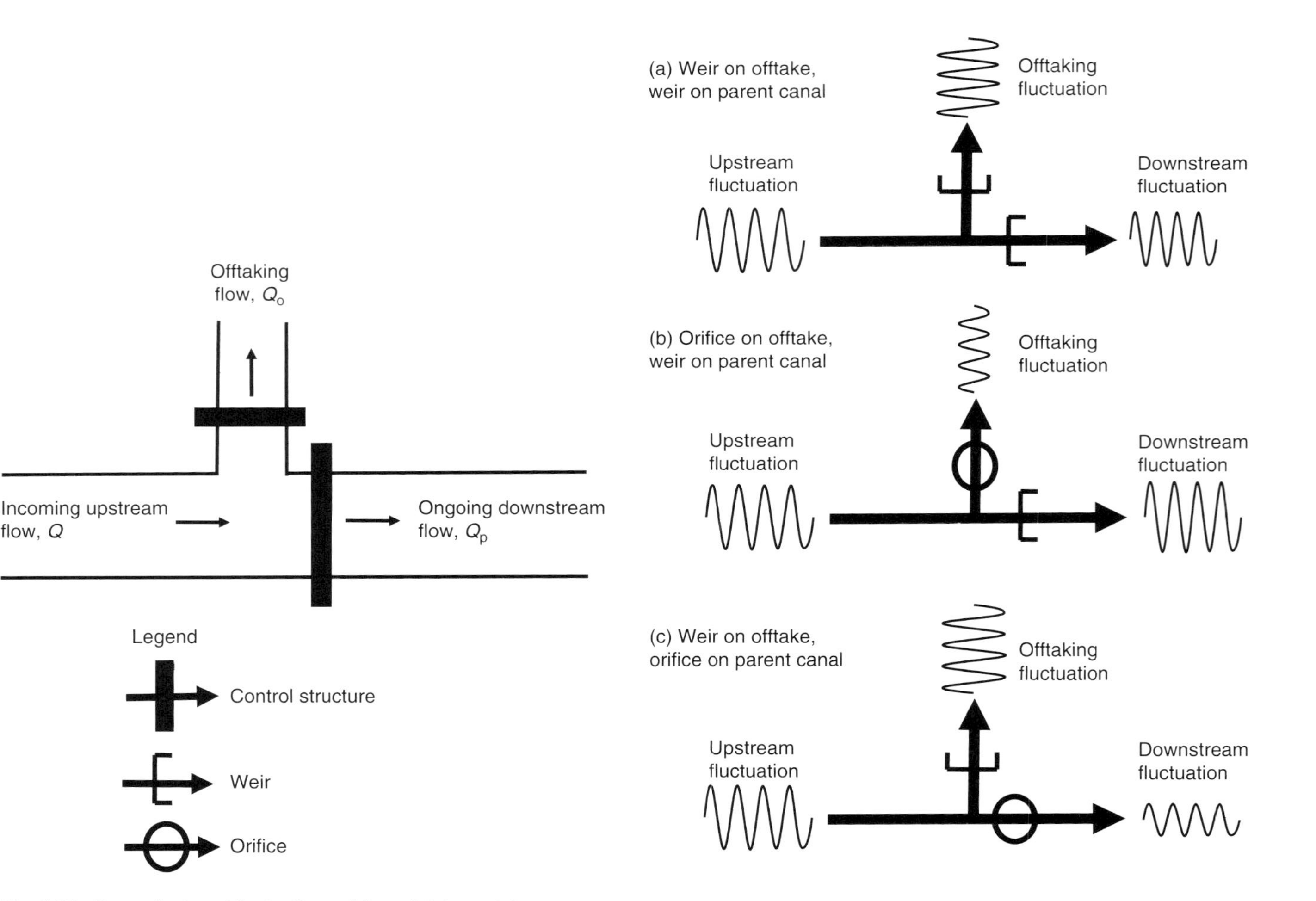

Fig. 4.31. Transmission of fluctuations at flow division points.

ing canal will be different from those in the ongoing parent canal. Thus the flexibility at a control point with an orifice offtake and a weir on the parent canal will be:

$$F = \frac{0.5h_p}{1.5h_o}$$

Note that, due to the derivation of the relationship to define flexibility, the power functions and upstream heads are inverted in relation to the power functions.

Therefore in this case the fluctuations in the offtaking canal will be less than those in the parent canal (Fig. 4.31b). This is a preferred relationship for water management as it means that the offtaking discharge will vary less with fluctuations in the main canal. Figure 4.32(a) shows such a configuration on a system in Sri Lanka where the long-crested weir keeps the water level variation to a minimum for fluctuations in the parent canal. The Neyrpic modules (Fig. 4.32(b), Fig. 4.23e) described earlier are often used in association with long-crested weirs in the parent canal.

If there is an orifice on the parent canal and a weir on the offtaking canal the fluctuations in the offtaking canal will be greater than those in the parent canal (Fig. 4.31c). This is not recommended as fluctuations in the parent canal will result in greater outflows in upstream offtakes and therefore reduced flow available downstream. One useful application of this configuration, however, is for a side escape spillway discharging into a drainage canal.

Examples of Main System Operation Procedures

Simple proportional distribution

The simplest main system water distribution method is proportional division. With this method water is divided automatically by the control structures located at division points in the irrigation network. The most common division is in proportion to area, so the width of an offtake serving 10 ha will be one-tenth of the width of the structure opening in the main channel serving 100 ha downstream (Fig. 4.33). Such systems are found in farmer-managed schemes in Nepal and on the slopes of Mount Kilimanjaro in East Africa.

Where these systems have been constructed on traditional irrigation systems the widths of the openings on the ongoing and offtaking structures are decided at the design and construction stage, and are thus fixed for life. There are some non-traditional irrigation systems that have been constructed with movable proportional dividing gates; these flow splitters are moved horizontally

(a)

(b)

Fig. 4.32. Examples of orifice and weir combinations for flow control. (a) Gated orifice offtake with a duck-bill weir in the parent canal. Due to the extended length of the weir in the parent canal the head over the offtake orifice varies very little for relatively large variations in upstream discharge (Sri Lanka). (b) Diagonal weir in the parent canal with a set of Neyrpic offtake gates on the right-hand side (Morocco).

Fig. 4.33. Automatic proportional distribution on the main channel of a farmer-managed irrigation system in Nepal. Note the relative widths of the offtake on the right and the main channel.

depending on the downstream requirements, and operated much as ordinary undershot gates might be in allocating discharges between the main channel and the offtaking channel, or splitting the discharge between two main channels.

There are no operation procedures for the fixed proportional systems, and maintenance is limited to ensuring that there are no obstructions to the flow through the structure.

Warabandi system of water allocation and distribution

Overview

The development of the Indo-Gangetic Plain for irrigated agriculture began in earnest in the 1850s. At the time there were two options available for development of the available resources of water, land and labour. The first was to limit the canal command area so that the available water supplies matched crop water demands. In this case production would have been maximized per unit of land irrigated. The second option was to extend the irrigation area to support as large an area as possible. In this case a greater number of farmers would benefit from the irrigation water and production would be at a maximum per unit of water.[10] The second option was chosen, with a method of water allocation and distribution that came to be known as Warabandi.

In its modern form Warabandi involves the rotation of water supplies between distributaries on the main system, and between farmers' fields within the watercourse.[11] Within the watercourse, allocation is based on time shares that are proportional to the area of a farmer's fields, as defined by Malhotra (1982):

> Warabandi is a system of equitable distribution of the water available in the scheme by turn according to a predetermined schedule specifying the day, time and duration of supply to each irrigator in proportion to their holding in the outlet command.

The cardinal principle is that available water, whatever its amount, is allocated to cultivators in equal proportion to their holdings, and not only to some to meet their total demand. It attempts to guarantee equity of distribution, it is low-cost to build, easy to operate and straightforward for farmers to understand. It is probably the best possible method of water management for the huge schemes that exist in the Indo-Gangetic plains, and has stood the test of time.

In the climatic and institutional conditions in which it is used, the Warabandi system, by obviating the need for data collection and regular setting of gates, may achieve a more stable and equitable pattern of water distribution than more sophisticated methods.

Details of the method

A typical layout of a distribution system where Warabandi is practised is given in Fig. 4.34.

The main canal feeds two or more branch[12] canals, which operate by rotation. This primary distribution system runs throughout the season with varying supply.

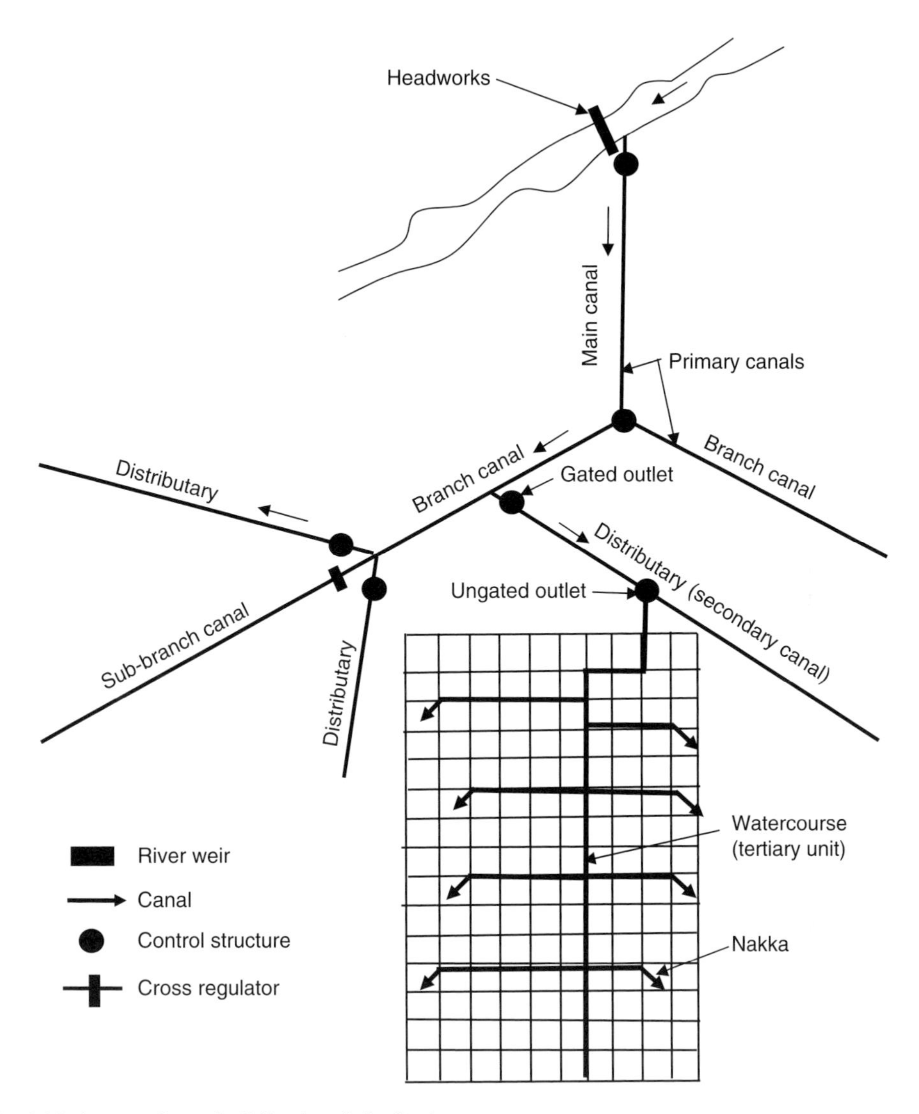

Fig. 4.34. Layout of a typical Warabandi distribution system.

Large numbers of distributaries[13] take off from the branch canals, these run at full supply level (FSL)[14] by rotation. The distributaries supply water to watercourses through ungated, fixed discharge outlets (adjustable proportional modules, APMs). Watercourses run at the design discharge when the distributary is running and water is allocated between farmers on a watercourse time roster. The Irrigation Department manages the main system down to the watercourse intake, below which the farmers manage the water.

Design of distributaries is based on the culturable command area (CCA), which is allocated a water allowance of about 2.4 cusecs per 1000 acres (0.168 cumecs per 1000 ha, or 0.17 l/s/ha).[15] For rice areas the duty is 7–10 cusecs per 1000 acres (0.49–0.7 cumecs per 1000 ha, or 0.5–0.7 l/s/ha).

The watercourse intake is designed for about 2.4 cusecs per 1000 acres (0.17 l/s/ha), with the supply into the watercourse being regulated by an APM (Fig. 4.35). The throat width and cross-sectional area in the throat of the APM is fixed in proportion to

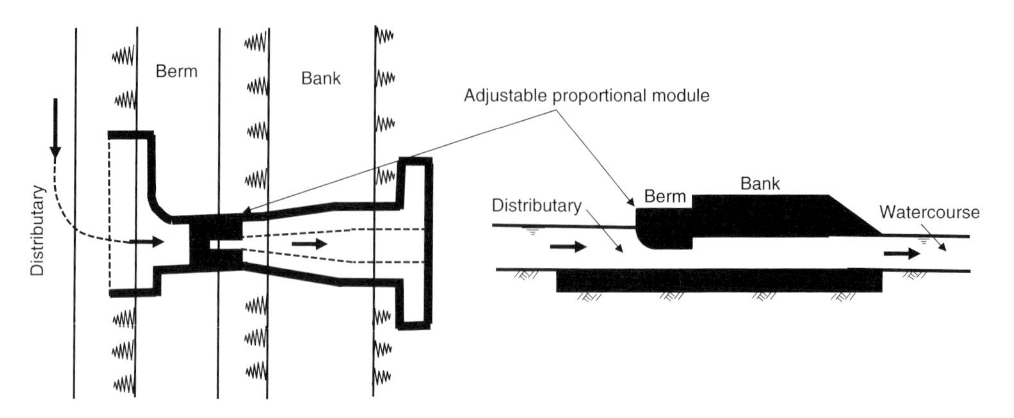

Fig. 4.35. Typical plan of an adjustable proportional module regulating discharge in proportion to culturable command area.

the CCA, assuming FSL is maintained in the parent canal.

No distributary operates for all the days of the growing period. The ratio of days operated to crop growth period is the *capacity factor*. This is 0.8 for Kharif (monsoon crop, July–October) and 0.72 for Rabi (winter crop, November–June) on some schemes. Thus each distributary may receive water supply for about 144 and 129 days out of each season, respectively.

Not all the land can be irrigated at once. The ratio of irrigated land to the CCA is called the *intensity of irrigation*, this is typically about 60% (see Equation (a) at the bottom of the page).

All distributaries are run at FSL for periods of 8 days. Each watercourse runs at full supply for 7 days, so that farmers receive their water at the same time, and for the same duration each week. The additional day is for filling the canal. The discharge in the watercourse generally varies between 30 and 85 l/s.

Roster of turns

The roster of turns or schedule (Table 4.4) is calculated based on the 168 hours available for irrigation during 1 week (see Equation (b) at the bottom of the page).

Bharai is the time a farmer must spend filling up the empty watercourse from the point of previous abstraction. Its value is 4–5 min per 220 ft (67 m) in good soils. This time is deducted from the common pool and added to the individual farmer's time. *Jharai* is a term related to the ponded water remaining in the watercourse when the supply has been cut off at the watercourse intake. This water can only be taken by tail-enders, so a deduction is made from their flow time to account for this additional water. It is difficult to determine the correct value of time to ascribe to this water as it does not flow at a constant rate.

No allowance is made in these calculations for losses due to seepage.[16] The calculation of the rosters is a formal procedure; once calculated and agreed it is posted for all farmers to follow.

$$\text{Water duty (l/s/ha)} = \frac{100\text{ ha}}{\text{Water allowance for 100 ha}} \times \text{Intensity (\%)} \qquad \text{(a)}$$

$$\text{Flow time per unit of area (FT)} = \frac{168 - \text{total Bharai} + \text{total Jharai}}{\text{Total area}} \qquad \text{(b)}$$

$$\text{Flow time per farmer} = (\text{FT for unit area} \times \text{farmer's area}) + (\text{farmer's Bharai}) - (\text{farmer's Jhara})$$

Table 4.4. Typical Warabandi schedule.

										Turn		Schedule A		Schedule B		
Sl. no.	Name of landowner	Name of tenant	Field nos. comprised in units	Total CCA	Time in preparation to CCA	Additional for Bharai	Deduction for Jhari	General deduction	Net length of time allocated	Take over from	Hand over to	Time from	Time to	Time from	Time to	Remarks
(1)	(2)	(3)	(4)	(5)	(6)	(7)	(8)	(9)	(10)	(11)		(12)				
32	Chander Bhan															
	Data Ram									181	181					
	Ram Saroop[a]			8.32	0.42	–	–	–	0.42	1–2	1–2	5.17 pm	5.59 pm	5.17 am	5.59 am	
	Mehar Chand															
33	Ram Saroop									181	181					
	Chhatar Singh[a]			4.21	0.23	0.15	0.05	–	0.33	1–2	2–3	5.59 pm	6.32 pm	5.59 am	6.32 am	
	Ramji Lal															
34	Hari Singh									181	181		Thursday			
	Ram Singh[a]			97.8	7.21	–	–	–	7.21	2–3	2–3	6.32 pm	1.53 am	6.32 am	1.53 pm	
35	Daya Nand											Thursday				
	Jiya Ram[a]			1.96	0.93	–	–	–	0.93			1.53 am	2.51 am	1.53 pm	2.51 pm	
36	Suraj Mal									181	Head				Thursday	
	Deep Chand[a]			54.87	4.69	–	1.3	–	3.39	2–3	outlet	2.51 am	6.30 am	2.51 pm	6.30 pm	
	Total			**2135**	**164.2**	**4.15**	**2.42**		**168**							

CCA, culturable command area.
Unit running time=164.2/2135.2=4.6 min.
[a] The farmers indicated are taking water.

Rotational running of distributaries

The running of the distributaries is fairly straightforward, when they are on it is for an 8-day period and the discharge is always at full supply. The only variation is the interval between being off and back on again.

The irrigation cycle (frequency) depends on the value of the branch canal's lowest probable supply. If the lowest probable river supply is one-third of the distributary canals' aggregate capacity the branch canal is divided into three about-equal parts (three groups). If the lowest probable supply is one-half of the aggregate capacity, the branch canal is divided into two about-equal parts (two groups). When supply matches demand all groups run at full supply, when the supply is less than demand supplies are rotated between each group. An analysis carried out for a Warabandi-type irrigation system in Nepal showed that considerable water savings could be made by rotating water supplies on the main canal when water supplies were short (Fig. 4.36). As the river water supply available at the intake to the scheme decreases, the main system operation is changed from all four zones operating concurrently to two zones operating, and finally to only one zone operating at a time. This procedure keeps the discharge in the canals close to design levels and prevents canals flowing with low flows.[17]

The rotational running of the canals attempts to take into account the crop growth stage. The lifespan of the crop is divided into three parts: sowing, growing and maturing (Table 4.5). Rotational operation of the canals attempts to equally divide the water available to all distributaries in each of these three growth stages.

Data collection, processing and analysis

Data collection, processing and analysis are limited in the Warabandi method of water allocation and distribution.

Irrigation Booking Clerks collect data each growing season on the area and type of

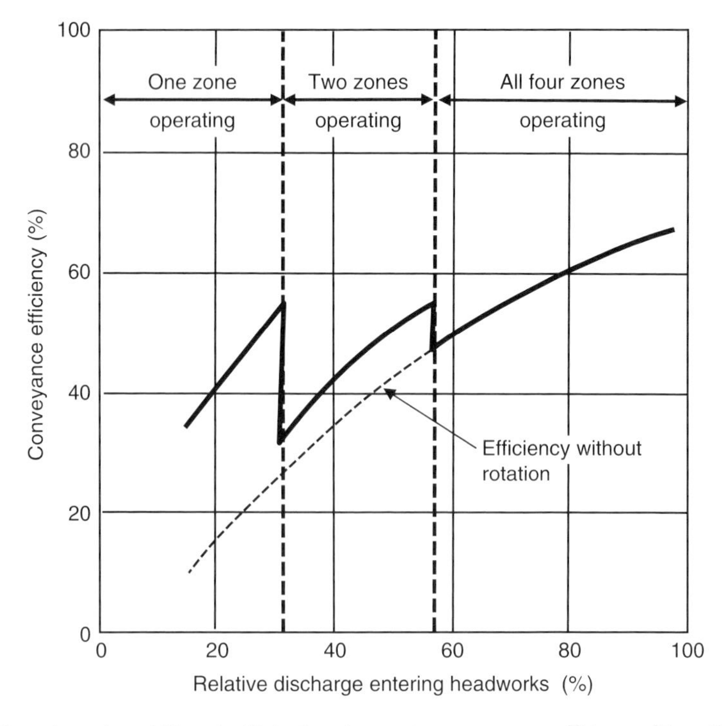

Fig. 4.36. Effect of rotation of Chandra Main Canal on water conveyance efficiency. (From Beadle *et al.*, 1988 with kind permission of Springer Science and Business Media.)

Table 4.5. Rotational water distribution schedule for distributaries.

Period		Cycle	Preferential order for groups		
			First	Second	Third
Sowing	3 Oct–10 Oct	1	B	A	C
	11 Oct–18 Oct	2	C	B	A
	19 Oct–26 Oct	3	A	C	B
	27 Oct–3 Nov	4	B	A	C
	4 Nov–11 Nov	5	C	B	A
	12 Nov–19 Nov	6	C	C	B
	20 Nov–27 Nov	Balancing period			
Growing	28 Nov–5 Dec	1	B	A	C
	6 Dec–13 Dec	2	C	B	A
	14 Dec–21 Dec	3	A	C	B
	22 Dec–29 Dec	4	B	A	C
	30 Dec–6 Jan	5	C	B	A
	7 Jan–14 Jan	6	C	C	B
	15 Jan–22 Jan	Balancing period			
Maturing	23 Jan–30 Jan	1	B	A	C
	31 Jan–7 Feb	2	C	B	A
	8 Feb–15 Feb	3	A	C	B
	16 Feb–23 Feb	4	B	A	C
	24 Feb–3 Mar	5	C	B	A
	4 Mar–11 Mar	6	C	C	B
	12 Mar–19 Mar	Balancing period			

crop cultivated under irrigation. The farmer is then taxed accordingly. These data are not used for organizing the irrigation rotation, except possibly at the start of the season to establish when cropping starts in earnest on a distributary.

There is no discharge measurement, except occasionally at a flume at the head of a distributary. Full supply discharge is determined and monitored using gauge boards at the head of each distributary. The basic principle is that if the canal is open, it must flow at full capacity (full supply level).

A check procedure has been designed at the tail of the system where the last cluster of watercourse intakes is weirs rather than orifices. If the distributary is working properly the depth over these weirs should be 1 ft (30 cm). A check is also kept on the watercourse intakes to ensure that they have not been tampered with in an attempt to increase the flow into the watercourse.

Some monitoring of system performance is done by keeping a check on the crop area cultivated in each watercourse. If there is a significant change from normal or design levels the cause for this can be investigated.

Relative area method

The relative area method of main system management has been developed for use on irrigation systems in Indonesia (Pasandaran, 1976; Khan, 1978; Burton, 1989b). Its simplicity of use makes it worthy of consideration elsewhere (though it would require adaption to local conditions). The method requires limited amounts of data, its calculation procedures are straightforward, and data can easily be analysed and monitored. It is designed to ensure equitable distribution of water.

The method is a compromise between the relatively complex water balance sheet approach and the relatively simple Warabandi method. It takes into account crop areas, mixed cropping patterns and crop water requirements, yet requires little calculation.

In the relative area method all crop areas are converted to a common equivalent crop area based on their relative crop water requirements. Typical conversion factors are:

Crop	Conversion factor
Maize	1
Groundnut	1
Soybean	1
Rice	4
Sugarcane	1.5

Thus a 1 ha field of maize would have a relative area of 1 (relative) ha; while a 1 ha field of rice would have a relative area of 4 (relative) ha. Thus if a unit discharge of say 1 l/s was allocated to each relative hectare, the 1 ha field of maize would get 1 l/s and the 1 ha field of rice would get 4 l/s. Having converted all the different crop areas to a relative area, a unit discharge per (relative) hectare is applied and the total demand calculated. If the total demand exceeds the supply available at the system intake, the amount to be supplied at each control point is reduced by the ratio of the supply available to the calculated demand.

The method of calculation and associated procedures greatly simplify the calculations of crop water requirements and water allocation for schemes with mixed cropping patterns.

With this method the main system is operated by the Irrigation Service while the tertiary unit is operated by the water users (usually formed into a WUA). The Irrigation Service collects data on the cropped area from the water users each time period (either weekly or 10-daily) and uses this to calculate the water allocation. Control structures comprise undershot gates with measuring structures, either weirs (Cipoletti, broad-crested) or flumes. Gates are adjusted to pass the required discharges at the start of each time period, with further adjustment during the time period in order to maintain the required discharge.

Details of the method

An example will help the explanation. Data for a typical irrigation system are presented in Fig. 4.37 and Table 4.6. The calculation in Table 4.6 shows how the required discharge is calculated at the river intake.

Figure 4.37 and Table 4.6 show how simple the system is to use. The crop areas are measured in the field, for Tertiary A the figures are 10 ha to rice and 30 ha to maize. These figures are converted to their relative area by multiplying by 4 and 1 respectively to obtain 40 ha and 30 ha, total 70 ha. A relative area water duty (RAWD) of 0.40 l/s/ha is applied to this relative area to obtain the discharge required at the tertiary intake, which gives a figure of 28 l/s. The tertiary unit discharges for

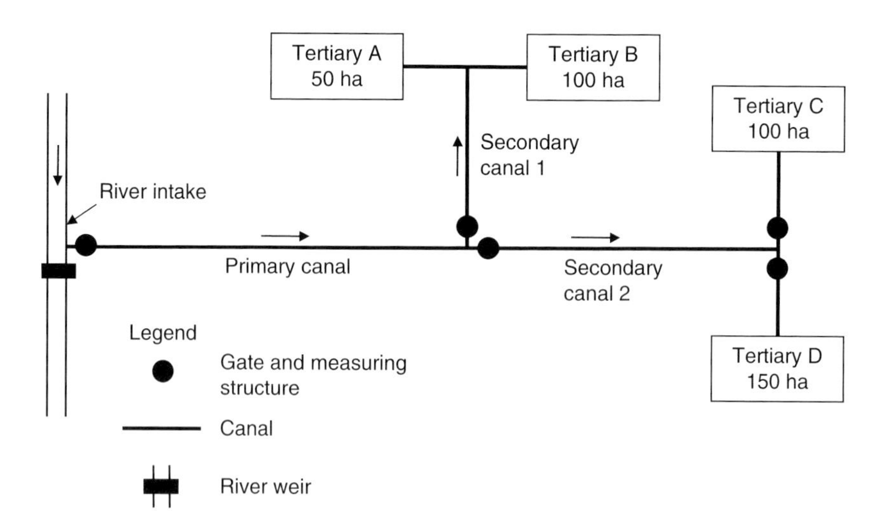

Fig. 4.37. Typical irrigation system layout.

Table 4.6. Data and calculations for water allocation using the relative area method.

		Tertiaries				Secondaries		Primary
Item	Units	A	B	C	D	S1	S2	P1
Command area	ha	50	100	100	150	150	250	400
Crop area – rice	ha	10	30	20	50	40	70	110
Crop area – maize	ha	30	60	60	80	90	140	230
Relative area – rice (area×4)	ha	40	120	80	200	160	280	440
Relative area – maize (area×1)	ha	30	60	60	80	90	140	230
Total relative area	ha rel.	70	180	140	280	250	420	670
RAWD at tertiary unit level	l/s/ha rel.	0.4	0.4	0.4	0.4	–	–	–
Discharge allocated to tertiary units (rel. area×water duty)	l/s	28	72	56	112	–	–	–
Estimated losses in secondary and primary canals	%	–	–	–	–	20	20	17
Discharges required in secondary and primary canals	l/s	–	–	–	–	125	210	404

RAWD, relative area water duty.

all offtakes on a secondary are summated and the discharge required at the head of the secondary canal calculated after allowing for losses.[18] Similarly the discharges required at the secondary canal intakes are summated and the discharge required at the primary canal intake calculated after allowing for the losses.

The RAWD has been determined from field measurements of crop water demands for different crops in different locations in Indonesia and is generally of the order of 0.35–0.45 l/s/ha rel. This gives an allocation 0.35–0.45 l/s to 1 ha of maize and 1.4–1.8 l/s to 1 ha of rice.

One of the great strengths of the relative area method is the ease with which it can be used to make water allocation in times of water shortage. Taking the above example, the desired RAWD at the tertiary gate is 0.40 l/s/ha rel., which gives a required discharge at the river intake of 404 l/s. However, there may only be a supply available of 300 l/s at the river intake.

The solution is outlined in Box 4.1, showing two approaches to determination of the RAWD, which is the basic figure required for determination of the discharges required at control points in the system.

In this case the RAWD at the tertiary unit intakes are the same, obviously Method 1 is quicker. The calculation of the discharges required at the various control point locations within the system are shown in Table 4.7. The calculation gives a discharge of 303 l/s at the primary canal intake, close enough to match the estimated 300 l/s available.

Data collection, processing and analysis

A major advantage of the relative area method is the very straightforward data collection, processing and analysis procedures. With the numerous irrigation systems that there are in Indonesia covering over 4 million ha, this is a significant consideration.

For *data collection*, the main data required are:

- canal discharges at all control points (primary, secondary and tertiary canal intakes);
- crop types and areas in each tertiary unit;

Box 4.1. Determination of the Relative Area Water Duty

- River discharge required, QR_r=404 l/s.
- River discharge available, QR_a=300 l/s.
- Water supply factor, WSF = QR_a/QR_r = 300/404 = 0.74.

Method 1

- RAWD at tertiary unit intakes = 0.4 × 0.74 = 0.30 l/s/ha rel.

Method 2

- Total losses in system = River intake discharge – Tertiary unit discharges = 404 – (28 + 72 + 56 + 112) = 136 l/s = 34%.
- Expected losses with lower discharge at intake = 300 × 34/100 = 102 l/s.
- Discharge available at tertiary unit intakes = 300 – 102 = 198 l/s.
- Relative area of tertiary units = 670 ha rel.
- RAWD at tertiary unit intakes = 198/670 = 0.30 l/s/ha rel.

Table 4.7. Recalculation of discharges based on water available at river intake.

		Tertiaries				Secondaries		Primary
Item	Units	A	B	C	D	S1	S2	P1
Command area	ha	50	100	100	150	150	250	400
Crop area – rice	ha	10	30	20	50	40	70	110
Crop area – maize	ha	30	60	60	80	90	140	230
Relative area – rice (area×4)	ha	40	120	80	200	160	280	440
Relative area – maize (area×1)	ha	30	60	60	80	90	140	230
Total relative area	ha rel.	70	180	140	280	250	420	670
RAWD at tertiary unit level	l/s/ha rel.	0.3	0.3	0.3	0.3	–	–	–
Discharge allocated to tertiary units (rel. area×water duty)	l/s	21	54	42	84	–	–	–
Estimated losses in secondary and primary canals	%	–	–	–	–	20	20	17
Discharges required in secondary and primary canals	l/s	–	–	–	–	93.75	157.5	303

RAWD, relative area water duty.

- river flows;
- drainage flows (if into canal or used for irrigation);
- abstractions for other uses from canals (industry, water supply, etc.).

Crop type and area are collected within the tertiary unit by village water masters or water users associations who pass the information onto the Irrigation Service water masters every 10 days.[19] All the other data are collected daily by the Irrigation Service water master.

For *data processing and analysis*, the Irrigation Service staff meet each 10 days in the office to evaluate the performance of the previous 10 days and plan the water supply for the coming 10-day time period. The crop area and type are recorded on one form (Form 04) and the crop area converted to relative area on that form. Daily canal discharge is recorded on another form (Form 01, see Fig. 4.38) and

FORM 01

DISCHARGE MEASUREMENT

Period: From 1 September to 10 September

Measurement location		Req. value	Day 1	2	3	4	5	6	7	8	9	10	11	Average discharge
Location: Penewon Type/width: ______	*H*													
Comm. area: 817 Comments: ______	*Q*		925	932	927	925	932	935	927	927	932	932		928
Location: Jambuwok I Type/width: ______	*H*													
Comm. area: 559 Comments: ______	*Q*		578	578	583	581	575	585	578	581	581	578		581
Location: Jambuwok II Type/width: ______	*H*													
Comm. area: 258 Comments: ______	*Q*		350	356	353	350	350	356	356	353	353	353		353
Location: Blendren I Type/width: ______	*H*													
Comm. area: 106 Comments: ______	*Q*		76	79	83	83	79	79	76	76	79	79		79
Location: Blendren II Type/width: ______	*H*													
Comm. area: 56 Comments: ______	*Q*		85	88	88	94	85	88	94	85	94	88		88
Location: Blendren III Type/width: ______	*H*													
Comm. area: 96 Comments: ______	*Q*		136	141	141	146	136	141	141	136	146	141		141
Location: Kedawung Type/width: ______	*H*													
Comm. area: 38 Comments: ______	*Q*		38	42	Fig	42	38	42	38	42	44	44		42
Location: Ked. Maling I Type/width: ______	*H*													
Comm. area: 64 Comments: ______	*Q*		68	64	72	68	68	72	64	68	68	68		68
Location: Ked. Maling II Type/width: ______	*H*													
Comm. area: 32 Comments: ______	*Q*		30	38	34	34	30	34	34	38	30	38		34
Location: Ked. Maling III Type/width: ______	*H*													
Comm. area: 69 Comments: ______	*Q*		49	54	54	59	54	54	49	54	54	59		54
Location: Ked. Maling IV Type/width: ______	*H*													
Comm. area: 25 Comments: ______	*Q*		17	14	17	20	17	17	14	17	17	20		17

Fig. 4.38. Form 01 used to record discharges at control points (*H* = gauge reading (cm); *Q* = discharge (l/s); Req. value = average required value of *H* and *Q* at the control structure for the 10-day period; Location = location of structure – canal name and structure number; Type = Cipoletti, broad-crested, V-notch, etc.; Width = width at control section in measuring structure; Comments = structure condition – good, broken, drowned out, silt upstream, etc.).

averaged for each 10-day time period. The relative area is transferred from Form 04 to Form 05 (see Fig. 4.39), where the average recorded discharge is divided by the relative area to give the RAWD (in l/s/ha rel.) over the recorded time period. The RAWD values of each control point in the system are then compared and anomalies investigated. For instance, all tertiary units should have a similar RAWD value. If one unit has an RAWD of

CROP- DISCHARGE DATA FORM

Period: From 1 September to 10 September

Canal (tertiary, secondary, primary)	Gross irrig. area (ha)	Authorized area for dry season paddy	PADDY									DRY-SEASON CROPS								Sugarcane		Tobacco	Fallow land	Total cultivated area (ha)	Total discharge (l/s)	Total relative area (ha rel.)	Relative factor (l/s/ha rel.)
			Wet season			Authorized dry season			Unauthorized dry season			Dry season I				Dry season II											
			Nursery crop	Land preparation	Main crop	Nursery crop	Land preparation	Main crop	Nursery crop	Land preparation	Main crop	Soyabean	Maize	Cassava	Other crops	Soyabean	Maize	Cassava	Other crops	Young cane	Old cane						
Penewon I	817																							817	928	1913	0.49
Jambuwok II	258																							258	353	600	0.59
Blendren I	106							32			10					26	38							106	79	232	0.34
Blendren II	56							26			8					13	9							56	88	158	0.56
Blendren II	96							30			8					19	39							96	141	210	0.67
Jambuwok I	559																							559	581	1313	0.44
Kedawung	38							24			4						10							38	42	122	0.34
Ked. Maling I	64							26			3						28		7					64	68	151	0.45
Ked. Maling II	32							23			3						3		3					32	34	110	0.31
Ked. Maling III	69																			69				69	54	104	0.52
Ked. Maling IV	25							23									2							25	17	94	0.18
Ked. Maling V	38							21									17							38	28	100	0.28
Sambiroti weir	293																							293	201	632	0.32
Sambiroto I	126							30									85		11					126	71	216	0.33
Sambiroto II	74							25									40		9					74	39	149	0.26
Sec. Sambirejo	93																							93	60	267	0.22
Sambirejo I	39							23									16							39	29	108	0.27
Sambirejo II	36							23									13							36	25	105	0.24
Sambirejo III	18							12									6							18	14	54	0.26

Fig. 4.39. Form 05 used to record and process crop areas and discharges.

0.41/s/ha rel. and another has 0.71/s/ha rel., the reason for the latter (higher) figure should be investigated.

For planning for the coming 10-day period the procedures shown in Tables 4.6 and 4.7 are followed, crop areas are converted to relative areas, an RAWD is applied to each tertiary unit, the tertiary unit discharges calculated, summated and increased to allow for losses to give the secondary canal discharge, and so on up the system. Canal losses can be determined by analysis of previous periods' data.

Schematic maps

Schematic maps for water distribution can be used in any irrigation system; they are particularly useful in the relative area method. A schematic map for the layout shown in Fig. 4.37 is given in Fig. 4.40.

The schematic map is a pictorial representation of the canal network showing the relative locations of the main canals, the tertiary units and control structures, together with key operational data. The schematic map can be used to display the actual discharges and RAWD values for the last 10-day period and then used to calculate and plan the allocations for the coming 10-day period.

The schematic map for the system represented in Figs 4.38 and 4.39 is presented in Fig. 4.41. Note that for ease of data processing and analysis, the canals in these data collection and processing forms (Figs 4.38 and 4.39) are listed in order from top to bottom of the canal system, and are also separated into primary and secondary canal reaches. From Fig. 4.41 it can be seen that the three tertiary units (Blendren I–III) on the Jambuwok II secondary canal are receiving higher water supplies per relative

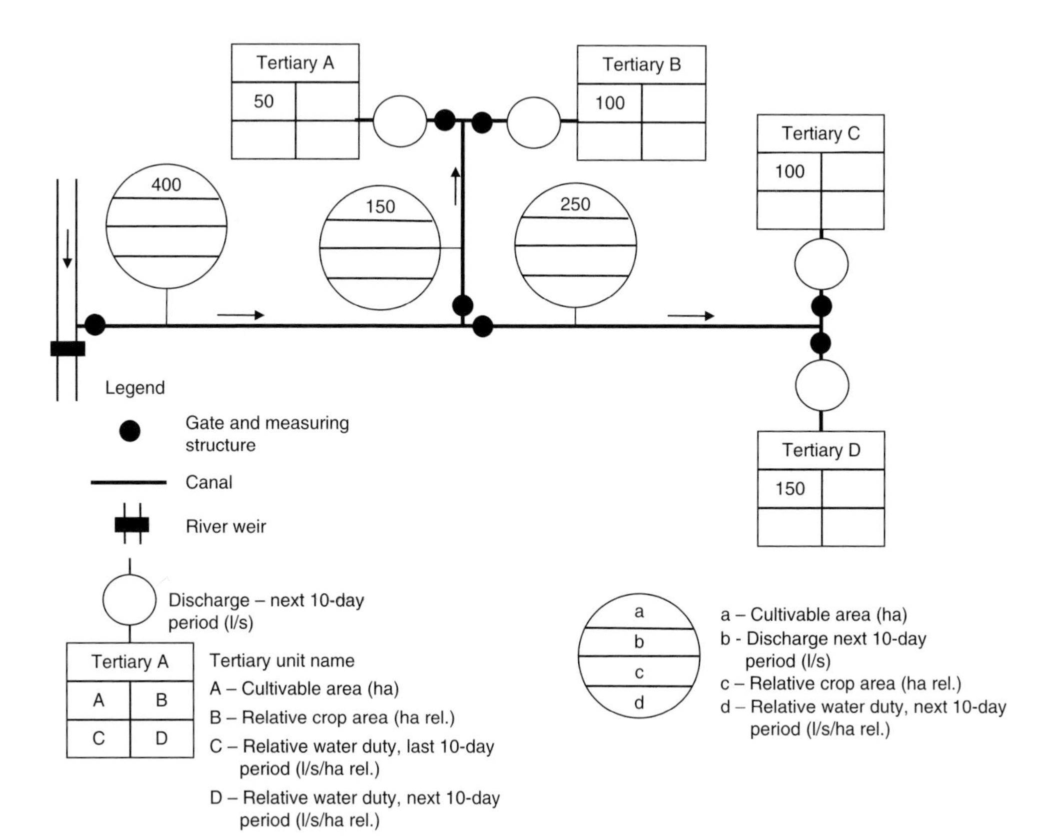

Fig. 4.40. Schematic map of irrigation system.

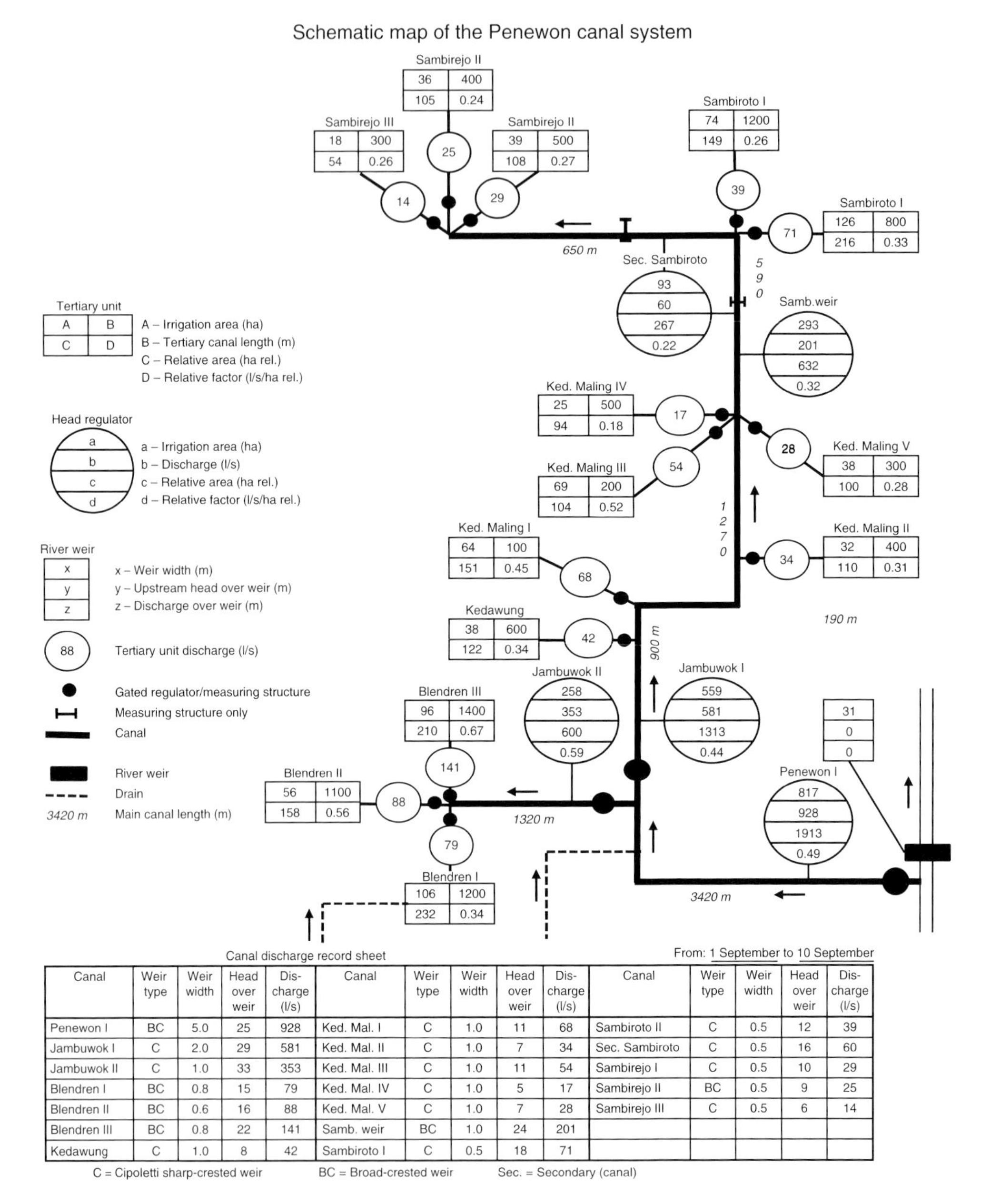

Canal discharge record sheet

From: 1 September to 10 September

Canal	Weir type	Weir width	Head over weir	Dis-charge (l/s)	Canal	Weir type	Weir width	Head over weir	Dis-charge (l/s)	Canal	Weir type	Weir width	Head over weir	Dis-charge (l/s)
Penewon I	BC	5.0	25	928	Ked. Mal. I	C	1.0	11	68	Sambiroto II	C	0.5	12	39
Jambuwok I	C	2.0	29	581	Ked. Mal. II	C	1.0	7	34	Sec. Sambiroto	C	0.5	16	60
Jambuwok II	C	1.0	33	353	Ked. Mal. III	C	1.0	11	54	Sambirejo I	C	0.5	10	29
Blendren I	BC	0.8	15	79	Ked. Mal. IV	C	1.0	5	17	Sambirejo II	BC	0.5	9	25
Blendren II	BC	0.6	16	88	Ked. Mal. V	C	1.0	7	28	Sambirejo III	C	0.5	6	14
Blendren III	BC	0.8	22	141	Samb. weir	BC	1.0	24	201					
Kedawung	C	1.0	8	42	Sambiroto I	C	0.5	18	71					

C = Cipoletti sharp-crested weir BC = Broad-crested weir Sec. = Secondary (canal)

Fig. 4.41. Schematic map used to display operational data (C = Cipoletti sharp-crested weir; BC = broad-crested weir; Sec. = secondary canal).

area (ranging from 0.34 to 0.67 l/s/ha rel.) than the tertiary units on Jambuwok I secondary canal, with the tail-end tertiaries Sambirejo I–III getting very low unit water supplies (0.24–0.27 l/s/ha rel.). In this case the system does not appear to be operated correctly, and further investigation is required to identify the cause of these differences in water supply to different locations within the command area.

In Fig. 4.41 colour coding the Relative Factor (unit water supply, l/s/ha rel.), as outlined in Fig. 4.4, can be used to highlight the variation in water supplies across the system.

The representation of data in a schematic diagram format is sometimes easier for operational staff to understand than data presented in tables, and is also useful where the operation planning and monitoring is computer-based.

Endnotes

[1] Maintenance planning is an integral part of operation planning. More details on maintenance planning are provided in Chapter 6.
[2] The exception to this is where storage is provided, such as with the on-line enlarged minor (secondary) canals in the Gezira Irrigation Scheme in Sudan, or with night-storage reservoirs. See later section on rotated flows.
[3] It is worth noting that there are detailed manuals on how to carry out accurate flow measurement using floats. In this case there is a range of float sizes for different channel depths and a range of reduction factors. These methods have generally been superseded by the advent of current meters.
[4] The head loss across the structure can be more in locations where there is sufficient head available.
[5] Defined as the average energy per unit of water at a channel section with respect to the channel bottom.
[6] Sediment removal upstream of the structure is thus required periodically to maintain the intended upstream cross-section and to reduce the approach velocity.
[7] The change in reach storage can be positive or negative.
[8] In these systems the check structures (cross regulators) have stop logs or bars, which are used to regulate the flow and maintain supply level.
[9] This is achieved by using walkie-talkies or mobile phones.
[10] An additional benefit to covering a larger area was that more income could be generated from water taxes.
[11] Tertiary unit.
[12] Primary canals.
[13] Secondary canals.
[14] Full supply level (FSL) means that when the canal flows the discharge is such that the canal runs full, at the design depth. This maintains the required head over the fixed discharge outlets to obtain the design flow through offtaking outlets.
[15] Cusecs=cubic feet per second (ft^3/s); cumecs=cubic metres per second (m^3/s).
[16] This is somewhat surprising, with those in the tail reaches being most affected, though the allowance for Jharai is some compensation.
[17] When canals flow with low discharges the wetted perimeter per unit discharge is high, and losses are correspondingly high.
[18] Secondary canal discharge = summation of tertiary canal discharges divided by the losses factor = (100 – % losses)/100.
[19] The 10-day time period breaks down into three time periods per month. In a 30-day month there are three 10-day time periods, in 31-day months there are two 10-day time periods and one 11-day time period.

5

Operation at the On-Farm Level

This chapter details the procedures for operation within the tertiary unit at the on-farm level. The building blocks for formulating irrigation schedules are presented, including identification of soil types and their water-holding capacity, estimation of crop evapotranspiration rates and determination of crop and irrigation water requirements. Losses of irrigation water within the tertiary unit are discussed together with measures to reduce these losses and improve irrigation efficiency and productivity. Procedures for organizing the rotation of irrigation water supplies are discussed together with worked examples.

Irrigated Soils

The physical and chemical composition of the soil is fundamental to irrigated agriculture. Soils are made up principally of solids, liquids and gaseous materials. The main solids are the soil or rock particles but there is also organic matter, such as plant roots, and mineral deposits, such as gypsum. The liquid content comprises water, dissolved minerals and soluble organic matter. The liquid fills the voids between the solid particles in the soil matrix. The gaseous or vapour portion is mainly air and occupies the space not occupied by the solid or liquid elements. It is important to have air in the root system as most crops, other than rice, require aeration of their roots. If there is no air, such as when the soil is waterlogged, most plants will perish, or suffer reductions in yield.

Soil texture

The *texture* of a soil depends on: (i) the size; and (ii) the distribution of the soil particles.

Thus soils can be described as:

General terminology	Soil terminology	Particle size
Coarse	Sand	0.05–2 mm
Medium	Silt	0.002–0.05 mm
Fine	Clay	<0.002 mm

The coarse, medium and fine terminologies are also sometimes referred to as light, medium and heavy, respectively.

There are different terminologies for classifying soils according to its texture, or mix of different soil particle sizes. Figure 5.1 shows one internationally recognized classification system, the US Department of Agriculture (USDA) soil textural triangle. Using this terminology the soil can be classified into 12 categories as:

©Martin Burton 2010. *Irrigation Management: Principles and Practices* (Martin Burton)

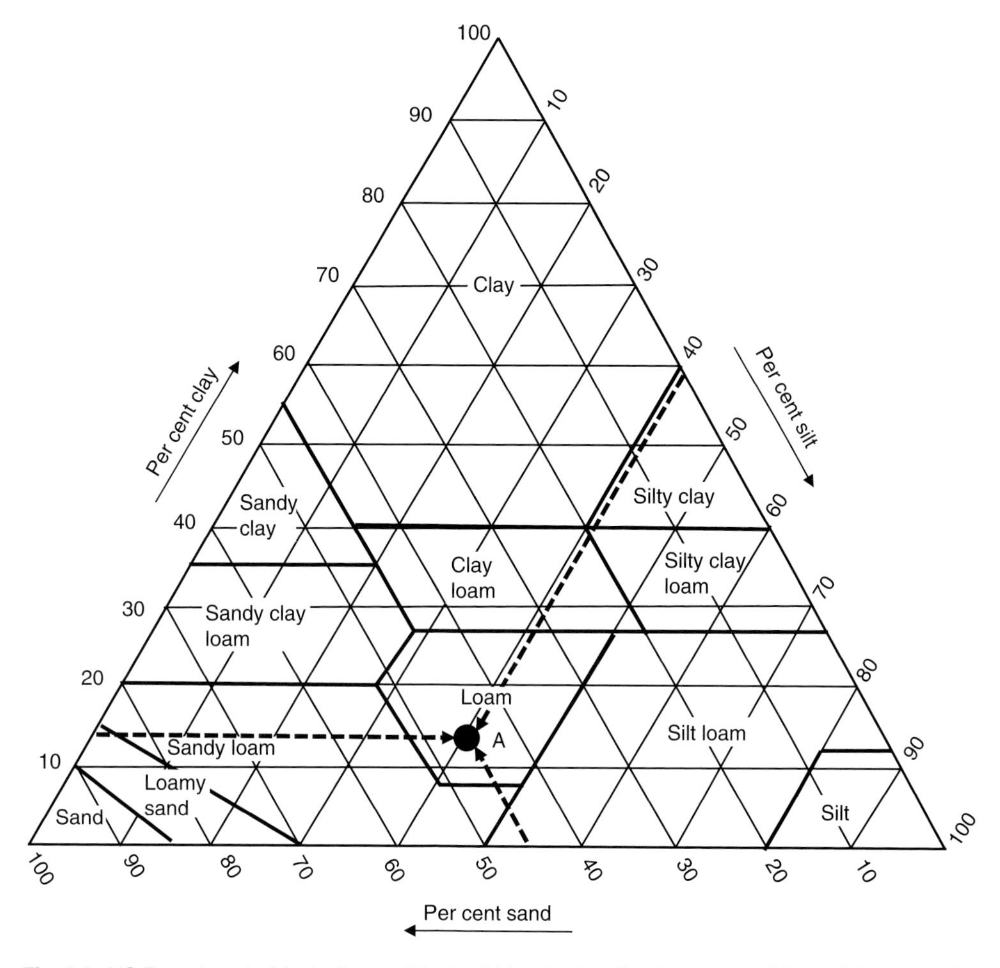

Fig. 5.1. US Department of Agriculture soil textural triangle classification system. (From FAO, 1984 with permission from the Food and Agriculture Organization of the United Nations.)

- Clay
- Sand
- Silt

- Sandy clay
- Silty clay
- Sandy clay loam

- Clay loam
- Silty clay loam
- Sandy loam

- Loam
- Silt loam
- Loamy sand

Thus soil containing 13% clay, 41% silt and 46% sand would be located at point A in the triangle and termed a 'loam'. An increase in the proportion of silt in this soil would classify it as a 'silt loam', while an increase in the proportion of sand would classify it as a 'sandy loam'.

The water-holding capacity of a soil is closely related to its texture. A clay soil, for example, can hold more water per unit of volume than a sandy soil. As a result a clay soil does not have to be irrigated as frequently as a sandy soil. Thus for scheduling purposes it is essential that the texture of the soil is known.

Determining the texture of a soil

The texture of a soil can be accurately determined by analysis in a soils laboratory. There are also several methods to determine the texture based on the 'feel' of the soil when rubbed in the hand (FAO, 1984).

1. *Sand* is free flowing with individual grains, which can be seen or felt when rubbed between thumb and forefinger. If squeezed in the hand when dry the soil will disintegrate when the hand is opened. If squeezed when moist it will form the shape of the clenched hand but will disintegrate when touched.
2. *Sandy loam* contains a large proportion of sand, but has sufficient silt and clay to make it slightly cohesive. The sand grains can be felt when the soil is rubbed between thumb and forefinger. If squeezed when dry it will form a shape that will easily fall apart when the hand is opened; if there is some moisture in the soil the shape will remain without disintegrating too easily.
3. *Loam* soil has a relatively equal mix of all sand, silt and clay. The sand particles can be felt when the soil is rubbed between thumb and forefinger, and the soil feels slightly plastic and malleable when moist. A dry soil when squeezed will hold its shape when the hand is opened, while a moist soil when squeezed can be handled quite freely and retain its shape after being squeezed.
4. *Silt loam* has a low percentage of sand particles and a low percentage of clay particles. When dry it forms into lumps, which can easily be broken up. The soil is smooth and soft when rubbed between thumb and forefinger, and feels rather like ground flour. When dry or moist it can be squeezed into a cylinder which can be handled a fair bit without breaking up. If moistened and rolled into a cylinder it will not, however, hold together, and will break up.
5. *Clay loam* is a fine-textured soil which breaks into lumps when it is dry. When moist it feels soft and silky when rubbed between thumb and forefinger. It is difficult to squeeze into a shape when dry, but when moist will easily take a shape and can be handled a fair amount without breaking up. If moistened and rolled out into a cylinder it will hold its shape but will break up if it is manipulated to form a circle.
6. *Clay* is fine-textured soil that forms into large hard lumps or clods when it is dry, and becomes difficult to work. When moist it feels soft and silky when rubbed between thumb and forefinger, and can be rolled into a thin cylinder, which can be formed into a circle without breaking up. Soils with a high percentage of clay are very plastic and sticky when wet.

Figure 5.2 shows a flowchart that can be used to identify the soil texture, while Box 5.1 shows another method for determination of soil texture. Using a mixture of the methods and descriptions outlined here enables a fairly good determination of soil texture in the field.

Soil Structure

The *structure* of the soil relates to the adhesion between the soil particles and the tendency to form larger blocks of soil. On the soil surface the structure is controlled by the ploughing and working of the soil, which results in breaking up the soil structure into finer blocks, or tilth. Below the plough layer the soil structure strongly influences the permeability of the soil, and the ability for water and air to move through the soil (Fig. 5.3). Single-grain soil structures, such as sand, are highly permeable; prismatic structures are moderately permeable; while plate-like structures, such as heavy clays, have poor permeability.

For growing rice the ploughing and working (often termed 'puddling') of clay soils when saturated is a deliberate action to destroy the soil structure and create an impermeable layer below the crop's root zone. This destruction of the soil structure can have an adverse impact on crops that follow on from rice as the movement of air in the soil is restricted.

Soil structure, unlike soil texture, can be changed by farming practices. It can be improved by rotation of crops, and good cultural practices. Cycles of wetting and drying, and freezing and thawing, can improve soil structure. Poor cultural practices can also destroy the soil structure; ploughing or working the soil when it is too wet will puddle the soil and compress it, while ploughing or working the soil when it is too dry may powder the soil. The chemical content of the soil can also affect its structure; heavy concentrations of alkali salts cause a deterioration of the structure, making it impermeable to water.

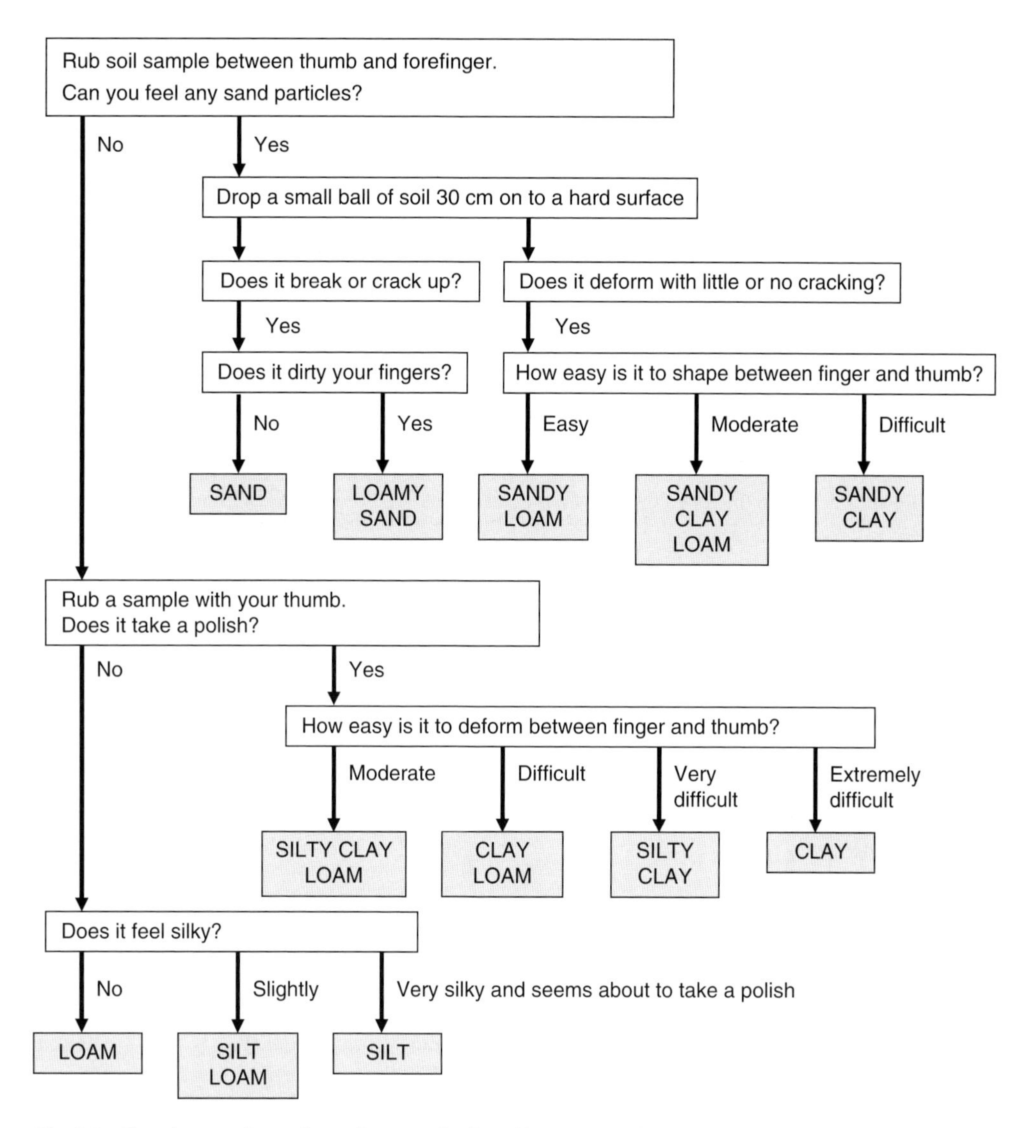

Fig. 5.2. Flowchart to determine soil texture by feel. (Courtesy of Cranfield University.)

Soil Moisture-holding Capacity

The ability of the soil to store water is central to irrigation scheduling. The more water that can be stored in the soil the less frequent the irrigations required by a crop and the less risk that the crop will suffer a shortage of water.

Soil comprises particles of soil and other matter, touching each other and leaving space in between the particles. This space is called 'pore space', and comprises between 40 and 60% of the total soil volume.

From an irrigation point of view there are different levels of water content in the soil, and four terms are used to identify these water content levels:

- saturation;
- field capacity;
- permanent wilting point;
- available soil water.

Each of these is discussed briefly below.

Box 5.1. Determination of Soil Texture by Manipulation

Instructions

1. Take a small sample of soil, about 2 cm in diameter, and moisten it until it starts to stick to the hand.
2. In the ball of your hand, or on a flat surface, try to roll the soil into a cylinder, about 5–7 mm in diameter.
3. The ability of the soil to be rolled into a cylinder is indicative of its soil texture, as set out below.

Classification

- *Sand*: The soil will not form into a cylinder. It remains loose and will only form a cone.
- *Sandy loam*: The soil has some cohesion and can be shaped into a ball, though it falls apart easily. It cannot be rolled into a cylinder.
- *Silt loam*: The soil has more cohesion than the sandy loam and can be rolled into a thick cylinder (6–9 mm). It breaks up if attempts are made to roll it into a thin cylinder.
- *Loam*: The soil can be rolled into a thin cylinder (4–6 mm) but will break when bent.
- *Clay loam*: The soil can be rolled into a cylinder and can be bent into a 'U' shape without breaking up.
- *Silty clay*: The soil can be rolled into a cylinder and can be bent into a circle, but with cracks.
- *Heavy clay*: The soil can be rolled into a cylinder and can be bent into a circle, without any cracks.

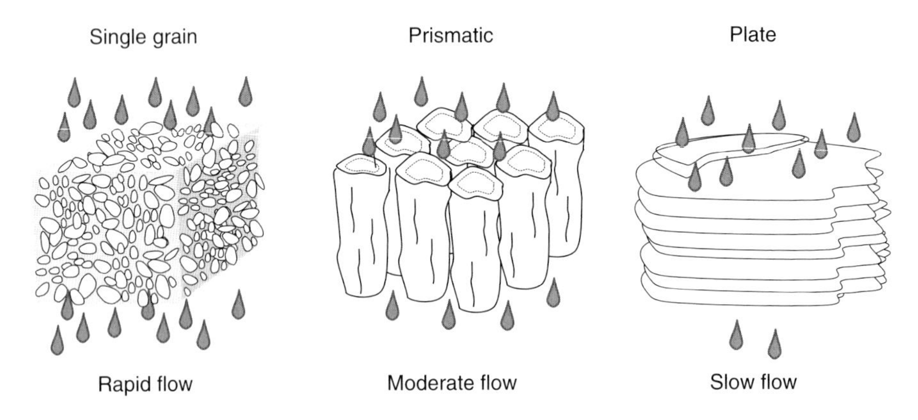

Fig. 5.3. Varieties of soil structure and the influence on permeability.

Saturation

During and immediately after irrigation all the pore space in the soil is filled with water and the soil is saturated (Fig. 5.4a). There is little air in the soil, and for most crops (other than rice) if the soil stays saturated the crop will be damaged due to this lack of air for the roots to breathe. If there are no drainage problems the water in the soil will drain away under gravity following irrigation, leaving space for air in the soil's pore space.

Field capacity

Field capacity is the quantity of water held in the soil once the water has drained away from the saturated soil (Fig. 5.4b). This water is held to the soil particles by surface tension forces, and much of it is available for taking up by the plant's roots.

The volume of water held by the soil at field capacity depends primarily on its texture and structure. The forces holding the water in the soil against the gravitational pull are surface tension forces. Soils with small particle size, such as silts and clays, have a large surface area and thus can hold more water. For example, 1 m^3 of loam soil has soil particles with an internal surface area of about 70,000 m^2, thus a maize plant rooting to 120 cm depth has an internal surface area of 7 ha from which to draw water and nutrients. At field capacity 1 m^3 of sandy soil will typically hold 135 l of water, a loam soil will hold

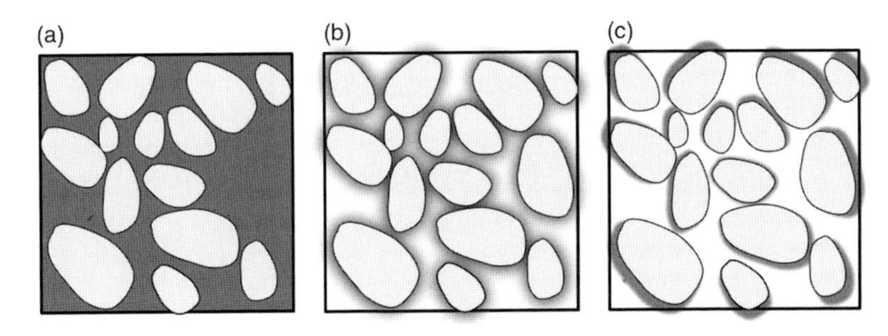

Fig. 5.4. Soil moisture at different stages of moisture content. (a) Saturation – all pores filled with water, little or no air; this situation occurs during and immediately following irrigation or rainfall. (b) Field capacity – water is held in the soil after surplus has drained away under gravitational forces. (c) Permanent wilting point – water attached by surface tension forces to soil particles, cannot be removed by plant root suction.

about 270 l of water and a clay soil will hold about 400 l of water.

Permanent wilting point

Water can be removed from the soil by the plant's roots exerting a greater pull or tension than the surface tension holding the water to the soil particle's surface. At some point, termed the permanent wilting point, the suction exerted by the plant's roots is not sufficient to remove the water from around the soil's particles (Fig. 5.4c). At this point, if additional water is not added by rainfall or irrigation, the crop will become stressed, the yield will be reduced and the crop may perish.

At permanent wilting point a crop's leaves may droop or wilt; in some crops, such as fruit trees, there will be a change in appearance in the leaf colour. Drooping or wilting of a crop's leaves does not always signify that the permanent wilting point has been reached; in some cases this is caused by the crop's inability to withdraw water quickly enough from the root zone. This is typified by drooping or wilting of a crop's leaves in the afternoon and recovery overnight, particularly on very hot days when evapotranspiration rates are high.

The permanent wilting point is affected by the soil texture in the same way as with field capacity, thus for fine-textured soils the moisture content at permanent wilting point is higher than for coarse-textured soils.

Available soil water

The water available to the plant is the difference between the moisture content at field capacity and that at the permanent wilting point. Though there may still be water in the soil at the permanent wilting point it cannot be removed by the plant, and is thus unavailable.

The objective of irrigation is to allow the soil moisture to reduce to a safe limit (above the permanent wilting point) and then to irrigate the soil to bring it back to field capacity. The interval between irrigation will thus depend on the available moisture in the soil and the rate at which the soil water is abstracted by the crop.

Table 5.1 summarizes the soil moisture situation for different soil types. It is worth noting that per metre depth clay has total available soil moisture content almost four times that of sand, while a loam soil has almost twice as much water available to the crop as a sandy soil.

Figure 5.5 presents the data from Table 5.1 for a 1 m depth of soil. The difference in the total available water for each soil type can clearly be seen.

Field Estimation of Soil Moisture Status

It is possible to make assessments of the soil moisture status in the field through taking samples of the soil at different depths in the

Table 5.1. Typical moisture content levels for different soil textures.

Soil texture	Bulk density (g/cm^3)	Soil moisture content: Saturation (mm/m)	Field capacity (mm/m)	Permanent wilting point (mm/m)	Total available soil water (mm/m)
Sand	1.65	380	150	70	80
Sandy loam	1.50	430	210	90	120
Loam	1.40	470	310	140	170
Clay loam	1.35	490	360	170	190
Silty clay	1.30	510	400	190	210
Clay	1.25	530	440	210	230

root zone. For this a simple soil auger can be used, with soil samples typically being taken for depths from 0 to 30 cm, 30 to 60 cm, 60 to 90 cm and 90 to 120 cm. The samples are then weighed, dried in an oven and then weighed again to determine the water content.

The moisture content can also be estimated by the feel of the soil in the hand. The soil is squeezed in the palm of the hand three or four times and the behaviour of the soil observed to see how it behaves when formed as a ball and tossed in the air, and when it is rolled into a cylinder. Table 5.2 can then be used to assess the status of the soil moisture for different soil textures.

The soil moisture deficit shows the amount of water needed (per metre depth of soil) to bring the soil back to field capacity. This thus represents the irrigation depth needed. Thus for a coarse-textured soil with 50% depletion of total available soil moisture, 45 mm depth of water is required to return 1 m depth of soil to field capacity. For a fine-textured soil at 50% depletion, 100 mm depth of water is required to return 1 m depth of the soil to field capacity.

In good irrigation practice the soil moisture content over the root depth is checked before irrigation and then again after irrigation to determine if the water has penetrated the full depth of the root zone but not deeper. Checking the soil moisture status before and after irrigation provides the farmer with valuable information on the amount of water to apply for each

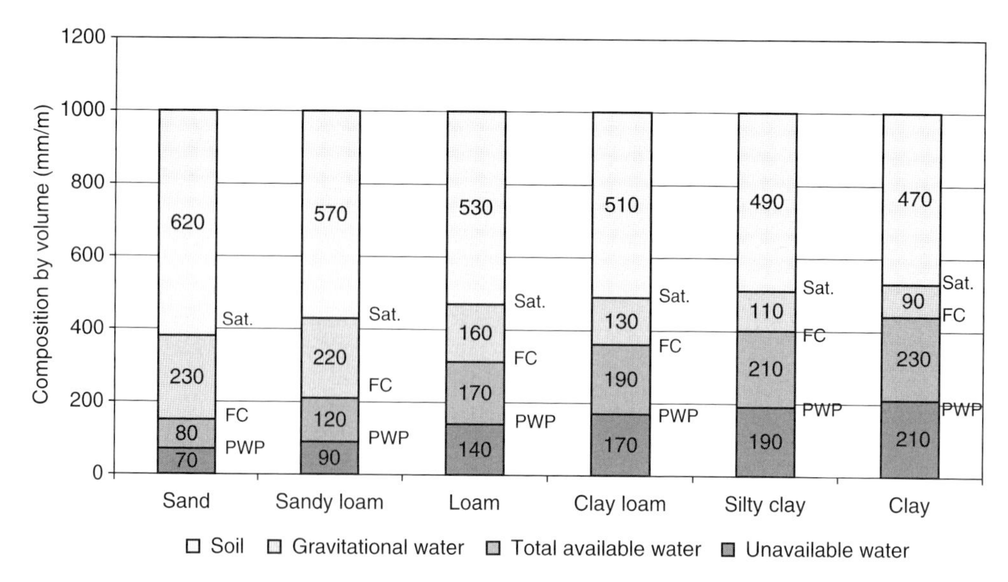

Fig. 5.5. Moisture categories for different soil textures (Sat., saturation; FC, field capacity; PWP, permanent wilting point).

Table 5.2. Field guide for estimating how much available water has been removed from the soil (adapted from Israelson and Hanson, 1962).

Level of depletion of available soil moisture (%)	Feel or appearance of the soil and moisture deficiency in millimetres per metre of soil depth (mm/m depth)			
	Coarse texture (sand)	Moderately coarse texture	Medium texture	Fine and very fine texture (Clay)
0 (Field capacity)	When squeezed: no free water appears on the soil; wet outline is left on the hand 0 mm/m	When squeezed: no free water apparent on the soil surface; wet outline of the soil ball is left on the hand 0 mm/m	When squeezed: no free water appears on the soil; wet outline of the ball is left on the hand 0 mm/m	When squeezed: no free water apparent on the soil surface; wet outline of the ball is left on the hand 0 mm/m
0–25	Sticks together slightly, may form a very weak ball under pressure 0–20 mm/m	Forms a weak ball, breaks easily 0–35 mm/m	Forms a ball, is very pliable, slicks readily if relatively high clay content 0–40 mm/m	Easily forms a ribbon between the fingers, has a smooth feeling when rubbed 0–50 mm/m
25–50	Appears to be dry, some cohesion, will not form a ball with pressure 20–45 mm/m	May form a ball under pressure, but will not hold together 35–70 mm/m	Forms a ball, somewhat plastic feel, will slick slightly with pressure 40–85 mm/m	Forms a ball, will form a ribbon between thumb and forefinger 50–100 mm/m
50–75	Appears to be dry, some cohesion, will not form a ball with pressure 40–70 mm/m	Appears to be dry, will not form a ball when squeezed hard 65–100 mm/m	Crumbles, but will hold together under pressure 80–125 mm/m	Pliable, will form a ball under pressure 100–160 mm/m
75–100 (100% represents the permanent wilting point)	Dry, loose, single grained, flows through the fingers 65–85 mm/m	Dry, loose, flows through fingers 100–125 mm/m	Powdery, dry, easily broken into powdery condition 125–170 mm/m	Hard, baked, cracked, sometimes has loose crumbs on the surface 160–210 mm/m

[a] 'Slick' means that the soil slides when rubbed between thumb and forefinger.
[b] The soil moisture deficit shows the amount of water needed (per metre depth of soil) to bring the soil back to field capacity. This thus represents the irrigation depth needed. Thus for a coarse textured soil with 50% depletion of total available soil moisture, 45 mm depth of water is required to return 1 m depth of soil to field capacity. For a fine textured soil at 50% depletion, 100 mm depth of water is required to return 1 m depth of the soil to field capacity.

irrigation, and helps to reduce wastage through over-irrigation. To check the depth of penetration the soil samples at different depths should be taken 1 (for sand) to 3 (for clays) days after irrigation, as the water takes time to percolate through the soil horizon.

Other methods for determining the soil moisture status include use of tensiometers, gypsum blocks, theta probes or neutron probes, which give a reading of the soil moisture tension in the soil, which can be converted to soil moisture content. These methods are not commonly found on smallholder irrigation systems.

Wetting Profiles for Different Soil Textures

The wetting profile varies with the soil texture. For sandy soils the water tends to move

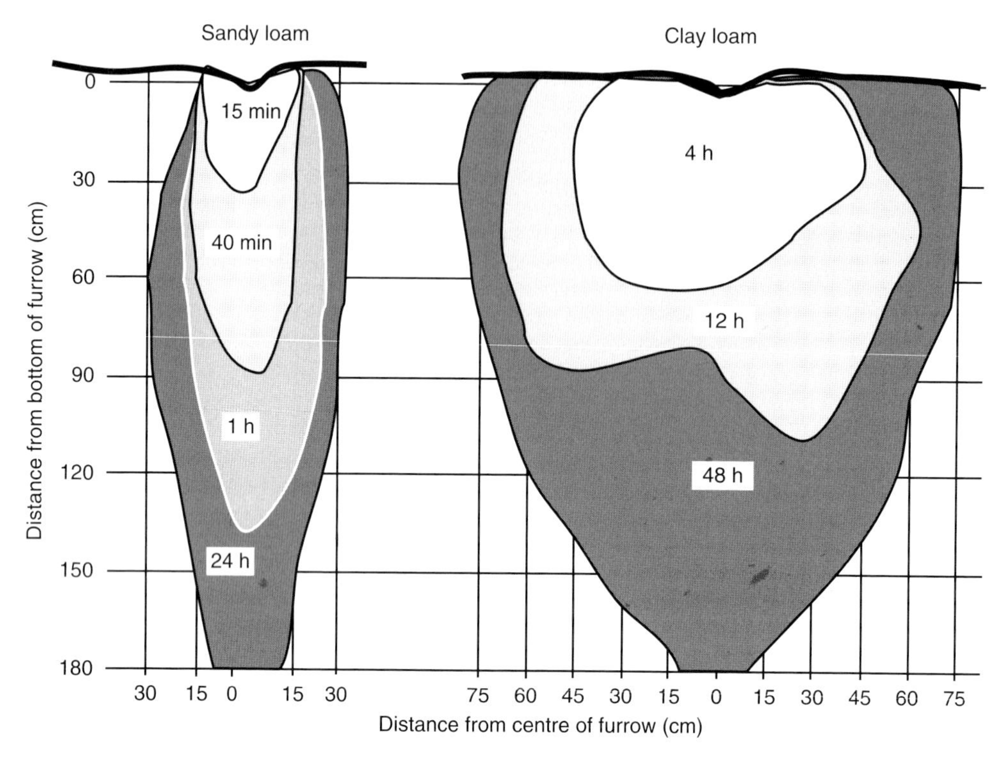

Fig. 5.6. Distribution of soil water in different soils using furrow irrigation. (From FAO, 1984 with permission from the Food and Agriculture Organization of the United Nations.)

quickly down vertically under gravity, for clay soils the vertical movement is less rapid and the water tends to spread horizontally as well as vertically. Figure 5.6 shows typical wetting profiles for a sandy loam and a clay loam, together with the depth the water has infiltrated over different timeframes.

Understanding how the water moves through different soils is important as it enables the irrigator to control the water application to suit the soil type. For example on coarse-textured (sandy) soils, furrows need to be closer in order to irrigate the upper root zone and the irrigation duration needs to be shorter to avoid deep percolation below the root zone.

Cropping Patterns

A cropping pattern diagram is used to show:

- the area under each crop type at any time;
- the percentage of the total area that is cropped;
- the cropping intensity.

Knowing the cropping pattern is an essential prerequisite for determining the water requirements for the irrigation season.

Preparing a cropping pattern

Figure 5.7 shows a typical cropping pattern diagram drawn using the data presented in Table 5.3.

As shown in Table 5.3, in order to plot the cropping pattern diagram the following information is required:

- The crop types.
- For each crop type:
  - the start date of planting;
  - the end date of planting;
  - the area planted;
  - the growth duration;

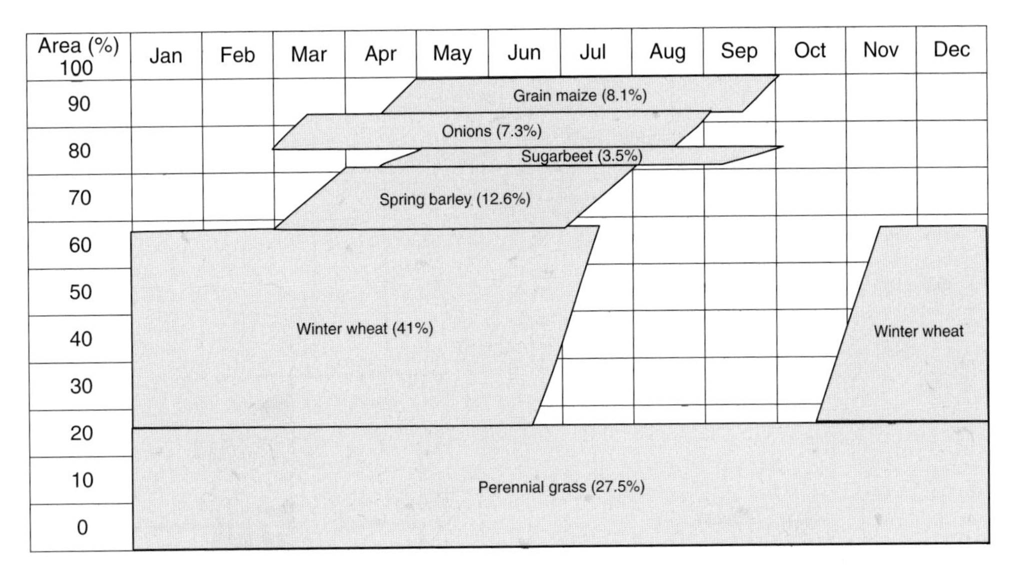

Fig. 5.7. Example of a cropping pattern showing percentage of area planted.

Table 5.3. Tabulation of cropping pattern data (for Fig. 5.7).

	Area planted		Planting period			Harvesting period	
Crop	(ha)	(%)	Start	End	Total crop growth period (days)	Start	End
Winter wheat	864	41.0	15 Oct	15 Nov	240	15 June	15 July
Spring barley	266	12.6	1 Mar	1 Apr	140	1 July	1 Aug
Sugarbeet	73	3.5	15 Apr	1 May	150	15 Sept	1 Oct
Grain maize	170	8.1	15 Apr	1 May	150	15 Sept	1 Oct
Perennial grass	581	27.5	15 Mar	1 Apr	365	15 Mar	1 Apr
Onions	155	7.3	1 Mar	15 Mar	180	15 Aug	1 Sept
Total	**2109**	**100**					

- the start date of harvesting;
- the end date of harvesting.

The difference between the start and end date of planting is called the 'crop stagger'. This spread of planting of the crop is due to farmers not all wanting to plant on the same date, and also to other factors such as shortage of labour or machinery for land preparation. The crop stagger is different for each crop type and can last several days to several weeks.

The area of each crop shown in Fig. 5.7 can be plotted either as the actual area with the maximum point on the *y*-axis being the total command area (2109 ha in this case) or as a percentage of the total command area. What is important is to show the relative scale of the different areas, and what fraction or percentage of the total command area is planted.

Cropping patterns are generally plotted on graph paper. To plot the data for each crop, mark the start and end dates of planting and the start and end dates of harvesting. The start and end date marks should be separated by a vertical distance equal to the crop area planted. Connect the four points in a parallelogram, as shown in Fig. 5.7. Repeat the process for each crop.

Plotting cropping patterns get more complicated when farmers plant more than one crop per year. If for the data above the farmer follows the winter wheat with 210 ha of carrots then the cropping pattern diagram will look as in Fig. 5.8.

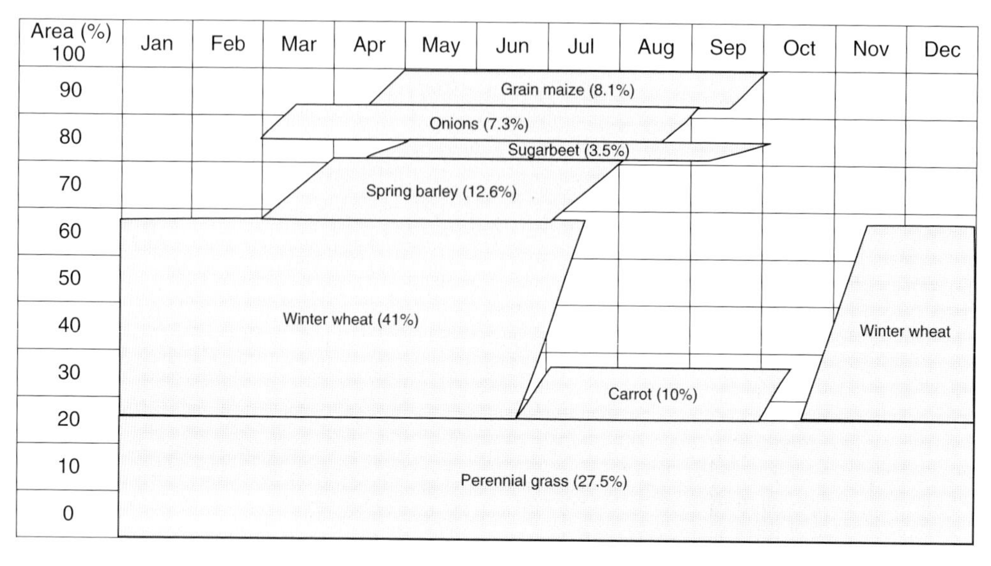

Fig. 5.8. Cropping pattern with double cropping.

In the first case the cropping intensity is 100%, as 2109 ha out of a total command area of 2109 ha has been cultivated in the year. In the second case the cropping intensity is 110% as an additional area of 210 ha of carrots has been planted immediately following the harvesting of the winter wheat.

Crop Rooting Depths

The rooting depth of the crop determines the size of the soil water reservoir available to the crop and thus the irrigation scheduling. The deeper the rooting depth the greater the soil water reservoir available to the crop, the greater the volume of water that can be applied during each irrigation and the greater the interval between irrigations. Table 5.4 gives typical rooting depths for selected crops.

The water extraction is not uniform over the rooting depth of the crop (Fig. 5.9), with the majority of the water being extracted from the top 50% of the root zone. Water extraction by the deeper roots becomes more important, however, as the moisture content in the soil decreases. It is thus important not to over-irrigate a crop during the early root development stage; it is better to stress it to some degree so that the roots develop and seek out the water in the lower profiles of the soil.

Infiltration

The infiltration rate of the soil determines how long it takes to fill the crop's root zone with water. The infiltration rate is high in the initial instance when the soil is wet, and then

Table 5.4. Typical rooting depths for selected crops. (From FAO, 1977 with permission from the Food and Agriculture Organization of the United Nations.)

Crop	Rooting depth (m)
Beans	0.5–0.7
Beets	0.6–1.0
Cabbage	0.4–0.5
Citrus	1.2–1.5
Cotton	1.0–1.7
Grapes	1.0–2.0
Grass	0.5–1.5
Groundnuts	0.5–1.0
Lucerne	1.0–2.0
Maize	1.0–1.7
Onions	0.3–0.5
Potatoes	0.4–0.6
Sorghum	1.0–2.0
Soybeans	0.6–1.3
Sugarbeet	0.7–1.2
Tobacco	0.5–1.0
Vegetables	0.3–0.6
Wheat	1.0–1.5

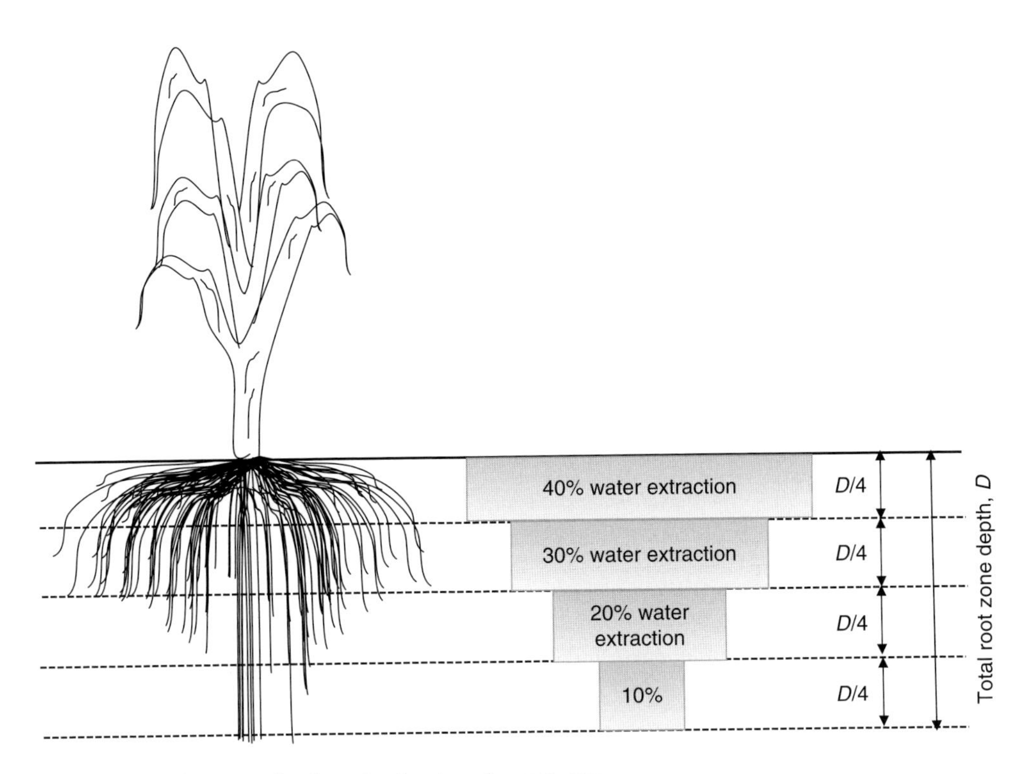

Fig. 5.9. Typical pattern of water extraction from the root zone.

slows down as the soil at the surface becomes saturated, after which the infiltration rate reaches a steady state.

Importance of determining the infiltration rate

The soil's infiltration rate is a key factor in obtaining an adequate and uniform irrigation application on a field. To obtain this it is important to match the horizontal speed of the irrigation streamflow over the surface of the soil with the vertical infiltration rate of the water into the soil. In a typical field of cotton irrigated by furrow irrigation, water has to travel vertically about 60 cm in the time it takes the surface flow to travel 300 m.

Figure 5.10 shows the consequences of good and bad irrigation at the field level. In Fig. 5.10a the surface flow rate and the infiltration rate are balanced, with a uniform wetting of the root zone and very little deep percolation or surface runoff at the bottom of the field. In Fig. 5.10b the surface flow rate and the infiltration rate are not balanced, there is excessive runoff at the tail end and excessive infiltration below the root zone, leading to inefficient irrigation and wastage of water.

Typical infiltration rates for different soil textures

The infiltration rate is determined by several factors as well as the soil texture. These factors include the cultivation practices, tillage operations and the movement of machinery over moist soil (which leads to compaction of the soil and a reduction in the infiltration rate). Figure 5.11 shows typical figures for infiltration rates for three types of soil: sand, loam and clay. It can be seen in all cases that the infiltration rate starts off high and then attenuates to a uniform or terminal rate. As expected sand has the fastest rate, starting above 125 mm/h and reaching its terminal rate of about 55 mm/h after 2 h.

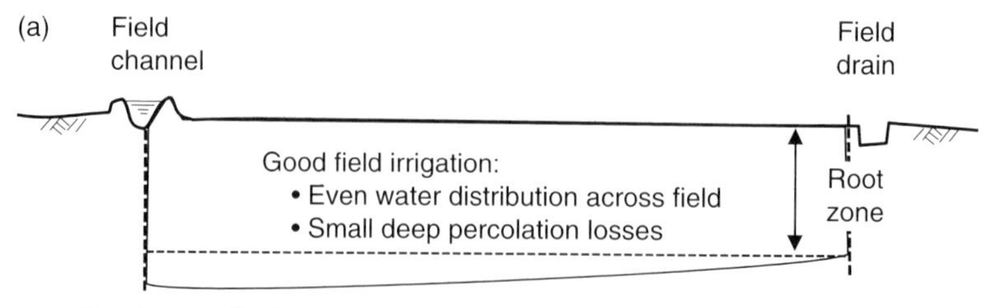

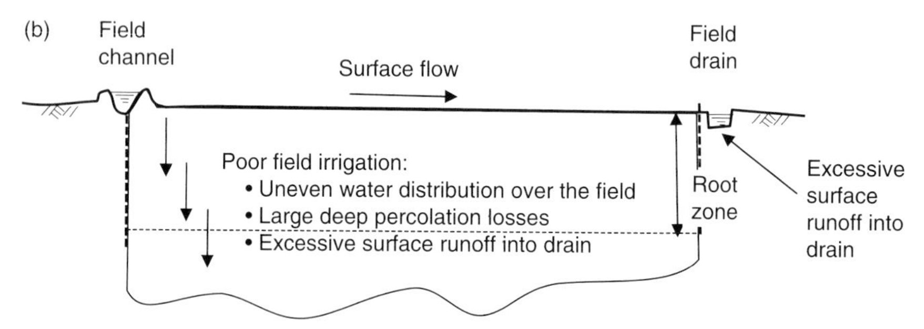

Fig. 5.10. (a) Good and (b) poor water distribution on a field as a consequence of the balance between surface flow rate and the soil's infiltration rate.

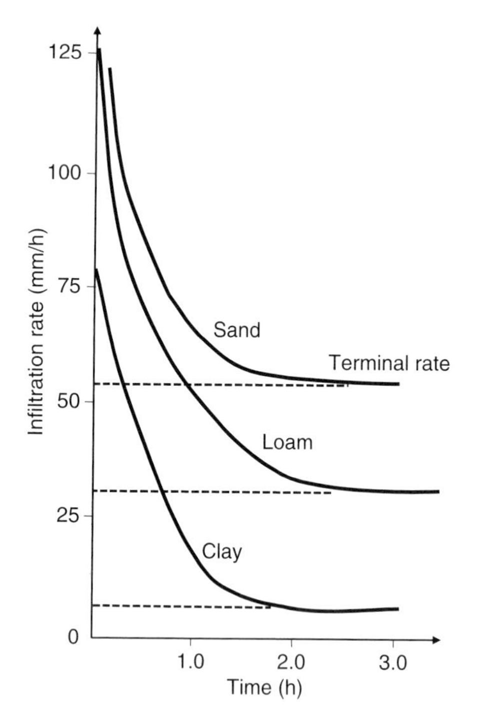

Fig. 5.11. Infiltration rates of different soils. (Modified from Withers and Vipond, 1974.)

Loam has a slower infiltration rate, starting at about 125 mm/h and reaching its terminal rate of about 30 mm/h after 2 h. Clay has the slowest infiltration rate, starting at about 75 mm/h and reaching its terminal rate of about 10 mm/h after 1.5 h (this is due to the clay particles swelling on application of water and closing the pore spaces). Table 5.5 gives the range of terminal infiltration rates for different soil textures.

Measuring the infiltration rate

Equipment

The infiltration rate can be measured using a simple set of equipment comprising two concentric rings (Fig. 5.12), one of 30 cm diameter, the other of 60 cm diameter. The rings are made of 3–4 mm thick and 25 cm wide mild steel. Other equipment required includes a stopwatch (an ordinary watch with a seconds hand can also be used), a measuring flask graduated in cubic centimetres, and a hook gauge. The hook gauge can be made

Table 5.5. Terminal infiltration rates for different soil textures. (From FAO, 1984 with permission from the Food and Agriculture Organization of the United Nations.)

Classification	Terminal infiltration rate (mm/h)	Soil texture
High	30 to >80	Sandy loam; sandy clay loam
Medium high	15 to 30	Loam, silt loam
Medium low	5 to 15	Clay loam, clay, silty clay loam
Low	2 to 5	Clay, heavy clay

from a 20 cm length of 1–2 mm diameter wire bent into an ‘S’ shape hooked over the side of the inner ring. The end measuring the water level should be sharpened to a point.

The purpose of the outer ring is to stop the infiltrated water under the inner ring from spreading sideways. This would increase the wetting front and thus increase the infiltration rate. It is important therefore to fill the inner and outer rings at the same time, and to keep the water levels in both rings approximately the same.

Procedure

The procedure for carrying out the test is as follows.

1. The 30 cm diameter ring is hammered 10 cm into the soil and then the 60 cm ring is hammered into the soil to the same depth around the smaller ring. The hook gauge is positioned securely on to the inner ring to measure the water level in the inner ring, with

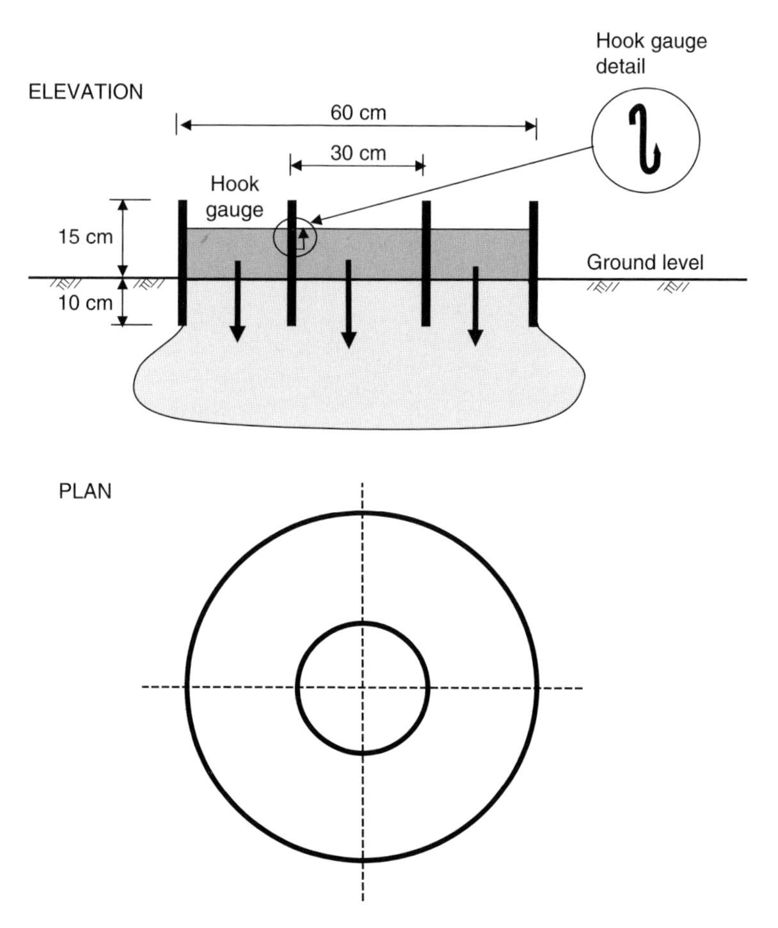

Fig. 5.12. Infiltration test equipment.

the tip of the gauge positioned approximately 5 cm below the rim of the ring (to give a water depth in the ring of approximately 10 cm).
2. Water is added carefully to the inner ring, taking care not to disturb the soil surface. To reduce the risk of disturbing the soil surface the water can be poured on to a cloth laid on the soil surface until an adequate depth of water has been achieved. The water is added to the outer ring at the same time, to about 10 cm depth. The water in the inner ring is brought up to the level of the tip of the hook gauge and the timing started.
3. At intervals a measured quantity of water is added to the inner ring to bring the water level back up to the level of the tip of the hook gauge. The usual timings are 5, 10, 15, 30, 45, 60 min etc., though these timings may need to be varied depending on the soil type (they will be more frequent on lighter soils). In order to maintain a relatively constant head the water level should not be allowed to fall more than 1–2 cm below the tip of the hook gauge. The interval between topping-up should be short at the start and can be lengthened as the test proceeds and the infiltration rate slows down. The volume of water added each time interval is recorded and the cumulative volume added since the start of the procedures determined (see Form Inf-01 in Fig. 5.13).
4. The water level in the outer ring should be maintained at a depth close to the depth in the inner ring. The quantity added to the outer ring does not, however, need to be measured. The point of the outer ring is to act as a buffer to ensure that the inner ring water goes down vertically and does not spread out laterally. In some soils, such as heavy cracking clays, a large outer ring filled with water (2 m diameter) can be formed with an earth bund to reduce the tendency for large lateral movement through soil cracks. In this case the outer ring of the infiltrometer can be used instead of the inner ring.
5. The measurements can be stopped when the infiltration rate has reached the terminal rate. This may take between 1 and 3 h depending on the soil type, though in heavy clays it may take 1–2 days to reach the terminal rate. When the volume of water added is the same for three consecutive measurements then it can be assumed that the terminal rate has been reached, and the measurements can

Form Inf-01		Infiltration test recording form		
Location:		Date:	7th June 2008	
Field number:	MA-23	Soil state at start of test (wet/dry):	Dry	
WUA name:	Maz-Aikal	Person conducting the test:	WUA Engineer	Approximate depth of water in inner ring (cm): 10 cm
Diameter of inner ring (cm):	30 cm	Surface area of inner ring (cm^2):	707 cm^2	

Elapsed time from start (min)	Volume of water added to return water level to zero point (cm^3)	Equivalent depth of water added (mm)	Infiltration rate (mm/h)	Cumulative infiltration amount (mm)
0	0	0	0	0
5	1072	15.2	182	15
10	907	12.8	154	28
15	695	9.8	118	38
30	1661	23.5	94	61
45	1573	22.2	89	84
60	1379	19.5	78	103
75	1361	19.2	77	122
90	1343	19.0	76	141
115	2180	30.8	74	172
130	1308	18.5	74	191

Fig. 5.13. Example of a completed infiltration test recording form.

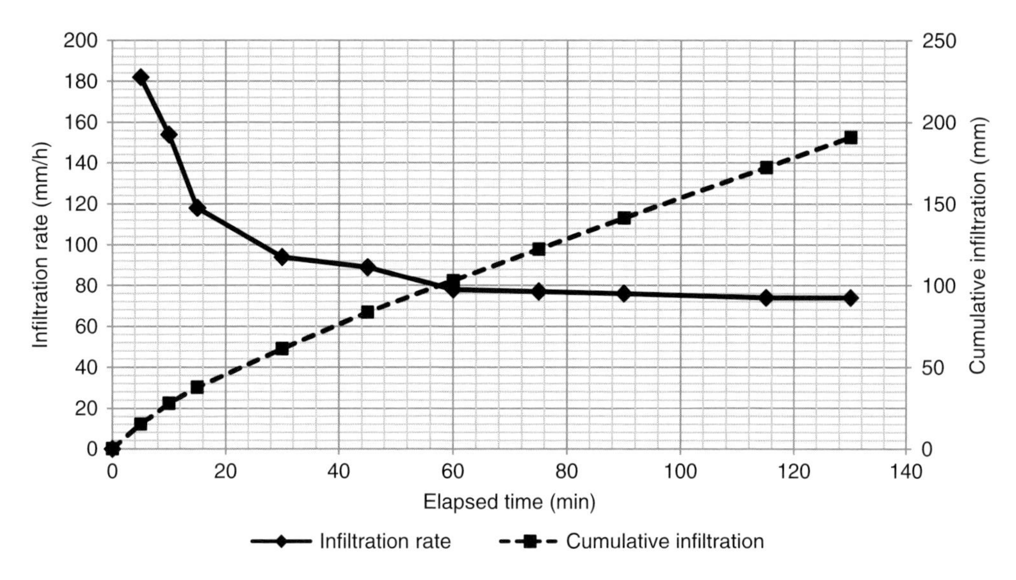

Fig. 5.14. Plot of data recorded from an infiltration test.

stop. Note that it is better to continue for a bit longer than to finish the measurements early.

6. It is advisable to do at least three infiltration tests in a given location in order to get an average set of figures.

7. The data are collected and calculated on a data collection form (Fig. 5.13), and the data then plotted to show the infiltration rate and the cumulative infiltration (Fig. 5.14).

8. Note that to get the full range of data the infiltration test should be carried out when the soil is relatively dry just prior to irrigation.

Rainfall

The quantity and timing of rainfall play a significant role in the quantity of irrigation water that needs to be provided to the crop. Not all the rainfall that falls on a field can be utilized by the crop. Some may run off and enter the drainage system, while rainfall that exceeds the storage capacity in the crop's root zone will not be available to crop, and will thus not be effective in contributing to the crop's water needs. For paddy irrigation, rainfall which exceeds the ponded water storage capacity will flow out of the bunded field and will not be effective.

Estimating the effectiveness of rainfall is an important part of the design of an irrigation system in order to determine the irrigation water requirements and therefore the design canal capacities. A number of approaches have been developed for calculating the effectiveness of rainfall; the Food and Agriculture Organization of the United Nations (FAO) details these well in FAO Irrigation and Drainage Paper No. 25, *Effective Rainfall* (FAO, 1978). Though somewhat dated, this publication still provides a comprehensive analysis of methods used for determining effective rainfall.

One of the approaches that the publication recommends is the USDA Soil Conservation Method based on analysis of long-term climatic and soil moisture data. The method provides a table of monthly mean rainfall against mean monthly consumptive use to determine the effective rainfall. Thus out of a monthly mean rainfall of 50 mm only 25 mm will be effective if the mean monthly consumptive use is 25 mm, but all 50 mm will be effective if the mean monthly consumptive use is 250 mm or more.

A simpler approach that is sometimes used is to apply a percentage (70–80%) to the mean monthly rainfall to allow for the fact that 20–30% of the rainfall may not be effective.

For day-to-day scheduling, account has to be taken of the actual rainfall that has taken place and is likely to take place. Keeping a water balance sheet for each irrigated field (or group of fields with the same crop and soil characteristics) is an effective way of monitoring the contribution from rainfall to a crop's water requirements. If it has rained the irrigation schedule should be amended to allow for the rainfall, and the irrigation water conserved. This is particularly important where irrigation water comes from a storage reservoir, or is pumped. In addition, if rain is forecast irrigation might be delayed to make the most use of the rainfall when it occurs. Very little of the rainfall occurring immediately following full irrigation of a crop will be effective.

The quantity and timing of rainfall have another impact that is often not fully appreciated. In more humid regions rainfall may be adequate and irrigation is provided to supplement rainfall. In more arid regions rainfall is inadequate and irrigation is essential if crops are to be grown. The perception of irrigation in these two situations is significantly different. In the arid region irrigation is highly valued as crops cannot be grown without it. In the more humid region farmers will endeavour to utilize rainfall as much as possible and avoid having to irrigate, especially if they have to pay for the irrigation water. These different perceptions can have significant repercussions on how irrigation water is managed and used by farmers.

Evapotranspiration

Water is consumed through evaporation from the soil surface and transpiration through the crop's leaves. The combined term is called *evapotranspiration*. The rate at which water evapotranspires governs the amount of irrigation water required and the frequency of irrigation. The evapotranspiration rate is usually measured in units of mm/day, representing the volume of water that has evapotranspired over a given area.

Factors influencing evapotranspiration

The factors influencing the rate of transpiration from the crop and evaporation from the soil surface include:

- sunlight;
- temperature;
- humidity;
- wind;
- degree of canopy development.

The influence of these different factors is summarized below.

- *Length of sunlight hours*: Long sunshine hours increase the water evapotranspired.
- *Intensity of sunlight*: Intense sunshine increases the evapotranspiration rate.
- *Temperature*: High temperatures increase the evapotranspiration rate.
- *Humidity*: Low humidity increases the evapotranspiration rate.
- *Wind speed*: High wind speeds increase the evapotranspiration rate.
- *Canopy development*: The more the canopy cover (leaf area) the greater the transpiration from the crop, and the less the evaporation from the soil surface. The greater the canopy cover the greater the amount of sunlight intercepted by the leaves and the greater the photosynthesis.

The behaviour of the plant is governed by the relationship between the soil, plant and the atmosphere. When there is sufficient water in the soil and the transpiration rate is low the stomata of the plant open at dawn and water is lost during the day by transpiration. This loss will be matched by the uptake of water from the soil through the roots and plant stem, without causing any stress to the plant. Photosynthesis will take place and the plant will grow. At night the stomata will close and there will be little water loss from the plant.

As the moisture is taken out, either by evaporation from the soil surface or by transpiration through the crop's leaves, the soil dries and the rate of water supply from the soil reduces and cannot match the demand placed on it by the climatic conditions. At this point the rate of leaf expansion is slowed and the stomata may close to conserve water, leading to the characteristic wilting of the crop. The rate of photosynthesis reduces and the crop's yield will be adversely affected. Under these conditions the actual evapotranspiration will be less than the potential evapotranspiration rate.[1]

Change in evapotranspiration during the year

In order to plan the schedule for delivery of irrigation water supplies it is necessary to make estimates of the evapotranspiration rates of crops at different times of the year. Table 5.6 and Fig. 5.15 show how the evapotranspiration rate changes as the temperature increases during the summer, and the demand for irrigation water increases. Note the close relationship in the shapes of the crop evapotranspiration rates for alfalfa and the fruit tree crop and the temperature, showing that the evapotranspiration rate is very closely linked to temperature. The change in the evapotranspiration rates can be seen most clearly from the daily rates shown in Table 5.6b. For lucerne, which grows throughout the year, the daily evapotranspiration rate changes from 1.19 mm/day in March to 6.45 mm/day in July, over five times higher.

Determining evapotranspiration rates

Crop evapotranspiration can be determined in a number of ways using measurements from:

- meteorological stations;
- evaporation pans;
- lysimeters;
- atmometers.

Meteorological stations

Data collected from meteorological stations can be used to estimate the potential evapotranspiration. The data are collected and then the potential evapotranspiration calculated using a mathematical formula linking all the

Table 5.6. Example evapotranspiration rates for selected crops in Central California, USA. (From FAO, 1984 with permission from the Food and Agriculture Organization of the United Nations.)

(a) Monthly rates (mm/month)

	Monthly evapotranspiration rate (mm/month)								
Crop	Mar	Apr	May	June	July	Aug	Sept	Oct	Total (mm)
Lucerne	37	97	132	162	200	165	125	50	968
Beans				37	75	162	100		374
Maize				80	155	152	100	12	500
Fruit (deciduous)	12	62	100	180	192	135	85	45	810
Tomatoes				80	112	175	130	102	600
Mean monthly temperature (°C)	12.7	15.1	18.3	21.8	24.0	23.3	21.8	17.7	

(b) Daily rates (mm/day)

	Daily evapotranspiration rate (mm/day)								
Crop	Mar	Apr	May	June	July	Aug	Sept	Oct	Average (mm/day)
Lucerne	1.19	3.23	4.26	5.40	6.45	5.32	4.17	1.61	3.95
Beans				1.23	2.42	5.23	3.33		3.07
Maize				2.67	5.00	4.90	3.33	0.39	4.10
Fruit (deciduous)	0.39	2.07	3.23	6.00	6.19	4.35	2.83	1.45	3.31
Tomatoes				2.67	3.61	5.65	4.33	3.29	3.92

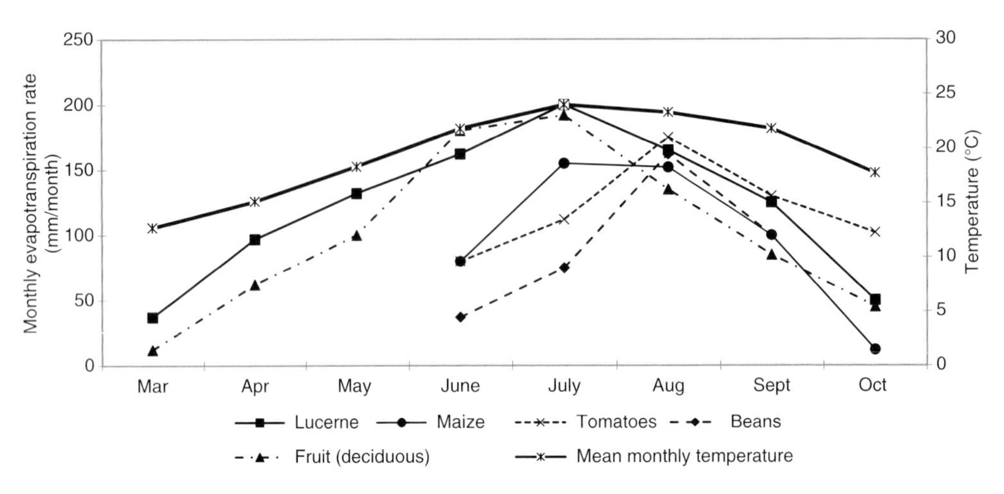

Fig. 5.15. Changes in the monthly evapotranspiration rates for different crops.

Table 5.7. Data requirements for measuring potential evapotranspiration.

	Data requirements					
Method	Sunshine hours	Temperature	Wind speed	Relative humidity	Incoming radiation rate	Crop albedo (reflectivity)
Thornthwaite		✓				
Blaney–Criddle	✓	✓				
Penman–Monteith	✓	✓	✓	✓	✓	✓

variables. As shown in Table 5.7, different mathematical formulas use different sets of data.

Depending on the data collected and the method used the estimates of potential evapotranspiration can be quite accurate. The FAO recognizes the Penman–Monteith method as the most accurate, though it does require the most data to be collected.

The Penman–Monteith equation (FAO, 1998) is shown in the Equation at the bottom of the page:

where

ET_0 = reference crop evapotranspiration (mm/day)
R_n = net radiation at the crop surface (MJ/m²/day)
G = soil heat flux density (MJ/m²/day)
T = mean daily air temperature at 2 m height (°C)
u_2 = wind speed at 2 m height (m/s)
e_s = saturation vapour pressure (kPa)
e_a = actual vapour pressure (kPa)
$e_s - e_a$ = saturation vapour pressure deficit (kPa)
Δ = slope of saturation vapour pressure curve at temperature T (kPa/°C)
ρ = psychometric constant (kPa/°C).

The equation requires standard meteorological data for solar radiation (sunshine), air temperature, wind speed and humidity to enable the potential crop evapotranspiration to be determined for given time periods (daily, weekly, 10-day, bi-monthly or monthly). In addition to the meteorological data the equation requires information related to

$$ET_0 = \frac{0.408\Delta(R_n - G) + \rho(900/T + 273)u_2(e_s - e_a)}{\Delta + \rho(1 + 0.34u_2)}$$

the location of the site (the altitude above sea level and latitude).

Reference should be made FAO Irrigation and Drainage Paper No. 56, *Crop Evapotranspiration: Guidelines for Computing Crop Water Requirements* (FAO, 1998; http://www.fao.org/docrep/X0490E/X0490E00.htm) for a full description of the calculation process and associated data tables.

Evaporation pans

Data collected at meteorological stations may be useful for research stations and large farms; it is not always available for individual farmers. A more practical approach is to measure the evaporation of water from an evaporation pan and then convert this to an estimate of the potential evapotranspiration.

For this the evaporation of water from a standard sized evaporation pan is measured. A fairly universal evaporation pan is the Class A pan (Fig. 5.16). This pan is made of galvanized iron, 120.5 cm in diameter, 25 cm deep, supported on a 122 cm×122 cm grid of 5 cm×10 cm timber. Water is maintained at a depth of 20 cm, with hook gauge being used to mark this depth. A measured amount of water is added each day to return the water level to this depth, and the evaporation rate determined by dividing the volume of water added by the area of the evaporation pan.

The potential evapotranspiration can be determined from the pan evaporation by means of the formula:

$$ET_0 = k_{pan} \times E_{pan}$$

where

ET_0 = potential evapotranspiration
k_{pan} = pan coefficient (as given in Table 5.8)
E_{pan} = evaporation in mm/day from a Class A pan.

An example calculation is given in Box 5.2.

Lysimeters

A lysimeter comprises a watertight cylinder or tank containing soil, sealed at the base, set into the ground and planted with a crop. A primary requirement is that the vegetation inside and immediately outside the lysimeter is similar, with the same height and leaf area. With precision weighing lysimeters the evapotranspiration from the crop is measured directly and very accurately at regular intervals, even down to 1 h. With non-weighing lysimeters drainage water below the root zone is collected and measured, and deducted from the total water added to the lysimeter to

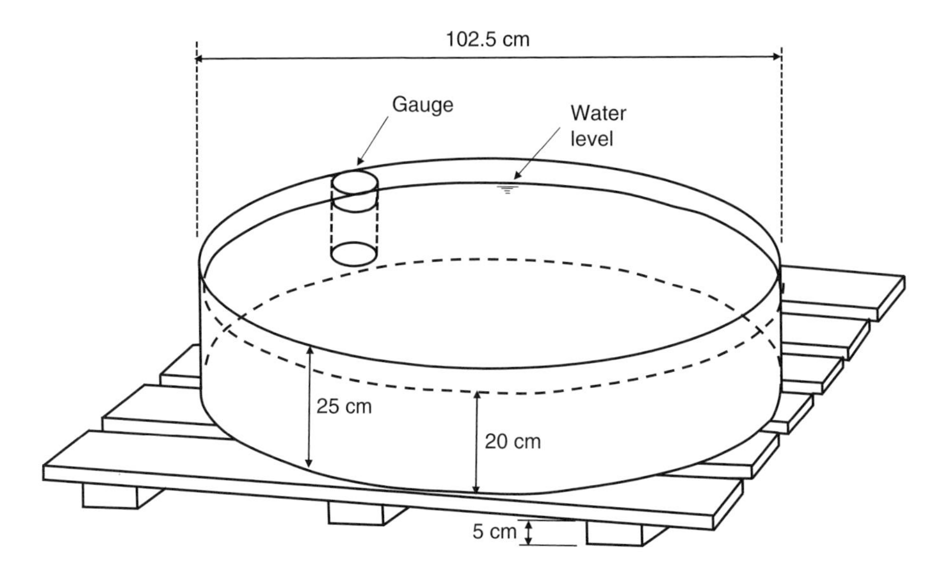

Fig. 5.16. Class A evaporation pan.

Table 5.8. Pan coefficients for Class A pan. (From FAO, 1977 with permission from the Food and Agriculture Organization of the United Nations.)

	For a pan placed in an area of short green crop				For a pan placed in a dry fallow area			
		Relative humidity (%)				Relative humidity (%)		
Wind speed (km/day)	Windward side distance of green crop (m)	Low <40%	Medium 40–70%	High >70%	Windward side distance of dry fallow (m)	Low <40%	Medium 40–70%	High >70%
Light (<175)	1	0.55	0.65	0.75	1	0.70	0.80	0.85
	10	0.65	0.75	0.85	10	0.60	0.70	0.80
	100	0.70	0.80	0.85	100	0.55	0.65	0.75
	1000	0.75	0.85	0.85	1000	0.50	0.60	0.70
Moderate (175–425)	1	0.50	0.60	0.65	1	0.65	0.75	0.80
	10	0.60	0.70	0.75	10	0.55	0.65	0.70
	100	0.65	0.75	0.80	100	0.50	0.60	0.65
	1000	0.70	0.80	0.80	1000	0.45	0.55	0.60
Strong (425–700)	1	0.45	0.50	0.60	1	0.60	0.65	0.70
	10	0.55	0.60	0.65	10	0.50	0.55	0.65
	100	0.60	0.65	0.70	100	0.45	0.50	0.60
	1000	0.65	0.70	0.75	1000	0.40	0.45	0.55
Very strong (>700)	1	0.40	0.45	0.50	1	0.50	0.60	0.65
	10	0.45	0.55	0.60	10	0.45	0.50	0.55
	100	0.50	0.60	0.65	100	0.40	0.45	0.50
	1000	0.55	0.60	0.65	1000	0.35	0.40	0.45

Box 5.2. Example of Using Pan Evaporation to Calculate Potential Evapotranspiration

- Month = July.
- E_{pan} = 6.7 mm/day.
- Conditions:
  - pan surrounded by cropped area of several hectares;
  - windward side distance of green crop = 100 m.
- Relative humidity = medium.
- Wind speed = moderate.
- From Table 5.8, k_{pan} = 0.75.
- Thus $ET_0 = k_{pan} \times E_{pan} = 0.75 \times 6.7 = 5.0$ mm/day.

determine the quantity of water lost due to evapotranspiration.

Atmometers

An atmometer (Fig. 5.17) is a specially designed evaporation tube that can be used to measure the daily evaporation rate. Its advantage is that it is relatively small and inexpensive and can be located close to the crop.

Atmometers comprise a wet, porous ceramic cup mounted on a cylindrical water reservoir. The ceramic cup is covered with a green fabric that simulates the canopy of the crop. The reservoir is filled with distilled water (to avoid the pores in the ceramic cup getting clogged by minerals in the water) and the water then evaporates from the surface of the cup, much like the transpiration from a crop's leaves. Underneath the green fabric the ceramic

Fig. 5.17. Atmometer.

cup is covered by a special membrane to stop rainwater entering, and a rigid wire frame above the cup stops birds perching on the cup.

Atmometers are usually mounted on a wooden post near the irrigated crops, with the top of the ceramic cup 1 m above ground level. The daily evaporation from the atmometer is measured on a gauge on the side of the cylinder. At full canopy the evaporation from the atmometer is close to the potential evapotranspiration from the crop; when the canopy is not fully developed tables are used to convert the atmometer readings to evapotranspiration of the crop.

Calculation of Crop Water Requirements

A crop's water requirements are dependent on:

- crop type;
- climatic conditions.

As discussed in previous sections the climatic conditions strongly influence the crop's evapotranspiration rate (ET_c), which is made up of *evaporation* from the soil surface and *transpiration* from the crop. As for most crops the evapotranspiration varies with the growth stage of the crop, a perennial crop such as grass or lucerne is used as the *reference crop* to measure the climatic conditions.

The *reference crop evapotranspiration* (ET_0) is defined as (FAO, 1977):

> The rate of evapotranspiration from an extended surface of 8–15 cm grass cover of uniform height, actively growing, completely shading the ground and not short of water.

Thus the evapotranspiration from this reference crop can be measured during the year as a means of determining the evapotranspiration caused by the climatic conditions. In the winter the reference crop evapotranspiration rate may be low, 1–2 mm/day, or zero if there is snow cover; in the summer the rate may increase to 5–10 mm/day depending on the climatic conditions.

To allow for the different crop water demands of different crops the reference crop evapotranspiration ET_0 is multiplied by a *crop coefficient* (K_c) to obtain the *crop evapotranspiration* (ET_c):

$$ET_c = ET_0 \times K_c$$

Typical values of the crop coefficient during the different crop growing phases are shown in Table 5.9. It can be seen that in general the crop coefficient is low to start with, rises to a peak and then falls again. Also in all cases, except for rice, the value for the total growing season is less than 1.0.

Knowing the duration of each of the crop's growth phases allows the crop coefficient diagram to be plotted (on graph paper), as shown in Fig. 5.18. In this example the diagram is constructed using the data presented in Table 5.10 for a maize crop.

From this plot the K_c values can be determined for any required time period, e.g. weekly, 10-daily, twice monthly, etc. These values can then be multiplied by the reference crop evapotranspiration figures to obtain the evapotranspiration figures for the crop, as shown in Fig. 5.19. It can be seen from Fig. 5.19 that the maize crop evapotranspiration rate starts off low and then increases to equal the reference crop evapotranspiration rate before dropping back again as the crop matures and becomes ready to harvest.

In order to keep the crop at its maximum potential yield the water supply has to match the crop's potential evapotranspiration needs

Table 5.9. Typical crop coefficient values. (From FAO, 1977, 1998 with permission from the Food and Agriculture Organization from the United Nations.)

Crop	Crop development phase					Total growing period
	Initial	Crop development	Mid-season	Late season	At harvest	
Bean – green	0.30	0.65	0.95	0.90	0.85	0.85
Bean – dry	0.30	0.70	1.05	0.65	0.25	0.70
Cabbage	0.40	0.70	0.95	0.90	0.80	0.70
Cotton	0.40	0.70	1.05	0.80	0.65	0.80
Maize – grain	0.30	0.70	1.05	0.80	0.55	0.75
Potato	0.40	0.70	1.05	0.85	0.70	0.75
Rice	1.10	1.10	1.10	0.95	0.95	1.05
Sunflower	0.30	0.70	1.05	0.70	0.35	0.75
Wheat	0.30	0.70	1.05	0.65	0.20	0.80

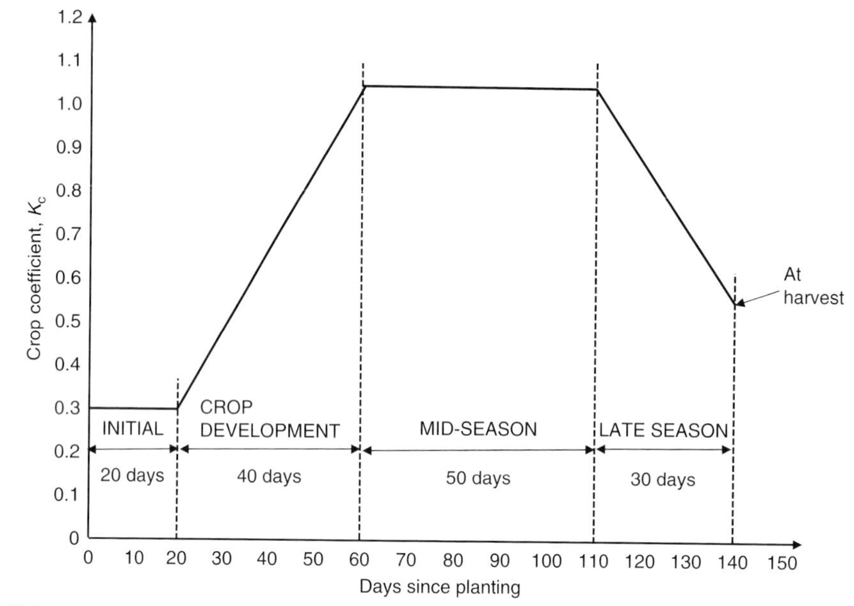

Fig. 5.18. Crop coefficient curve for a maize crop.

Table 5.10. Key data for plotting a crop coefficient (K_c) diagram.

Crop phase	K_c value for maize	Duration of phase (days)
Initial	0.30	20
Crop development	–	40
Mid-season	1.05	50
Late season	–	30
At harvest	0.55	
Total	–	**140**

during the growing season, as shown in Fig. 5.19. Where the crop is irrigated, the soil reservoir must be refilled at intervals to compensate for the loss of water through evapotranspiration.

Calculation of Irrigation Water Requirements

The calculation of the *crop* water requirements is only part of calculation of the

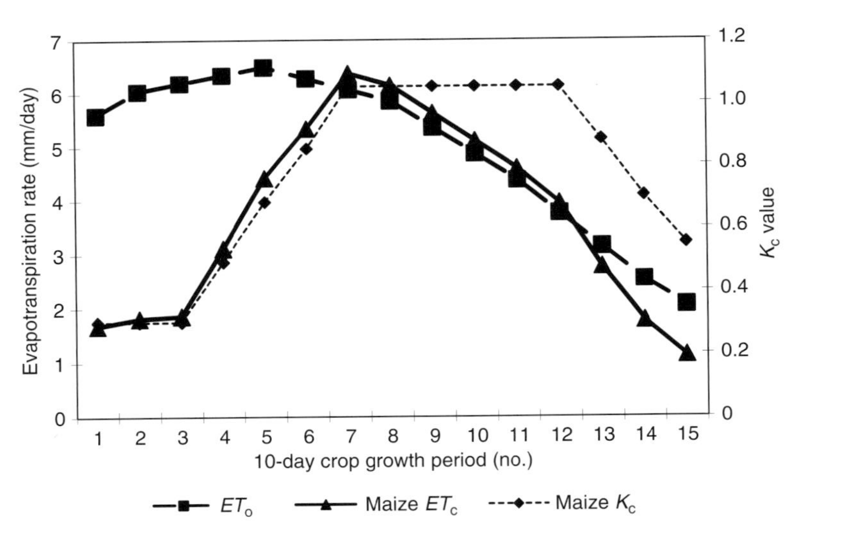

	June			July			August			September			October		
10-day period	1–10	11–20	21–end	1–10	11–20	21–end	1–10	11–20	21–end	1–10	11–20	21–end	1–10	11–20	21–end
Growing period	1	2	3	4	5	6	7	8	9	10	11	12	13	14	15
ET_o	5.60	6.04	6.19	6.34	6.49	6.28	6.06	5.85	5.36	4.86	4.37	3.76	3.14	2.53	2.05
Maize K_c	0.30	0.30	0.30	0.49	0.68	0.85	1.05	1.05	1.05	1.05	1.05	1.05	0.88	0.70	0.55
Maize ET_c	1.68	1.81	1.86	3.11	4.41	5.34	6.37	6.14	5.62	5.11	4.59	3.94	2.77	1.77	1.13

Fig. 5.19. Determination of 10-daily maize crop potential evapotranspiration (ET_c) from reference crop potential evapotranspiration (ET_o).

irrigation water requirements. To calculate the irrigation water requirements information is required on the:

- crop water requirement;
- effective rainfall;
- build-up of the crop area (the crop stagger);
- water requirement for land preparation;
- contribution, if any, to crop water needs from groundwater;
- leaching requirement, if any required;
- application efficiency;
- distribution efficiency;
- for rice, the soil percolation rates and the depth of water to be ponded.

For dry foot crops

For 'dry foot' crops (crops other than rice) the equation for determining the irrigation water requirements is:

$$I = ET_c + L - R_e - G_w$$

where

I = irrigation requirement (mm/day)
ET_c = crop potential evapotranspiration (mm/day)
L = leaching requirement (mm/day)
R_e = effective rainfall (mm/day)
G_w = groundwater contribution (by upward capillary rise, mm/day).

If there is no groundwater contribution or leaching requirement the equation reduces to:

$$I = ET_c - R_e$$

Thus if a farmer is using a Class A evaporation pan to measure the pan evaporation, the calculation of the daily irrigation water demand is as set out in Table 5.11.

For rice crops

For rice crops allowance has to be made for the ponding depth of water on the soil surface, and for deep percolation of water below the crop's root zone. The equation for determining the irrigation water requirements is:

$$I = ET_c - R_e + P + (D_2 - D_1)$$

where

I = irrigation requirement (mm/day)
ET_c = crop potential evapotranspiration (mm/day)
R_e = effective rainfall (mm/day)
P = percolation water seeping below the crop root zone (mm/day)
D_2 = required ponded depth of water (mm/day)
D_1 = ponded depth after last irrigation (mm/day).

If the ponded depth is to be kept the same the equation becomes:

Table 5.11. Estimation of the daily irrigation demand for a maize crop using a Class A pan.

		Day						
	Units	1	2	3	4	5	6	7
Pan evaporation, E_{pan}	mm/day	4	4	5	4	5	5	4
Pan coefficient, k_{pan}		0.8	0.8	0.8	0.8	0.8	0.8	0.8
Reference crop potential evapotranspiration, ET_0	mm/day	3.2	3.2	4.0	3.2	4.0	4.0	3.2
Crop coefficient, K_c (for maize, 11–20 September, Fig. 5.19)		1.05	1.05	1.05	1.05	1.05	1.05	1.05
Crop potential evapotranspiration, ET_c ($ET_0 \times K_c$)	mm/day	3.4	3.4	4.2	3.4	4.2	4.2	3.4
Effective rainfall, R_e	mm/day	0	0	4	2	0	0	0
Irrigation demand, ET_c–R_e	mm/day	3.4	3.4	0.2	1.4	4.2	4.2	3.4
Cumulative irrigation demand	mm	3.4	6.8	7.0	8.4	12.6	16.8	20.2

$$I = ET_c - R_e + P$$

The percolation rate is a key variable in determination of irrigation water requirements for paddy rice. The percolation rate is dependent on the soil type and the effort put into puddling the soil during land preparation, and can vary from 1 mm/day to 10 mm/day or more. It is worth noting that there is little point in worrying about canal distribution losses (see following section) in a paddy rice scheme if the percolation rates in the paddy fields are high. The total volume of water loss from the paddy fields will far outweigh the losses from the canals.

Application and distribution efficiencies

In order to determine the irrigation demand at the tertiary unit intake, it is necessary to allow for the efficiency of *application* of the water to the field, and the efficiency of *distribution* in moving the water from the tertiary unit intake to the field.

The *application efficiency* takes account of the following possible losses in irrigating the crop:

- over-irrigation, leading to losses of water below the root zone;
- over-irrigation, leading to losses of water from runoff;
- lateral seepage from the field (in the case of rice fields).

The application efficiency varies depending on the capability of the farmer and the irrigation method (Table 5.12).

The *distribution efficiency* (Table 5.13) takes account of the following possible losses in conveying the water from the intake to the field:

- seepage through the canal bed and banks;
- spillage through the canal banks;
- management losses due to emptying and filling the canals.

The combination of the application and distribution efficiencies gives the on-farm efficiency to be applied to the demand at the field. A typical calculation is given in Box 5.3.

Note that in this example the irrigation depth required is 75 mm. Using the information from Table 5.11, the average demand is about 3.5 mm/day (not allowing for rainfall), which means that the irrigation interval will be approximately 75/3.5 = 21 days.

Tabulating crop and irrigation water requirement calculations

The above calculations can be put into a tabular format and the calculations carried out for the whole season, and for a mix of crops. If a computer is available then the tables can be put into a spreadsheet and the calculations done very quickly. Figures 5.20 and 5.21 are proformas giving examples of such tables, while Fig. 5.22 shows the plots of the field irrigation and tertiary unit requirements calculated in Fig. 5.21.

Table 5.12. Typical values of application efficiency, E_a. (From Bos and Nugteren, 1974 with permission.)

	Application efficiency, E_a	
Application method	Light soil Small fields	Heavy soil Large fields
Graded border	0.60	0.75
Basin	0.60	0.80
Furrows	0.55	0.70
Sprinklers		
Hot dry climate	0.60	0.80
Moderate climate	0.70	0.85
Humid	0.80	0.85
Drip	1.00	1.00

Table 5.13. Typical values of distribution efficiency, E_d. (From Bos and Nugteren, 1974 with permission.)

Application method	Distribution efficiency, E_d
Blocks of 20 ha or more	
Unlined	0.8
Lined or piped	0.9–0.95
Blocks of 1–20 ha	
Unlined	0.6–0.75
Lined or piped	0.7–0.9

Box 5.3. Calculation of Irrigation Demand at the Field

For a 1 ha field of maize in a large field on heavy soils irrigated by furrow irrigation:

- depth of water needed to fill the root zone = 75 mm;
- application efficiency (from Table 5.12 – furrow irrigation, large field, heavy soils) = 0.70;
- total depth of application = 75/0.70 = 107 mm;
- flow required to irrigate 1 ha in 1 day = 107 × 0.1157[a] = 12.4 l/s;
- distribution efficiency (from Table 5.13 – 1 ha block, unlined canal in poor condition) = 0.6;
- requirement at tertiary unit intake = 12.4/0.6 = 20.63 l/s.

[a]Note: An irrigation depth of 1 mm applied in 1 day (24 h) = 0.1157 l/s/ha.

The details of the calculations carried out in each cell of the table in Figs 5.20 and 5.21 are provided in the Source/Calculation column (col. (5)). Using such proformas the irrigation requirements for different cropping patterns and their pattern of demand can be calculated for different time periods (7, 10, 15 days or monthly).

Computer Applications

CROPWAT, developed by the FAO, is one of the best known computer applications in irrigation management. The program enables the determination of crop and irrigation water requirements and can produce indicative irrigation schedules. The program can be downloaded from the FAO website (http://www.fao.org/waicent/faoinfo/agricult/agl/aglw/CROPWAT.stm) together with guidance details and other relevant information. There is also a CROPWAT for Windows version written in Visual Basic, which provides useful visual representations of the data. The program uses monthly climatic data (temperature, relative humidity, wind speed, sunshine hours, rainfall) for the calculation of reference crop potential evapotranspiration. Through the input of crop data (growth stages, K_c factors, root zone depth and allowable soil moisture depletion factor), the program calculates the crop and irrigation water requirements for the selected cropping pattern on a decade (10-day) basis. A proforma for entering data into the program is presented in Fig. 5.23, while typical output is provided in Table 5.14.

CRIWAR (Bos *et al.*, 2009) is another item of crop and irrigation water requirement software, which can be downloaded free of charge (www.bos-water.nl). Like CROPWAT, the program calculates the crop and irrigation water requirements for specified cropping patterns, and has the additional facility of analysing alternative water management strategies.

Combining Crop, Soil and Water Relationships to Schedule Irrigation

Previous sections have discussed the water-holding capacity of the soil and methods used to determine the crop and irrigation requirements. This section outlines how these factors are brought together to schedule the supply of irrigation water to match crop needs based on procedures detailed in FAO Irrigation and Drainage Papers No. 24 and 56 (FAO, 1977, 1998). Some of the terms used in previous sections are summarized below for convenience.

- *Deep percolation*: Water percolating below the root zone of the crop, and effectively lost.
- *Easily available soil water* (EAW): Water that can be abstracted from the soil without causing any stress to the crop.
- *Saturation*: All the pore space in the soil is filled with water, with little or no air.
- *Field capacity* (S_{fc}): Maximum moisture content of the soil where soil tension forces are balancing the gravitational pull on the water.
- *Permanent wilting point* (S_w): Moisture content of the soil at which the crop can no longer abstract water.
- *Root depth* (D): The depth of soil profile from which the crop can abstract water.

IRRIGATION WATER REQUIREMENT CALCULATION TABLE

Calculations for :- [ ] Date :- [ ]

Line no.	Item			Source/ Calculation	Unit	1	2	3	4	5	6	7	8	9	10	11	12	13	14	15	16	17	18
1	Month (optional)																						
2	Period			No. days in period:-																			
3	Potential evapotranspiration, ET_0				mm/day																		
4	Rainfall				mm/day																		
5	Effective rainfall, R_e				mm/day																		
6	Number of blocks (max 5):-		a.Type	b. Area (ha)		c. Crop coefficients, K_c																	
7	Crop, crop area and	1.			K_c																		
8	crop coefficients	2.			K_c																		
9		3.			K_c																		
10		4.			K_c																		
11		5.			K_c																		
12	Crop consumptive use	1.		3*7c	mm/day																		
13	(ET_c)	2.		3*8c	mm/day																		
14		3.		3*9c	mm/day																		
15		4.		3*10c	mm/day																		
16		5.		3*11c	mm/day																		
17	Land preparation/pre-	1.			mm/day																		
18	irrigation/leaching/	2.			mm/day																		
19	groundwater	3.			mm/day																		
20	contribution	4.			mm/day																		
21		5.			mm/day																		
22	Field irrigation	1.		17-5 or 12-5	mm/day																		
23	requirement (mm/day)	2.		18-5 or 13-5	mm/day																		
24		3.		19-5 or 14-5	mm/day																		
25		4.		20-5 or 15-5	mm/day																		
26		5.		21-5 or 16-5	mm/day																		
				Block no.		1	2	3	4	5													
27	Application efficiency, E_a																						
28	Distribution efficiency, E_d																						
29	Conveyance efficiency, E_c					(Note: 1 mm/day = 0.1157 l/s/ha)																	
30	Tertiary head irrigation	1.		22/(27*28)	l/s/ha																		
31	requirement (l/s/ha)	2.		23/(27*28)	l/s/ha																		
32		3.		24/(27*28)	l/s/ha																		
33		4.		25/(27*28)	l/s/ha																		
34		5.		26/(27*28)	l/s/ha																		
35	Tertiary head irrigation	1.		30*7b	l/s																		
36	requirement (l/s)	2.		31*8b	l/s																		
37		3.		32*9b	l/s																		
38		4.		33*10b	l/s																		
39		5.		34*11b	l/s																		
40	Total tertiary head irrigation requirement (l/s)			sum(35..39)	l/s																		
41	Scheme irrigation requirement (l/s)			40/29	l/s																		

Fig. 5.20. Blank proforma for calculating irrigation water requirements.

IRRIGATION WATER REQUIREMENT CALCULATION TABLE

Calculations for :- Sample Calculation Date :- 8 February 2007

Line no.	Item			Source/ Calculation	Unit	1	2	3	4	5	6	7	8	9	10	11	12	13	14	15	16	17	18
1	Month (optional)																						
2	Period			No. days in period :-	15	1	2	3	4	5	6	7	8	9	10	11	12	13	14	15	16	17	18
3	Potential evapotranspiration ET_0				mm/day	4.00	4.00	5.00	5.00	6.00	6.00	5.00	5.00	4.00	4.00								
4	Rainfall				mm/day																		
5	Effective rainfall, R_e				mm/day	2	3	4	4	3	2	1	0	0	0								
6	Number of blocks (max 5):		a. Type	b. Area (ha)		c. Crop coefficients, K_c																	
7	Crop, crop area and crop coefficients	1.	Onion	5	K_c		0.60	0.70	0.90	1.00	1.00	0.90											
8		2.	Beetroot	5	K_c		0.60	0.65	0.80	1.00	1.05	1.05	1.00										
9		3.	Maize	10	K_c		0.40	0.50	0.70	1.00	1.10	1.10	1.05	0.75									
10		4.	Pepper	5	K_c		0.60	0.65	0.80	0.95	1.00	1.00	1.00	0.90									
11		5.	Tomato	5	K_c		0.60	0.70	1.00	1.25	1.25	1.15	0.80										
12	Crop consumptive use ET_c	1.		3*7c	mm/day	0.0	2.4	3.5	4.5	6.0	6.0	4.5	0.0	0.0	0.0	0.0	0.0	0.0	0.0	0.0	0.0	0.0	0.0
13		2.		3*8c	mm/day	0.0	2.4	3.3	4.0	6.0	6.3	5.3	5.0	0.0	0.0	0.0	0.0	0.0	0.0	0.0	0.0	0.0	0.0
14		3.		3*9c	mm/day	0.0	1.6	2.5	3.5	6.0	6.6	5.5	5.3	3.0	0.0	0.0	0.0	0.0	0.0	0.0	0.0	0.0	0.0
15		4.		3*10c	mm/day	0.0	2.4	3.3	4.0	5.7	6.0	5.0	5.0	3.6	0.0	0.0	0.0	0.0	0.0	0.0	0.0	0.0	0.0
16		5.		3*11c	mm/day	0.0	2.4	3.5	5.0	7.5	7.5	5.8	4.0	0.0	0.0	0.0	0.0	0.0	0.0	0.0	0.0	0.0	0.0
17	Land preparation/pre-irrigation/ leaching/ groundwater contribution	1.			mm/day	3																	
18		2.			mm/day	3																	
19		3.			mm/day	5																	
20		4.			mm/day	3																	
21		5.			mm/day	4																	
22	Field irrigation requirement (mm/day)	1.		17-5 or 12-5	mm/day	1.00	0.00	0.00	0.50	3.00	4.00	3.50	0.00	0.00	0.00	0.00	0.00	0.00	0.00	0.00	0.00	0.00	0.00
23		2.		18-5 or 13-5	mm/day	1.00	0.00	0.00	0.00	3.00	4.30	4.25	5.00	0.00	0.00	0.00	0.00	0.00	0.00	0.00	0.00	0.00	0.00
24		3.		19-5 or 14-5	mm/day	3.00	0.00	0.00	0.00	3.00	4.60	4.50	5.25	3.00	0.00	0.00	0.00	0.00	0.00	0.00	0.00	0.00	0.00
25		4.		20-5 or 15-5	mm/day	1.00	0.00	0.00	0.00	2.70	4.00	4.00	5.00	3.60	0.00	0.00	0.00	0.00	0.00	0.00	0.00	0.00	0.00
26		5.		21-5 or 16-5	mm/day	2.00	0.00	0.00	1.00	4.50	5.50	4.75	4.00	0.00	0.00	0.00	0.00	0.00	0.00	0.00	0.00	0.00	0.00
				Block no.		1	2	3	4	5													
27	Application efficiency, E_a					0.6	0.6	0.6	0.6	0.6													
28	Distribution efficiency, E_d					0.7	0.7	0.7	0.7	0.7													
29	Conveyance efficiency, E_c				0.8																		
						(Note: 1 mm/day = 0.1157 l/s/ha)																	
30	Tertiary head irrigation requirement (l/s/ha)	1.		22/(27*28)	l/s/ha	0.28	0.00	0.00	0.14	0.83	1.10	0.96	0.00	0.00	0.00	0.00	0.00	0.00	0.00	0.00	0.00	0.00	0.00
31		2.		23/(27*28)	l/s/ha	0.28	0.00	0.00	0.00	0.83	1.18	1.17	1.38	0.00	0.00	0.00	0.00	0.00	0.00	0.00	0.00	0.00	0.00
32		3.		24/(27*28)	l/s/ha	0.83	0.00	0.00	0.00	0.83	1.27	1.24	1.45	0.83	0.00	0.00	0.00	0.00	0.00	0.00	0.00	0.00	0.00
33		4.		25/(27*28)	l/s/ha	0.28	0.00	0.00	0.00	0.74	1.10	1.10	1.38	0.99	0.00	0.00	0.00	0.00	0.00	0.00	0.00	0.00	0.00
34		5.		26/(27*28)	l/s/ha	0.55	0.00	0.00	0.28	1.24	1.52	1.31	1.10	0.00	0.00	0.00	0.00	0.00	0.00	0.00	0.00	0.00	0.00
35	Tertiary head irrigation requirement (l/s)	1.		30*7b	l/s	1.38	0.00	0.00	0.69	4.13	5.51	4.82	0.00	0.00	0.00	0.00	0.00	0.00	0.00	0.00	0.00	0.00	0.00
36		2.		31*8b	l/s	1.38	0.00	0.00	0.00	4.13	5.92	5.85	6.89	0.00	0.00	0.00	0.00	0.00	0.00	0.00	0.00	0.00	0.00
37		3.		32*9b	l/s	8.26	0.00	0.00	0.00	8.26	12.67	12.40	14.46	8.26	0.00	0.00	0.00	0.00	0.00	0.00	0.00	0.00	0.00
38		4.		33*10b	l/s	1.38	0.00	0.00	0.00	3.72	5.51	5.51	6.89	4.96	0.00	0.00	0.00	0.00	0.00	0.00	0.00	0.00	0.00
39		5.		34*11b	l/s	2.75	0.00	0.00	1.38	6.20	7.58	6.54	5.51	0.00	0.00	0.00	0.00	0.00	0.00	0.00	0.00	0.00	0.00
40	Total tertiary head irrigation requirement (l/s)			sum(35..39)	l/s	15.15	0.00	0.00	2.07	26.45	37.19	35.12	33.75	13.22	0.00	0.00	0.00	0.00	0.00	0.00	0.00	0.00	0.00
41	Scheme irrigation requirement (l/s)			40/29	l/s	18.94	0.00	0.00	2.58	33.06	46.49	43.90	42.18	16.53	0.00	0.00	0.00	0.00	0.00	0.00	0.00	0.00	0.00

Fig. 5.21. Proforma for calculating irrigation water requirements.

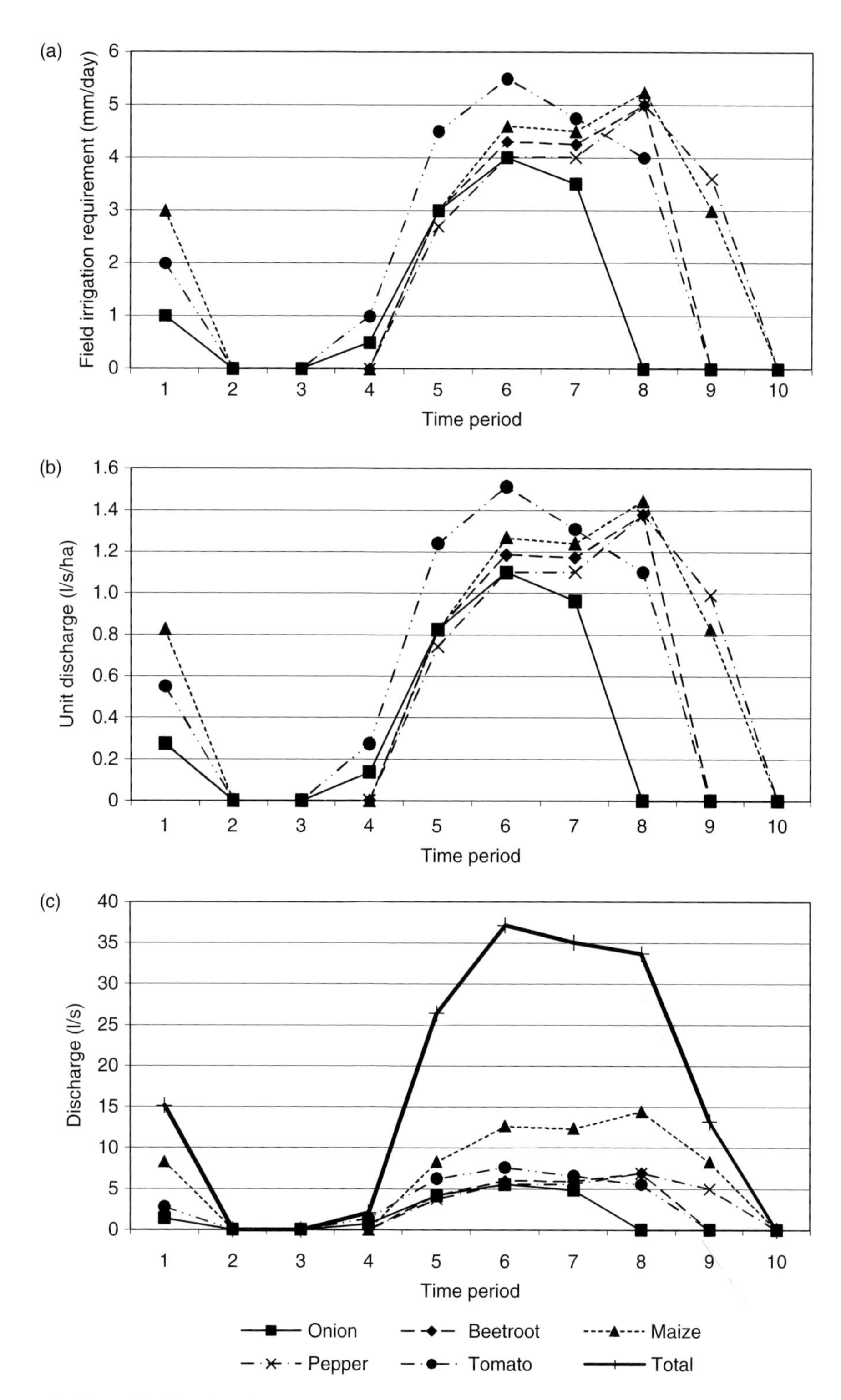

Fig. 5.22. Plots of field and tertiary unit irrigation requirements: (a) field irrigation requirement (mm/day); (b) tertiary head irrigation requirement (l/s/ha); (c) tertiary head irrigation requirement (l/s).

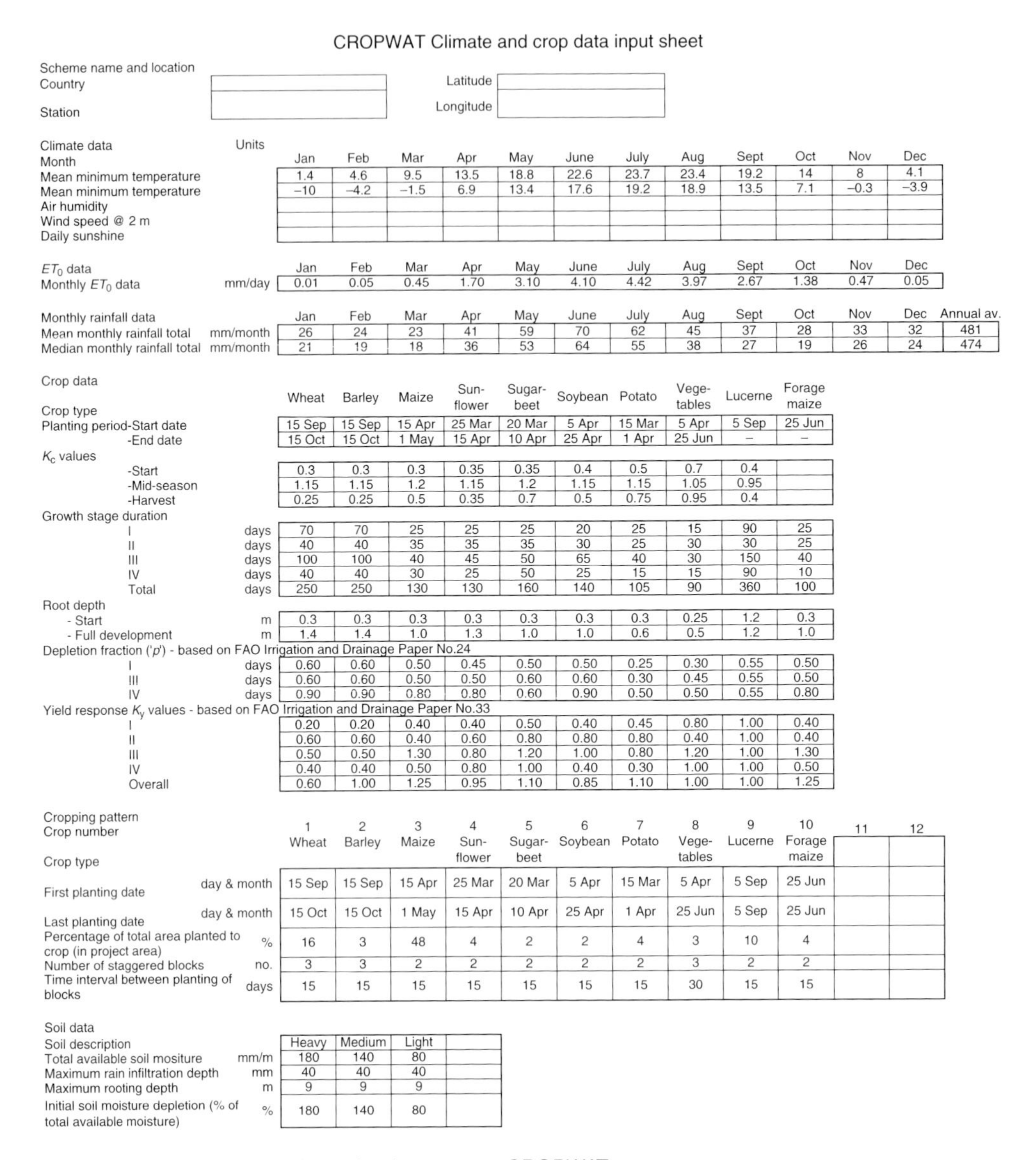

CROPWAT Climate and crop data input sheet

Scheme name and location			
Country		Latitude	
Station		Longitude	

Climate data	Units	Jan	Feb	Mar	Apr	May	June	July	Aug	Sept	Oct	Nov	Dec
Mean minimum temperature		1.4	4.6	9.5	13.5	18.8	22.6	23.7	23.4	19.2	14	8	4.1
Mean minimum temperature		–10	–4.2	–1.5	6.9	13.4	17.6	19.2	18.9	13.5	7.1	–0.3	–3.9
Air humidity													
Wind speed @ 2 m													
Daily sunshine													

ET_0 data		Jan	Feb	Mar	Apr	May	June	July	Aug	Sept	Oct	Nov	Dec
Monthly ET_0 data	mm/day	0.01	0.05	0.45	1.70	3.10	4.10	4.42	3.97	2.67	1.38	0.47	0.05

Monthly rainfall data		Jan	Feb	Mar	Apr	May	June	July	Aug	Sept	Oct	Nov	Dec	Annual av.
Mean monthly rainfall total	mm/month	26	24	23	41	59	70	62	45	37	28	33	32	481
Median monthly rainfall total	mm/month	21	19	18	36	53	64	55	38	27	19	26	24	474

Crop data		Wheat	Barley	Maize	Sun-flower	Sugar-beet	Soybean	Potato	Vege-tables	Lucerne	Forage maize
Crop type											
Planting period-Start date		15 Sep	15 Sep	15 Apr	25 Mar	20 Mar	5 Apr	15 Mar	5 Apr	5 Sep	25 Jun
-End date		15 Oct	15 Oct	1 May	15 Apr	10 Apr	25 Apr	1 Apr	25 Jun	–	–
K_c values											
-Start		0.3	0.3	0.3	0.35	0.35	0.4	0.5	0.7	0.4	
-Mid-season		1.15	1.15	1.2	1.15	1.2	1.15	1.15	1.05	0.95	
-Harvest		0.25	0.25	0.5	0.35	0.7	0.5	0.75	0.95	0.4	
Growth stage duration											
I	days	70	70	25	25	25	20	25	15	90	25
II	days	40	40	35	35	35	30	25	30	30	25
III	days	100	100	40	45	50	65	40	30	150	40
IV	days	40	40	30	25	50	25	15	15	90	10
Total	days	250	250	130	130	160	140	105	90	360	100
Root depth											
- Start	m	0.3	0.3	0.3	0.3	0.3	0.3	0.3	0.25	1.2	0.3
- Full development	m	1.4	1.4	1.0	1.3	1.0	1.0	0.6	0.5	1.2	1.0
Depletion fraction ('p') - based on FAO Irrigation and Drainage Paper No.24											
I	days	0.60	0.60	0.50	0.45	0.50	0.50	0.25	0.30	0.55	0.50
III	days	0.60	0.60	0.50	0.50	0.60	0.60	0.30	0.45	0.55	0.50
IV	days	0.90	0.90	0.80	0.80	0.60	0.90	0.50	0.50	0.55	0.80
Yield response K_y values - based on FAO Irrigation and Drainage Paper No.33											
I		0.20	0.20	0.40	0.40	0.50	0.40	0.45	0.80	1.00	0.40
II		0.60	0.60	0.40	0.60	0.80	0.80	0.80	0.40	1.00	0.40
III		0.50	0.50	1.30	0.80	1.20	1.00	0.80	1.20	1.00	1.30
IV		0.40	0.40	0.50	0.80	1.00	0.40	0.30	1.00	1.00	0.50
Overall		0.60	1.00	1.25	0.95	1.10	0.85	1.10	1.00	1.00	1.25

Cropping pattern													
Crop number		1	2	3	4	5	6	7	8	9	10	11	12
Crop type		Wheat	Barley	Maize	Sun-flower	Sugar-beet	Soybean	Potato	Vege-tables	Lucerne	Forage maize		
First planting date	day & month	15 Sep	15 Sep	15 Apr	25 Mar	20 Mar	5 Apr	15 Mar	5 Apr	5 Sep	25 Jun		
Last planting date	day & month	15 Oct	15 Oct	1 May	15 Apr	10 Apr	25 Apr	1 Apr	25 Jun	5 Sep	25 Jun		
Percentage of total area planted to crop (in project area)	%	16	3	48	4	2	2	4	3	10	4		
Number of staggered blocks	no.	3	3	2	2	2	2	2	3	2	2		
Time interval between planting of blocks	days	15	15	15	15	15	15	15	30	15	15		

Soil data					
Soil description		Heavy	Medium	Light	
Total available soil mositure	mm/m	180	140	80	
Maximum rain infiltration depth	mm	40	40	40	
Maximum rooting depth	m	9	9	9	
Initial soil moisture depletion (% of total available moisture)	%	180	140	80	

Fig. 5.23. Standardized proforma for data entry to CROPWAT.

- *Total available soil water* (S_a): The quantity of water between field capacity and the permanent wilting point that can be abstracted for use by the crop.

As discussed in the previous section on soils, water is normally available to the plant at moisture contents between field capacity and permanent wilting point. Above field capacity, water will drain by gravity out of the soil's pores; below permanent wilting point, the plant cannot extract enough water from the soil and will die. The difference between the soil moisture content at field capacity (S_{fc}) and the soil moisture content at permanent wilting point (S_w) is described as the total available soil water (S_a):

$$\text{Total available soil water, } S_a = S_{fc} - S_w$$

For example, for clay loam with field capacity of 360 mm/m and a permanent wilting point of 170 mm/m (see Table 5.1):

Table 5.14. Example of data output from CROPWAT.

(a) Irrigated cropping area (ha) and percentage of total irrigated area

	Wheat	Maize	Sunflower	Vegetables	Lucerne	Forage maize	Total
Irrigated cropping area (ha)	290	768	15	865	97	69	2105
Percentage of total irrigated area	13.8	36.5	0.7	41.1	4.6	3.3	100

(b) Basic data

Rainfall	50% probability of exceedence
Irrigation efficiency (crop to TU intake)	78%
Calculation period	10 days
Soil type	Medium

(c) Irrigation water requirements per period[a,b,c]

Date	ET_0 (mm/period)	Area (%)	Crop K_c	ET_m (mm/period)	Total rain (mm/period)	Effective rainfall (mm/period)	Net irrigation (mm/period)	TU irrigation demand (l/s/ha)
01 Jan	0.10	19.0	0.18	0.02	1.11	1.09	0	0
11 Jan	0.10	19.0	0.18	0.02	1.08	1.03	0	0
21 Jan	0.10	19.0	0.19	0.02	1.09	0.99	0	0
31 Jan	0.46	19.0	0.19	0.09	1.16	1.00	0	0
10 Feb	0.50	19.0	0.20	0.10	1.26	1.03	0	0
20 Feb	0.90	19.0	0.20	0.18	1.40	1.02	0	0
02 Mar	4.50	19.0	0.20	0.92	1.58	0.37	0.55	0.01
12 Mar	4.50	19.0	0.21	0.93	1.78	0	0.93	0.02
22 Mar	5.59	19.7	0.21	1.19	2.06	0.66	0.52	0.01
01 Apr	11.65	24.8	0.25	2.93	2.90	1.92	1.00	0.02
11 Apr	17.35	38.8	0.30	5.22	4.99	3.75	1.47	0.02
21 Apr	22.33	52.6	0.36	8.03	7.34	5.87	2.16	0.04
01 May	26.85	72.0	0.42	11.38	10.76	9.05	2.33	0.04
11 May	31.02	76.8	0.48	15.01	12.17	10.75	4.26	0.07
21 May	34.81	76.0	0.53	18.63	12.58	11.58	7.05	0.12

Continued

Table 5.14. Continued

Date	ET_0 (mm/period)	Area (%)	Crop K_c	ET_m (mm/period)	Total rain (mm/period)	Effective rainfall (mm/period)	Net irrigation (mm/period)	TU irrigation demand (l/s/ha)
31 May	38.14	78.3	0.65	24.64	13.39	12.67	11.97	0.20
10 Jun	40.90	78.0	0.74	30.18	13.59	12.99	17.19	0.28
20 Jun	42.94	80.8	0.82	35.03	14.17	13.36	21.67	0.36
30 Jun	44.15	80.2	0.82	36.15	13.98	12.72	23.43	0.39
10 Jul	44.44	77.0	0.80	35.44	13.16	11.27	24.17	0.40
20 Jul	43.77	72.2	0.76	33.15	11.95	9.45	23.71	0.39
30 Jul	42.14	68.3	0.71	29.83	10.80	7.76	22.06	0.36
09 Aug	39.64	63.2	0.61	24.34	9.44	6.21	18.13	0.30
19 Aug	36.37	49.2	0.45	16.46	6.85	4.24	12.21	0.20
29 Aug	32.49	34.4	0.30	9.72	4.41	2.69	7.03	0.12
08 Sep	28.20	16.9	0.16	4.50	1.95	1.25	3.25	0.05
18 Sep	23.69	15.8	0.12	2.99	1.65	1.09	1.90	0.03
28 Sep	19.15	15.0	0.09	1.79	1.38	0.69	1.10	0.02
08 Oct	14.70	15.5	0.10	1.43	1.25	0	1.43	0.02
18 Oct	10.42	19.0	0.13	1.32	1.36	0.19	1.12	0.02
28 Oct	6.26	19.0	0.13	0.83	1.21	1.12	0	0
07 Nov	4.70	19.0	0.14	0.65	1.10	1.10	0	0
17 Nov	4.70	19.0	0.15	0.69	1.04	1.04	0	0
27 Nov	2.18	19.0	0.16	0.34	1.02	1.02	0	0
07 Dec	0.50	19.0	0.16	0.08	1.04	1.04	0	0
Total	**680.98**			**354.34**	**189.71**	**153.65**	**210.67**	

TU, tertiary unit; ET_0, reference crop evapotranspiration; K_c, crop coefficient; ET_m, maximum rate of evapotranspiration.
[a] ET_0 data are distributed using polynomial curve fitting.
[b] Rainfall data are distributed using polynomial curve fitting.
[c] Total rainfall is over the cropped area only (i.e. rainfall not accounted for when no crop in the ground).

Total available soil water, $S_a = S_{fc} - S_w$
$= 360 - 170 = 190$mm/m

Soil moisture contents are conventionally expressed as millimetres of water per metre depth of soil (mm/m). Thus a figure of 190 mm/m of total available soil water signifies that in a 1 m depth of soil there is a total of 190 mm of water (19% by volume). This is a relatively small amount compared with the volume occupied by the soil.

To calculate how much water can be made available for use by the crop it is also necessary to consider the depth of soil from which the plant roots can draw water. Typical ranges of rooting depth (D) are given in Table 5.15 for various crops. More detailed data can often be provided by agricultural research institutes within a locality. The total available water to the crop is thus:

Total available soil water in the root zone, TAW = $S_a \times D$

In the initial crop growth stages the root depth is not fully developed and will be less than that shown in Table 5.15. As mentioned previously the root development can be encouraged by not over-irrigating in the initial growth stages and allowing the roots to develop into the lower levels of soil moisture. In some cases this soil moisture is present from rainfall or snow melt that has occurred prior to the start of the irrigation season, in other cases pre-irrigation may be required to fill the root zone.

For example, for maize crop grown on a clay loam with total available soil water, S_a=190 mm/m:

Table 5.15. Typical values of crop rooting depth.

Crop	Effective rooting depth, D (m)
Pasture grasses, potatoes, vegetables	0.3 to 0.4
Small grains, wheat, maize, cotton, tobacco, most field crops	0.6 to 0.8
Sugarcane, lucerne, most tree crops	0.9 to 1.1

Total available soil water in root zone = $S_a \times D = 190 \times 0.8 = 152$ mm

When the soil is at field capacity, the plant can extract water easily and thus maintain the maximum rate of evapotranspiration (ET_m). As the plant removes water from the soil moisture reservoir, the suction it has to exert increases as the moisture content decreases. After a certain level of depletion, the crop cannot extract the water fast enough to maintain ET_m, and the evapotranspiration drops to a lower rate, termed the actual evapotranspiration rate (ET_a). The plant is then under (moisture) stress and the potential yield of the crop may be adversely affected. The relationship of ET_a to ET_m and the impact on the crop yield has been studied on some detail and approaches developed to predict the resultant crop yield (see later section on 'Yield Response to Water').

Easily available soil water (EAW) is defined as the fraction (p) to which the total available soil moisture can be depleted without causing the evapotranspiration to drop below ET_m. The depth of freely available soil moisture in a soil with rooting depth D is given by:

Easily available water, EAW = $p \ S_a \ D$

A reasonable general figure to take for the value of the 'p' fraction is 0.5, but more precise figures for different types of crop are provided from Tables 5.16 and 5.17. The value of the 'p' fraction of the total available soil water S_a depends on:

- the crop;
- the magnitude of ET_m;
- the soil.

The crop

Some crops like cotton and sorghum are better than others (e.g. vegetables) at abstracting water from the soil before ET_a falls below ET_m, as shown in Table 5.16.

The magnitude of ET_m

As the evapotranspiration rate ET_m increases, the crop finds it increasingly difficult to extract water from the soil, and the 'p' fraction falls. The values of the 'p' fraction for different values of ET_m are given in Table 5.17.

Table 5.16. Crop groups according to soil water depletion. (From FAO, 1979 with permission from the Food and Agriculture Organization of the United Nations.)

	Group	Crops
Increasing ability (Groups 1–4) to extract water from the soil ⇩	1	Onion, pepper, potato
	2	Banana, cabbage, grape, pea, tomato
	3	Bean, citrus, groundnut, lucerne, pineapple, sunflower, watermelon, wheat
	4	Cotton, maize, olive, safflower, sorghum, soybean, sugarbeet, sugarcane, tobacco

Table 5.17. Soil water depletion fraction (p) for crop groups at different values of maximum evapotranspiration rate (ET_m). (From FAO, 1979 with permission from the Food and Agriculture Organization of the United Nations.)

Crop group	ET_m (mm/day)								
	2	3	4	5	6	7	8	9	10
	Soil water depletion factor, p								
1	0.50	0.225	0.35	0.30	0.25	0.225	0.20	0.20	0.175
2	0.675	0.575	0.475	0.40	0.35	0.325	0.275	0.25	0.225
3	0.80	0.70	0.60	0.50	0.45	0.425	0.375	0.35	0.30
4	0.875	0.80	**0.70**	0.60	0.55	0.50	**0.45**	0.425	0.40

For example, for cotton (Crop Group 4) the 'p' fraction reduces from 0.70 at 4mm/day down to only 0.45 at 8mm/day.

The soil

Soil water is more easily transmitted and taken up by plant roots in light textured soil. However, this factor does not significantly improve accuracy of factor 'p'.

Tables 5.18 and 5.19 provide examples of calculations to determine the irrigation demand and irrigation interval for different crops on two different soil types, a silty clay loam and a sandy loam. It is apparent from the calculations provided in these tables that the irrigation interval is longer for the silty clay loam soils compared with the sandy loams. The silty clay loams have a water-holding capacity (S_a=210mm/m depth) almost double that of the sandy loams (S_a=120mm/m depth). This relationship carries through such that the irrigation intervals for the same crops on the silty clay loam soils are almost twice those on the sandy loam soil. Note that the total volume of water consumed by the crop does not change in the two cases, just the irrigation interval.

The influence of the crop group is also clear from the examples provided. Potatoes are in Crop Group 1, and though they have a similar rooting depth to cabbage they require irrigating twice as frequently if the yield potential is to be maintained. Crop Group 1 crops are not good at abstracting water from the soil, as shown by the 'p' fraction figures in Table 5.17.

The impact of a deeper rooting depth is also clear from the calculations, with cotton having twice the effective rooting depth of potato and cabbage, and therefore larger easily available water capacity EAW and a longer irrigation interval. The quantity of irrigation water that can be applied during each irrigation event for cotton is thus three to six times the amount that can be applied for cabbage or potato. With surface irrigation systems it is more likely that over-irrigation will occur for shallow rooting crops requiring small irrigation amounts, whereas with deep rooted crops there is a reduced likelihood of over-irrigation as each irrigation event takes longer and farmers are more likely to cut off the supply and move on to irrigate the next plot or area.

Table 5.18. Example calculation to determine the irrigation amount and interval for crops on a silty clay loam; total available water, S_a=210 mm/m.

Crop	Crop group	ET_0 (mm/day)	K_c	ET_m (mm/day)	D (m)	TAW (mm)	p	EAW (mm) – irrigation amount	Irrigation interval (days)
		(1)	(2)	(3)	(4)	(5)	(6)	(7)	(8)
	from tables	from climate station	from tables	col. (1) × col. (2)	from tables	col. (4) × S_a	from tables	col. (5) × col. (6)	col. (7)/ col. (3)
Cotton	4	3	1.05	3.15	0.8	168	0.78	131	42
Potato	1	3	1.05	3.15	0.4	84	0.25	21	7
Cabbage	2	3	0.95	2.85	0.3	63	0.6	38	13

ET_0, reference crop potential evapotranspiration; K_c, crop coefficient; ET_m, crop potential evapotranspiration; D, root depth; TAW, total available water; p, soil water depletion fraction; EAW, easily available water.

Table 5.19. Example calculation to determine the irrigation amount and interval for crops on a sandy loam; total available water, S_a=120 mm/m.

Crop	Crop group	ET_0 (mm/day)	K_c	ET_m (mm/day)	D (m)	TAW (mm)	p	EAW (mm) – irrigation amount	Irrigation interval (days)
		(1)	(2)	(3)	(4)	(5)	(6)	(7)	(8)
	from tables	from climate station	from tables	col. (1) × col. (2)	from tables	col. (4) × S_a	from tables	col. (5) × col. (6)	col. (7)/ col. (3)
Cotton	4	3	1.05	3.15	0.8	96	0.78	75	24
Potato	1	3	1.05	3.15	0.4	48	0.25	12	4
Cabbage	2	3	0.95	2.85	0.3	36	0.6	22	8

ET_0, reference crop potential evapotranspiration; K_c, crop coefficient; ET_m, crop potential evapotranspiration; D, root depth; TAW, total available water; p, soil water depletion fraction; EAW, easily available water.

The longer the irrigation interval the more efficient irrigation can be, as each time the crop is irrigated water is lost, either through distribution or application losses. Additional water is also lost due to evaporation from the wet soil surface during and immediately following irrigation.

Soil water and paddy rice

The description above applies to dry foot crops such as wheat, maize, legumes, vegetable and fruit crops. Paddy rice is a special case as its roots can tolerate waterlogging and so the soil is usually kept saturated with a ponded layer of water on the surface to provide water storage for crop growth, and to reduce weed growth. As mentioned previously, in order to reduce the percolation losses the structure of a paddy soil is manipulated by puddling, with the aim of eliminating the larger pores through which water flows by gravity. At the same time the number of smaller pores is also reduced, thus reducing the moisture content at field capacity, and the available moisture-holding capacity of the soil.

If there is a shortage of water and the ponded water is not replenished, the rice crop is dependent on the water stored in the soil moisture reservoir, as described above. This is a comparatively small amount however, because of both the low moisture-holding capacity of the puddled soil and the shallow rooting depth of the rice plant.

Scheduling using the water balance sheet

The soil moisture content can be tracked using an accounting process called the water balance sheet (Fig. 5.24). Following a full irrigation the soil moisture content is returned to field capacity. At field capacity the soil moisture deficit is zero, as each day passes the deficit increases by the amount of the crop's evapotranspiration. If the crop's daily evapotranspiration is estimated (using a Class A evaporation pan, for example) then the water balance deficit can be tracked. When the deficit reaches the easily available water limit it is time to irrigate the crop.

For example, Figure 5.24 shows the water balance sheet calculations for a field of cotton and a field of cabbages, while Fig. 5.25 provides a graphical plot of the soil moisture deficit over the period. The impact that the rooting depth and the '*p*' fraction have on the irrigation schedule can be seen in Fig. 5.25. For the cotton crop the field is irrigated once in the 15-day period, but with a large volume to fill the soil root zone. In the same period the cabbage crop is irrigated twice, and would have been ready for a third irrigation at the end of the period if it had not rained.

Yield Response to Water

The potential yield of a crop will be reduced if the crop suffers stress as a result of a shortage of water. The degree of water stress can be quantified by the rate of actual evapotranspiration ET_a in relation to the optimum or maximum evapotranspiration ET_m. Research has shown that the relative yield loss Y_a/Y_m (where Y_a is the actual crop yield and Y_m is the maximum potential yield) can be related to the relative evapotranspiration ET_a/ET_m. The FAO have produced a valuable summary and explanation in their Irrigation and Drainage Paper No. 33, *Yield Response to Water* (FAO, 1979).

A number of empirical relationships have been developed to relate the crop evapotranspiration to the crop yield. A widely used relationship is:

$$1-(Y_a / Y_m) = k_y[1-(ET_a / ET_m)]$$

where

Y_a = actual harvested yield
Y_m = maximum harvested yield
k_y = yield response factor
ET_a = actual evapotranspiration
ET_m = maximum evapotranspiration.

This relationship can be applied where:

- water shortages occur over the total growing period;
- water shortages occur for individual growth periods.

The relative sensitivities of different crops to water shortage, either over the total growing period or in individual growth stages, are summarized in the values of the yield response factor k_y (Table 5.20). The higher the value of k_y the greater the sensitivity to water shortage. Thus maize is more sensitive to water shortage than sorghum, and is most sensitive to water shortage in the flowering stage. In general the flowering stage is the most sensitive period to water shortage, though in the case of soybean the yield formation stage is the most sensitive.

Measuring and Improving Irrigation Application

Significant savings can be achieved through improving the application of irrigation water at the field level. The key is to improve the efficiency of irrigation by providing just enough irrigation water to match the available storage in the root zone (Fig. 5.26), thus reducing the amount of water lost to the crop through deep percolation and surface runoff.

The main variables influencing the irrigation application efficiency are:

- the soil type and intake rate of the soil;
- the field configuration (width and length);
- the slope, both in the direction of irrigation and across the field;
- the uniformity of the field slope;
- the irrigation method (basin, border, furrow, etc.).

For any given physical configuration of the field the following variables then need to

WATER BALANCE SHEET						WATER BALANCE SHEET					
Field No.: F1 Crop: Cotton Planting date: 1st April				Irrigation Plan Apply at	75 mm 75 mm SMD	Field No.: F2 Crop: Cabbage Planting date: 1st April				Irrigation Plan Apply at	22 mm 22 mm SMD
Date	Cropwater demand, ET_c (mm)	Irrigation, I (mm)	Effective rainfall, R_e (mm)	$ET_c - I - R_e$ (mm)	Soil moisture deficit, SMD (mm)	Date	Cropwater demand, ET_c (mm)	Irrigation, I (mm)	Effective rainfall, R_e (mm)	$ET_c - I - R_e$ (mm)	Soil moisture deficit, SMD (mm)
15 July	3			3	68	15 July	3			3	18
16	3			3	71	16	3			3	21
17	3			3	74	17	3	22		–19	2
18	3	75		–72	2	18	3			3	5
19	3			3	5	19	3			3	8
20	3			3	8	20	3			3	11
21	3			3	11	21	3			3	14
22	3			3	14	22	3			3	17
23	3			3	17	23	3			3	20
24	3			3	20	24	3	22		–19	1
25	3			3	23	25	3			3	4
26	3		10	–7	16	26	3		10	–7	0
27	4			4	20	27	3			3	3
28	4			4	24	28	3			3	6
29	4			4	28	29	3			3	9
30	4			4	32	30	3			3	12
31	4			4	36	31	3			3	15

Fig. 5.24. Monitoring soil moisture deficit using the water balance sheet method.

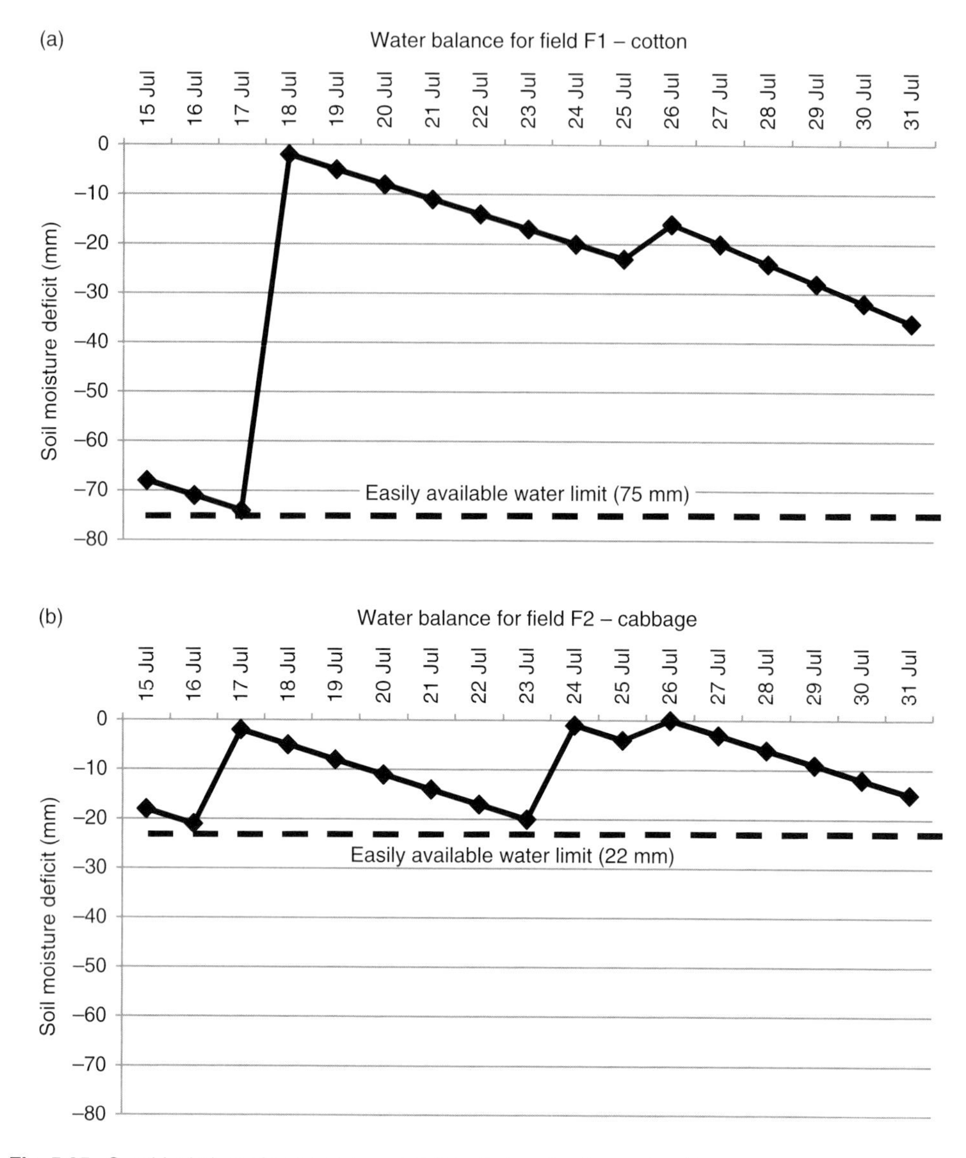

Fig. 5.25. Graphical plots of soil moisture deficit for fields of cotton and cabbage.

be controlled to achieve optimum irrigation application:

- the stream size;
- the contact time (of the water on the soil surface);
- advance, recession and shut-off times;
- surface runoff and deep percolation.

The overriding aim is to balance the horizontal speed of flow over the soil surface with the vertical speed of flow of water into the soil profile.

Table 5.21 outlines some of the areas where the efficiency of irrigation application can be improved by balancing these variables, while Fig. 5.27 provides some examples of the application of these measures.

The situation shown in Fig. 5.27e and f, and schematically in Fig. 5.28, was modelled using the SIRMOD computer program developed by Wynn Walker (FAO, 1989; Walker, 1993). The program can be used for the design and evaluation of basin, border strip and furrow irrigation for either continuous or surge

Table 5.20. Values of the yield response factor (k_y) for different crops. (From FAO, 1979 with permission from the Food and Agriculture Organization of the United Nations.)

Crop	Vegetative period			Flowering period	Yield formation	Ripening	Total growing period
	Early	Late	Total				
Bean			0.2	1.1	0.75	0.2	1.15
Cabbage	0.2				0.45	0.6	0.95
Cotton		0.2	0.5			0.25	0.85
Groundnut			0.2	0.8	0.6	0.2	0.7
Lucerne			0.7 –1.1				0.7–1.1
Maize			0.4	1.5	0.5	0.2	1.25
Onion			0.45		0.8	0.3	1.1
Potato	0.45	0.8			0.7	0.2	1.1
Safflower		0.3		0.55	0.6		0.8
Sorghum			0.2	0.55	0.45	0.2	0.9
Soybean			0.2	0.8	1.0		0.85
Sugarbeet							
Beet							0.6–1.0
Sugar							0.7–1.1
Sugarcane			0.75		0.5	0.1	1.2
Sunflower	0.25	0.5		1.0	0.8		0.95
Tobacco	0.2	1.0			0.5		0.9
Tomato			0.4	1.1	0.8	0.4	1.05
Wheat							
Winter			0.2	0.6	0.5		1.0
Spring			0.2	0.65	0.55		1.15

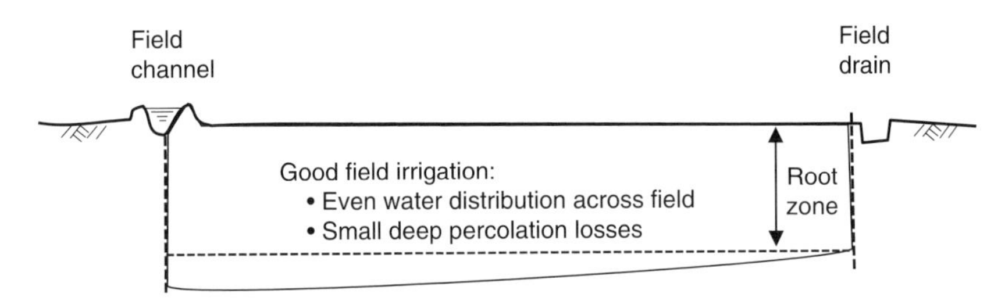

Fig 5.26. Efficient irrigation – matching water application to storage available in the crop's root zone.

flow conditions. The program utilizes the following equation:

$$Z = kr^a + f_0 r$$

where

Z = cumulative infiltration in units of volume per unit length per unit width
r = intake opportunity time
f_0 = long-term steady or basic infiltration rate in units of volume per unit length per unit time and width
k, a = empirical constants derived from field tests.

The data and results are presented in Table 5.22. As can be seen the performance of the observed irrigation method is very poor, with an application efficiency of only 6% and a storage efficiency of only 24%. Using the model to try different cut-off times for the same discharge the application efficiency is increased to 96% and the storage efficiency to 87%. In addition the total runoff is significantly reduced. Due to the slower surface flow rate the distribution uniformity is reduced from 96% to 75%. A further issue is the time and energy spent by the farmer in forming the earth checks and diverting the

Table 5.21. Measures to improve irrigation application efficiency.

No.	Measure	Why	What to do
1	Keep the irrigation interval as long as possible	Long intervals between irrigations reduce the number of irrigations. During each irrigation water is lost, so reducing the number of irrigations reduces the losses	• Irrigate when the soil moisture deficit reaches the easily available water limit, not before
2	Know how much water is required	If the farmer doesn't know how much water is required then he may over- or under-irrigate. Over-irrigation wastes water; under-irrigation means more frequent irrigations, and more wastage (see 1 above)	• Monitor the soil moisture deficit (by physical inspection or water balance sheet) and irrigate when the easily available water limit is reached • Monitor soil moisture status using a soil auger or moisture probe
3	Apply the correct amount of water	Over-irrigation results in wastage of water	• Know how much water needs to be applied • Measure and time the water application
4	Measure the depth of infiltration and uniformity of irrigation following irrigation	To know how far the water has infiltrated. If this is done several times following irrigation events the farmer will become familiar with the amount of water needed to fill a given depth of root zone on each plot of land	• Record the duration of irrigation for the field • 2 or 3 days after irrigation, take auger samples at five locations in the field • Sample the soil at regular intervals (by feel or oven-drying method) to ascertain the depth to which the water has penetrated
5	Improve the uniformity of the field slope	The more uniform the field slope the more uniform the flow and infiltration into the soil. Better uniformity gives more efficient irrigation and better and more uniform crop yields within the field	• Observe/inspect the field looking for high and low spots (these will show in differential crop growth) • Survey the field and measure the uniformity down the irrigation slope • If the uniformity is poor carry out land planing • If the uniformity is very poor consider land levelling
6	Use an appropriate irrigation method for the crop type	Farmers sometimes use inappropriate irrigation methods for the slope, soil or crop type	• Observe the irrigation method and advise on a more appropriate method if required
7	Set out the field to match the soils, slope, crop type and irrigation method	In many smallholder irrigation schemes the fields are not laid out in the ideal configuration for irrigation	• Observe and measure irrigation application and recommend changes to the method and/or irrigation practices • Computer modelling of irrigation application can be useful to prepare norms for varying field dimensions

Fig. 5.27. Examples of approaches to improving irrigation application. (a, b) A farmer uses furrow-in-basin to irrigate a crop of beans on a small plot with light soils. The farmer irrigates several furrows at a time and moves to the next set of furrows when the water reaches the end of the furrows. Water ponds at the end of the furrows and balances the longer contact time at the head of the furrows. This approach is a significant improvement on irrigating the plot as a single basin. (c, d) The farmer has split this long field into two halves, irrigating both halves at the same time. The furrows are well formed and the flow rate in each furrow relatively uniform. The main variables to control here are the flow rates in each furrow and the application time. To ascertain the depth irrigated and uniformity of irrigation the farmer should take auger samples 1 or 2 days after irrigating. (e, f) This farmer is irrigating wheat using a method on steeply sloping land (1.5% slope) with silty clay loam soils, which is used in flatter parts of the scheme for irrigating sugarbeet in flat basins. The 38 l/s discharge is diverted on to strips 3–4 m wide in turn (e) with a lot of surface runoff at the tail end of the field on to the farm road (f). The surface runoff is significant and infiltration minimal. Measurements were made and modelled using SIRMOD (see discussion in text). The modelling showed application efficiencies of only 6%, which could be increased to 96% if border strip irrigation is used. Changing to border strip irrigation spreads the 38 l/s flow at the field intake to give a flow rate of 0.5 l/s/m width compared with the 10 l/s/m width with the farmer's method.

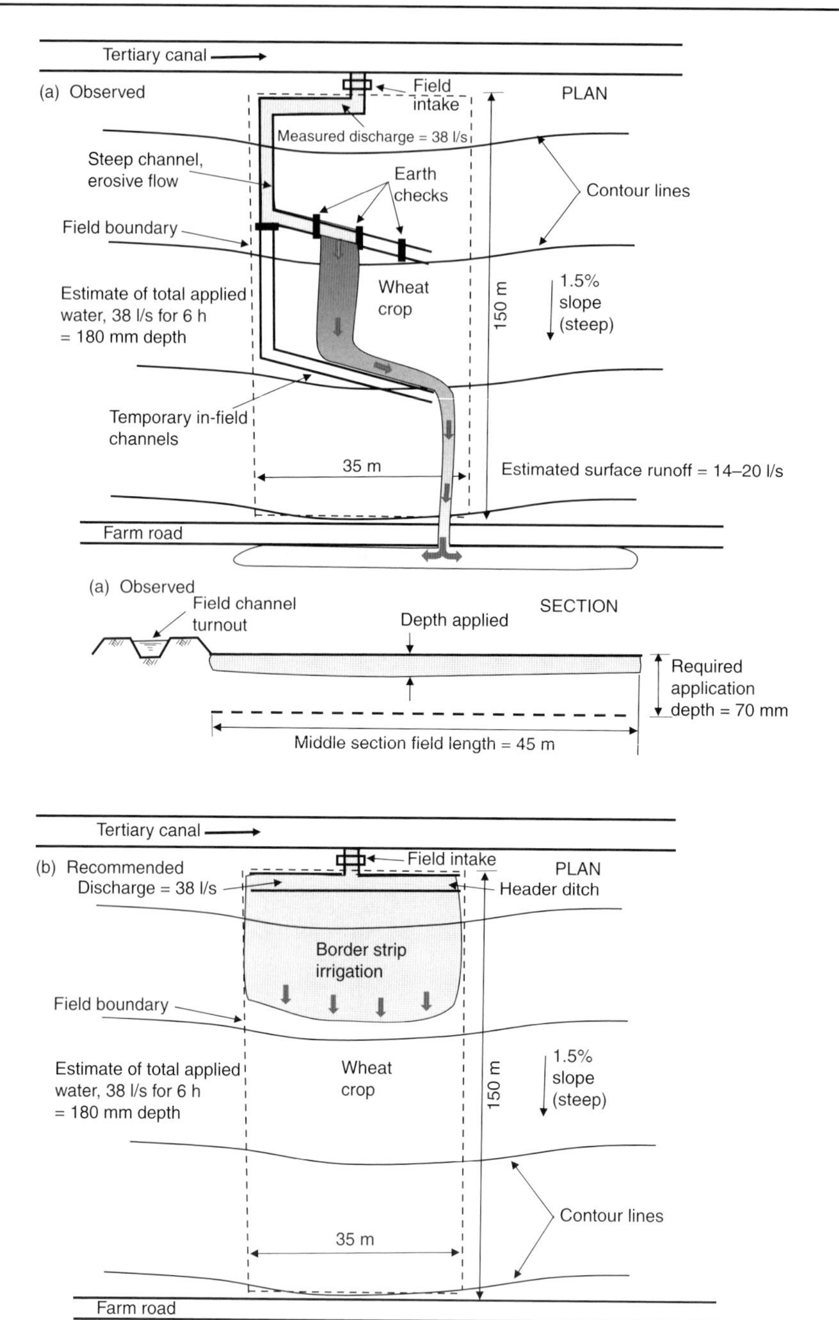

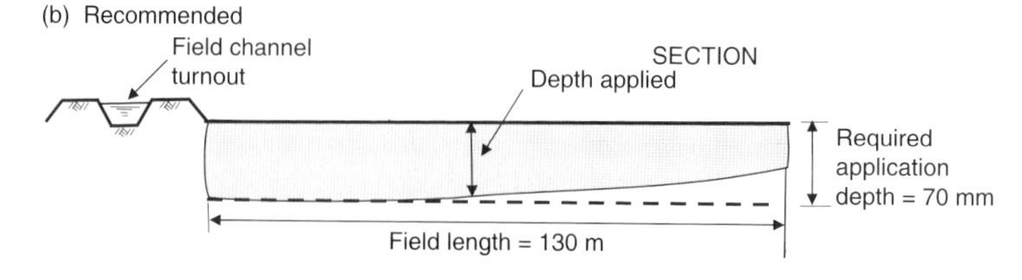

Fig. 5.28. Results of the SIRMOD modelling exercise. (a) Diagrammatic representation of the situation shown in Fig. 5.27e and f, with the observed irrigation method. (b) The recommended approach with border strip irrigation.

Table 5.22. Data for example of surface irrigation methods.

Data		Results	
(a) Observed			
Unit discharge	10 l/s/m	Total inflow	12 m^3
Field slope	0.015	Total infiltration	0.75 m^3
Manning's '*n*'	0.04	Total runoff	11.25 m^3
Length	130/3=45 m	Application efficiency	6%
Time to cut-off	20 min	Storage efficiency	24%
		Distribution uniformity	96%
(b) Recommended (border strip irrigation)			
Unit discharge	0.5 l/s/m	Total inflow	8.25 m^3
Field slope	0.015	Total infiltration	7.90 m^3
Manning's '*n*'	0.04	Total runoff	0.35 m^3
Length	130 m	Application efficiency	96%
Time to cut-off	250 min	Storage efficiency	87%
		Distribution uniformity	75%

water over new sections of the field compared with forming the header ditch for the border strip irrigation and ensuring a uniform flow over the field.

The exercise shows the benefits of using such a computer model to simulate the field application method, with significant savings in irrigation water and irrigation time.

The relationships between the key variables influencing irrigation application have been determined through empirical formulae, mathematical models and field trials. Tables 5.23 to 5.25 provide a summary of suggested field configurations for basin, border and furrow irrigation. They are useful in designing the field layout on new irrigation schemes, but are also helpful in determining suitable stream sizes where the field configuration is already set (as is often the case on smallholder irrigation schemes where landholding sizes are often determined by factors other than irrigation application efficiency). For a detailed and practical description of surface irrigation methods see Kay (1986).

For basin irrigation, especially on sandy soils, it is important to fill the basin as quickly as possible. A rule-of-thumb (the 'Quarter Time Rule') is that the stream size should be large enough such that it advances across the basin in one-quarter of the time needed to infiltrate the required irrigation depth into the soil (e.g. if the required irrigation depth is 75 mm and the required contact time is 60 min, the water should advance across the basin in 15 min). As can be interpolated from Table 5.23 basin size needs to be reduced on sandy soils in comparison to clay soils, and when the stream size is small.

For border irrigation it is important that there is no cross slope. If there is a cross slope then furrow irrigation should be used, or the basin width restricted. As with basin irrigation, the basin width and length need to be restricted if the stream size is small, and will be less on sandy soils than clay soils. The flow rate over the soil surface can be controlled either by matching the stream size to the border width or by matching the border width to the available stream size to give the unit stream sizes suggested in Table 5.24. If the available stream size is small in relation to the border width, temporary bunds can be formed down the border and the border irrigated in stages.

The suggested furrow lengths for different soil types, slopes and irrigation depths are shown in Table 5.25. Another important factor in furrow irrigation is the spacing of the furrows. The spacing should be more on clay soils than on sandy soils as the lateral infiltration is greater in the clay soils. As with basin and border irrigation the furrow length can be longer on clay rather than sandy soils, and can be longer when the streamflow in the furrow is greater. In order to prevent soil erosion the streamflow should be less and the furrow

Table 5.23. Suggested basin sizes, in hectares, for different soil types and stream sizes. (From Kay, 1986 with permission.)

	Soil type			
Stream size (l/s)	Sand	Sandy loam	Clay loam	Clay
	Field size (ha)			
15	0.01	0.03	0.06	0.1
30	0.02	0.06	0.12	0.2
60	0.04	0.12	0.24	0.4
90	0.06	0.19	0.36	0.6
120	0.08	0.24	0.48	0.8
150	0.10	0.30	0.60	1.0
180	0.12	0.36	0.72	1.2
210	0.14	0.42	0.84	1.6

Table 5.24. Suggested border sizes for different soil types, irrigation depths and slope. (From Kay, 1986 with permission.)

Soil type	Irrigation depth (mm)	Slope (%)	Width (m)	Length (m)	Unit stream (l/s/m)[a]
Sand	100	0.2	12–30	60–100	10–15
		0.4	10–12	60–100	8–10
		0.8	5–10	75	5–7
Loam	150	0.2	15–30	90–300	4–6
		0.4	10–12	90–180	3–5
		0.8	5–10	90	2–4
Clay	200	0.2	15–30	350+	3–6
		0.4	10–12	180–300	2–4

[a]Unit stream refers to flow per metre width of border (l/s/m).

Table 5.25. Suggested furrow lengths for different soil types, slopes and irrigation depths. (From Kay, 1986 with permission.)

		Clay		Loam			Sand		
		Average irrigation depth (mm)							
Slope (%)	Maximum stream size (l/s)	75	150	50	100	150	50	75	100
		Furrow length (m)							
0.05	3.0	300	400	120	270	400	60	90	150
0.1	3.0	340	440	180	340	440	90	120	190
0.2	2.5	370	470	220	370	470	120	190	250
0.3	2.0	400	500	280	400	500	150	220	280
0.5	1.2	400	500	280	370	470	120	190	250
1.0	0.6	280	400	250	300	370	90	150	190
1.5	0.5	250	340	220	280	340	80	120	190
2.0	0.3	220	270	180	250	300	60	90	150

length shorter on steeper slopes. The available flow can be divided into equal furrow flow sizes by use of a header ditch, such that a flow of, say, 15 l/s is divided into a furrow flow of 3 l/s for five furrows.

Determining irrigation application requirements

The following section outlines how the irrigation application can be measured in the field. The technique is very straightforward. Measurements can be carried out by one team on two to three plots in 1 day.

The following equipment is required:

- small measuring weir or flume;
- measuring tape (30 or 50 m);
- soil auger;
- 5 l bucket;
- infiltrometer rings;
- measuring flask;
- stopwatch, or a watch with a seconds hand;
- wooden pegs;
- hammer.

Approach

There are three parts to the process:

- determination of the soil texture;
- determination of the soil's infiltration characteristics;
- determination of the contact time and infiltration amount on the field.

Part 1: Determine the soil texture

The starting point for irrigation is the soil texture. The steps to be followed to determine soil texture are outlined below.

1. Take a sample of the soil and ascertain its textural class following the guidelines given in Fig. 5.2 or Box 5.1. Alternatively, send samples of the soil to the laboratory for determination of the textural class.
2. Knowing the soil textural class determines the soil moisture characteristics – field capacity, permanent wilting point and total available soil water (see Table 5.1). Alternatively, these characteristics can be determined in the laboratory.
3. Knowing the soil textural class ascertains the likely infiltration profile and rates (see Fig. 5.11 and Table 5.5).

Part 2: Determine the soil's infiltration characteristics

The rate at which the water infiltrates into the soil governs the duration of irrigation and the irrigation stream size. The infiltration rate is relatively easy to ascertain as follows.

1. Set up the infiltration rings in the field. Locate the rings away from the top and tail ends of the field where the soil might have been compacted by farm equipment.
2. Measure the infiltration rate and obtain the infiltration curve and the terminal infiltration rate.
3. Repeat the test in at least two other locations in the field to obtain an average for the field.
4. Check that the figures are in the range predicted for the soil texture (see Part 1 above).

Part 3: Determine the contact time and infiltration amount within the field

It is a relatively simple exercise to determine the contact time and amount of irrigation water infiltrated into the soil at different locations in the field. The steps are outlined below and illustrated in Figure 5.29.

1. Locate a typical field that is ready for irrigation.
2. Before the irrigation commences place pegs at regular measured intervals down the length of the field. Five measurement points are generally sufficient in the direction of irrigation, so if the field is 200 m long then the pegs should be spaced at 0, 50, 100, 150, 200 m. At least three lines of pegs should be set out in the field, and measurements taken along each line to get the average for the field. With furrow irrigation fields are often irrigated in blocks of five to ten furrows at a time, therefore set up lines of pegs in three separate blocks.

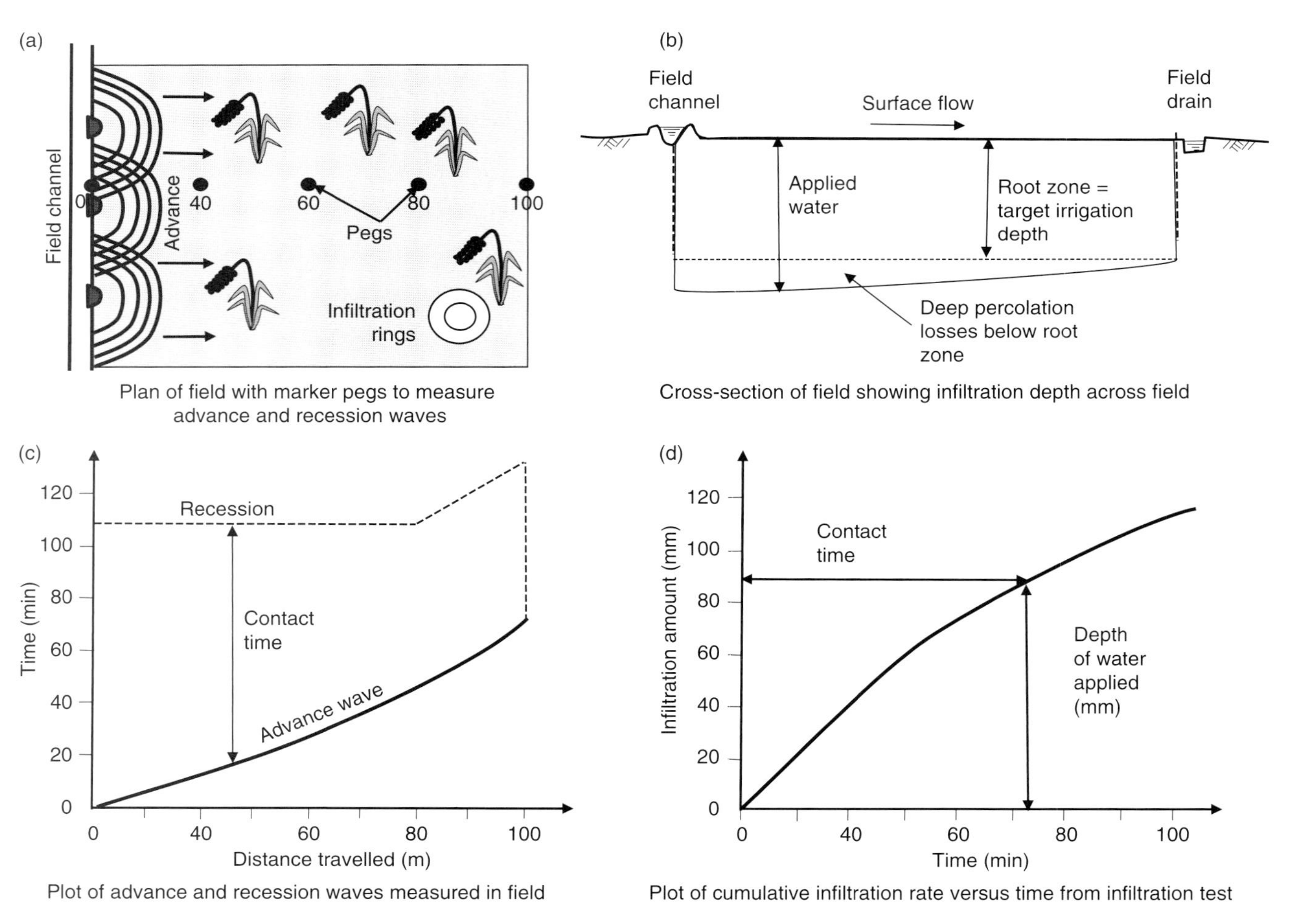

Fig. 5.29. Graphical representation of steps involved in determining water application for basin irrigation.

3. Set up a table for recording the data as set out in Fig. 5.30.
4. Start the irrigation and record the time that the advance wave of the flow reaches each peg.
5. Measure the streamflow size using the portable measuring weir or flume. For furrow irrigation measure the streamflow in one sample furrow in each block. Take care to measure the flow at regular intervals during the test in case the stream size changes.
6. Measure the time when the farmer cuts off the supply to the field (or furrow) and then measure the recession at each peg. The recession is seen when there is no more water on the soil's surface. The total contact time is the time that the water is on the soil's surface between the advance wave and the recession.
7. Measure the volume of any runoff at the tail end of the field. This can be done by placing a portable measuring flume in the drain at the end of the field. Take care to record the flow and the duration of flow, in order to obtain the total volume of runoff.
8. Calculate the contact time and using the infiltration curves obtained from Part 2 above determine the total (cumulative) amount of

Distance (m)	Advance time		Recession time		Contact time, (5) – (3) (min)	Water applied (mm)
	Clock	Elapsed (min)	Clock	Elapsed (min)		
(1)	(2)	(3)	(4)	(5)	(6)	(7)
0 (start)						
					Average	
DISCHARGE MEASUREMENT AT HEAD AND TAIL OF FIELD						
Head (Inflow)			Tail (Outflow)			
Time	Discharge (l/s)	Flow volume (l)	Time	Discharge (l/s)	Flow volume (l)	
(8)	(9)	(10)	(11)	(12)	(13)	
0			0			
20			20			
40			40			
60			60			
80			80			
100			100			
120			120			
140			140			
160			160			
180			180			
	Total			Total		

Notes:
Col. (1) = distance measured from quaternary/field channel.
Col. (2) = clock time taken for water to advance from quaternary/field channel.
Col. (3) = elapsed time in minutes from the start of the test.
Col. (4) = clock time measured when the recession occurs (water disappears from the soil surface).
Col. (5) = elapsed time in minutes from start of the test.
Col. (6) = contact time at different parts of basin found by subtracting (3) from (5).
Col. (7) = water applied at different parts of the basin found from contact time and graph of cumulative infiltration rate versus contact time (determined from 'Part 2: Determine the soil's infiltration characteristics').
Cols (8), (11) = time of discharge measurement, can be any interval, 20 min intervals shown here.
Cols (9), (12) = discharge measurement from portable weir or flume.
Cols (10), (13) = volume of flow – elapsed time multiplied by the average of the discharge at the start and end of each interval. For example: time = 20 min, discharge = 4.2 l/s; time = 40 min, discharge = 4.5 l/s; therefore flow volume = (40 – 20) × 60 × [(4.2 + 4.5)/2] = 20 × 60 × 4.35 = 5220 l = 5.22 m^3.

Fig. 5.30. Surface irrigation evaluation data collection form.

water infiltrated into the soil at each measurement point.

9. Plot the cumulative depth of irrigation against the location in the field and the target irrigation depth to fill the root zone.

10. Calculate the efficiency of application from the equation at the bottom of this page.

Example

The following example illustrates the evaluation procedure for a basin 40 m long. The required depth of irrigation is 75 mm. Table 5.26 shows the data collected from the irrigation and Fig. 5.31 shows the results of the infiltration test carried out before the irrigation.

The calculations in Table 5.26 show that the average application is 93.2 mm of water when the required irrigation depth was 75 mm. Percolation losses are thus 93.2–75=18.2 mm and the application efficiency is:

$$\text{Application efficiency, } E_a\ (\%) = \frac{75 \times 100}{93.2} = 80.4\%$$

Figure 5.32 shows the plot of the amount of water infiltrated at each point in the field. All parts of the field receive the target amount of 75 mm depth, but the upper ends of the field are over-irrigated due to the extra contact time. The area between the target amount and actual amount infiltrated represents the losses to deep percolation below the root zone.

Carrying out the above exercise has the following benefits.

- It quantifies the infiltration rate of the soil and makes it possible to set irrigation durations for individual crops and crop growth stages (for stated stream sizes).
- It quantifies the irrigation practices of the farmer. If the farmer is applying too much water during each irrigation, the process outlined above will identify this situation and a new irrigation regime can

Table 5.26. Data collected while irrigating the basin.

Distance (m)	Advance time		Recession time		Contact time, (5) – (3) (min)	Water applied (mm)
	Clock	Elapsed (min)	Clock	Elapsed (min)		
(1)	(2)	(3)	(4)	(5)	(6)	(7)
0	10.00	0	11.39	99	99	103
10	10.07	7	11.39	99	92	101
20	10.17	17	11.39	99	82	95
30	10.31	31	11.39	99	68	90
40	10.49	49	11.50	110	61	77
					Average	93.2
					Target	75.0

Col. (1) = distance measured from quaternary/field channel (see Fig. 5.29).
Col. (2) = clock time taken for water to advance from quaternary/field channel.
Col. (3) = elapsed time in minutes from the start of the test.
Col. (4) = clock time measured when the recession occurs.
Col. (5) = elapsed time in minutes from start of the test.
Col. (6) = contact time at different parts of basin found by subtracting (3) from (5).
Col. (7) = water applied at different parts of the basin found from contact time and infiltration data (Fig. 5.31).

$$\text{Efficiency of application } (\%) = \frac{\text{Total volume of water stored in the root zone}}{\text{Total volume applied to the field}}$$

Time (min)	Depth of water added (mm)	Infiltration rate (mm/h)	Cumulative depth infiltrated (mm)
0	0	0	0
5	15	180	15
10	10	120	25
20	15	90	40
30	12	72	56
50	22	66	74
70	22	66	90
100	33	66	103

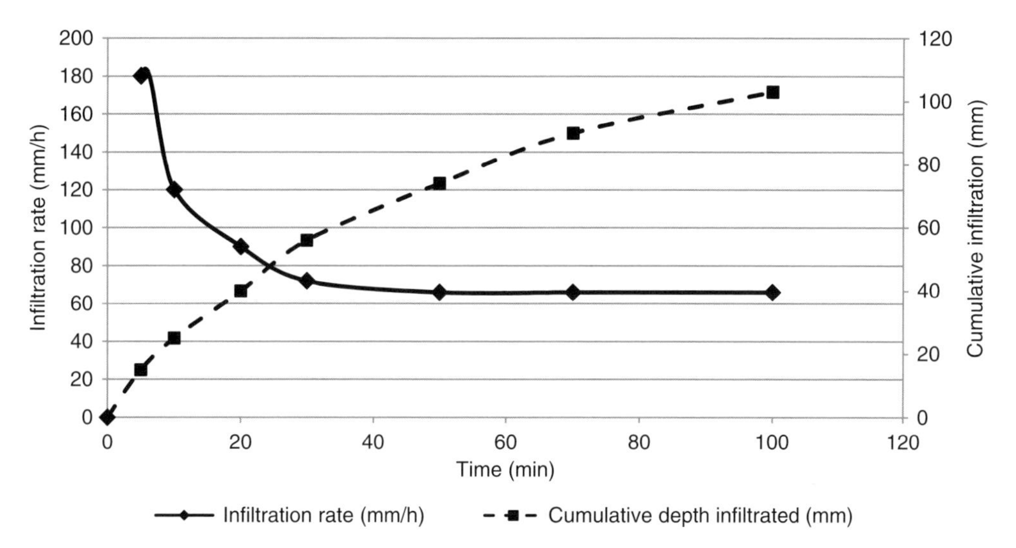

Fig. 5.31. Data collected from infiltration test (table, top) and of infiltration rate and cumulative infiltration (plot, bottom).

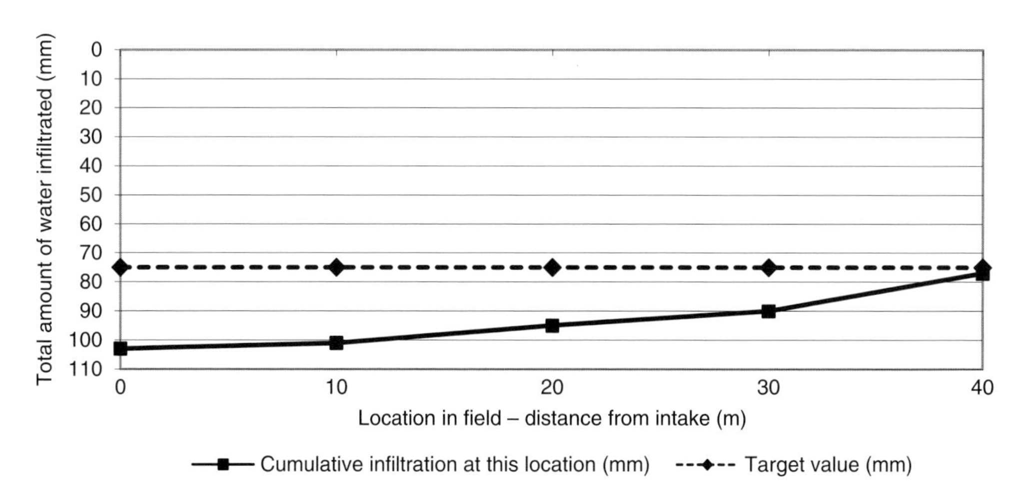

Fig. 5.32. Plot of the total cumulative amount of water infiltrated at each location in the field.

be developed, which better matches the supply to the target application depth and root zone capacity.

- It gives the irrigation duration for the measured stream size. If the process is carried out on several fields with similar characteristics over a period of time, it will be possible to quantify the time required to efficiently irrigate each field. Where the water allocation within the tertiary unit is controlled by a water users association these times can then be incorporated in the water master's irrigation schedule.

Discharge measurement at the tertiary unit level

Discharge measurement at the tertiary unit (on-farm) level can be used to determine the discharge:

- at the tertiary unit intake from the secondary canal;
- in tertiary and quaternary canals;
- in the field.

Discharge measurement at the tertiary unit intake is required to know how much water is being delivered to this level of the system. This is particularly important where the tertiary unit is managed and operated by a water users association. In this case daily measurement at this point forms the basis of the fee payment to the main system service provider.

Discharge measurement in tertiary and quaternary canals is used to know how much water is being distributed to different locations within the tertiary unit (Fig. 5.33). It can be used intermittently in order that the water masters gain experience in quantifying discharges.

The positioning of the portable weir in the channel is shown in Fig. 5.34, and the discharge tables in Table 5.27. Note that the values in Table 5.27 are per metre weir crest width; these need to be multiplied by the actual crest width to obtain the discharge.

Discharge measurement in the field can be used on occasion in order to establish how much water is being applied to a field, and how this application compares with the demand. As discussed in the previous section measurement at this level can be helpful in reducing over-application of irrigation water at the field level. Figure 5.35 shows the use of a small portable flume (Fig. 5.36) for discharge measurement in furrows. The flumes can be easily positioned in the furrow and can measure accurately discharges in the 0–6 l/s range.

An alternative to the portable flume for measuring flows in furrows is the sharp-crested V-notch weir (Fig. 5.37). This weir can be cut from a sheet of 4–5 mm mild steel, with a 1–2 mm sharp edge created by filing down the metal edge. The discharge range is from 0 to 100 l/s depending on the size of the weir. A small weir of 15 cm height in the V-section can accurately measure discharges in the range of 0–12 l/s (Table 5.28). The disadvantage with

(a)

(b)

Fig. 5.33. (a) Portable sharp-crested rectangular weir for on-farm discharge measurement and (b) a typical location for flow measurement where the tertiary canal bifurcates.

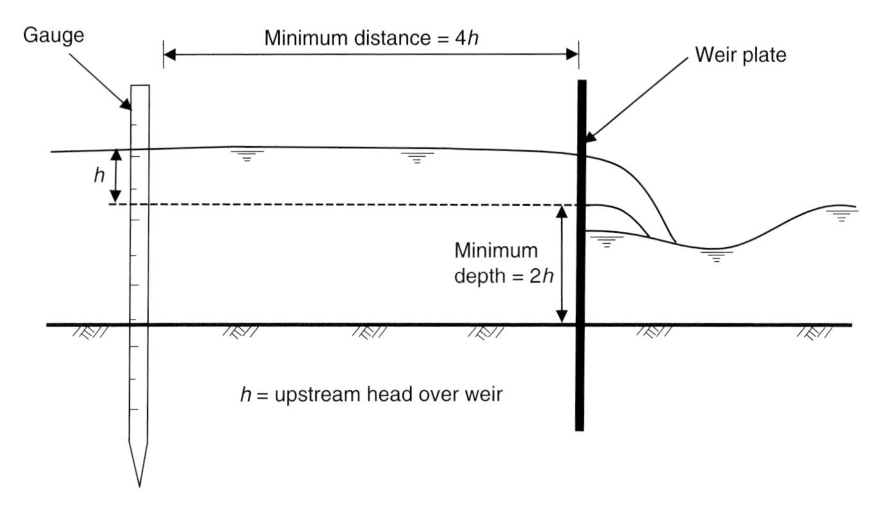

Fig. 5.34. In-channel arrangement of the portable weir for flow measurement.

Table 5.27. Flow rates for a rectangular weir, per metre crest width.

Head (mm)	Flow rate per metre crest width (l/s/m width)	Head (mm)	Flow rate per metre crest width (l/s/m width)	Head (mm)	Flow rate per metre crest width (l/s/m width)	Head (mm)	Flow rate per metre crest width (l/s/m width)
–	–	110	65.6	210	169.5	310	298.0
–	–	120	74.7	220	181.5	320	311.5
30	9.5	130	84.0	230	193.5	330	326.0
40	14.6	140	93.7	240	205.5	340	340.0
50	20.4	150	103.8	250	218.5	350	354.0
60	26.7	160	114.0	260	231.0	360	368.5
70	33.6	170	124.5	270	244.0	370	383.5
80	40.9	180	136.0	280	257.5	380	398.0
90	48.9	190	146.0	290	271.0		
100	57.0	200	158.5	300	284.0		

Fig. 5.35. Measuring the flow in a furrow using a portable flume.

the V-notch weir is the head loss required across the weir. For accurate measurement the downstream water level should be below the bottom of the 'V' in order that the flow over the weir is aerated and not hindered by the water downstream. It can be seen from Table 5.28 that if the flow in the channel is 4.4 l/s the head loss required is at least 10 cm. This is a significant head loss in a small channel and may result in overtopping of the channel section upstream.

A further alternative is the cut-throat flume. The discharge relationships for cut-throat flumes are based on empirical measurements, and require a series of tables for

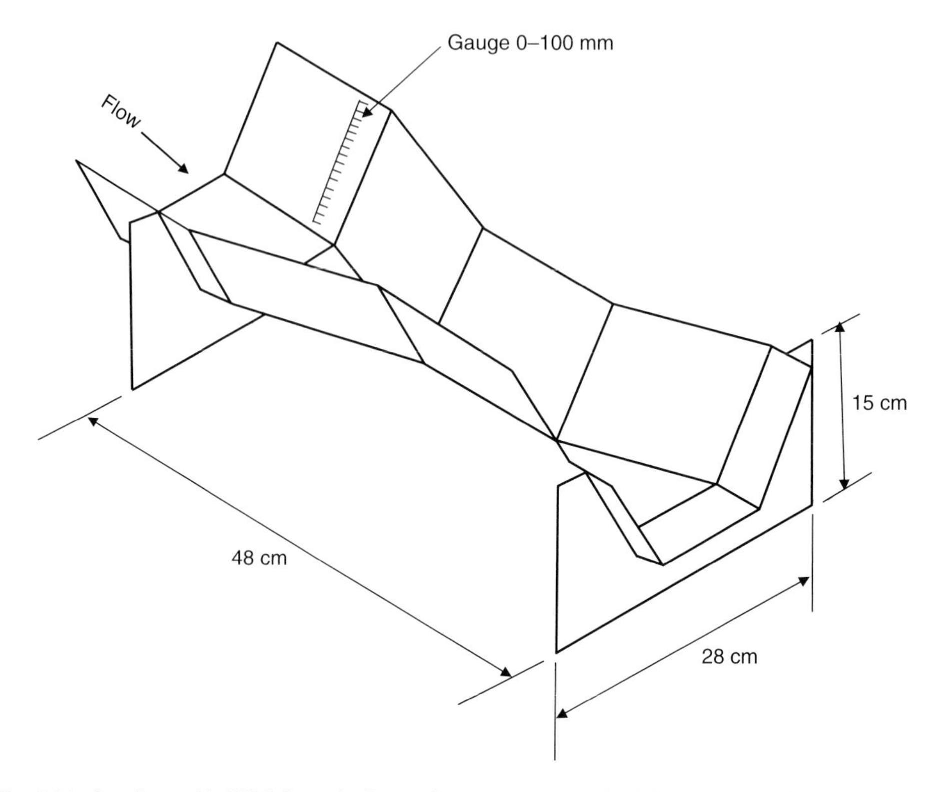

Fig. 5.36. Small portable WSC flume for furrow flow measurement (WSC, Washington State College).

the different sizes of flume (Skogerboe *et al.*, 1972). Reference should be made to the publication by Skogerboe and Merkley (1996) for more information on cut-throat flumes.

Losses of Irrigation Water Within the Tertiary Unit

Losses can occur in a variety of locations within the tertiary unit:

- through seepage from the tertiary, quaternary and field channels;
- by over-topping of the tertiary, quaternary or field channels;
- through breaches or holes in the channels;
- by over-application of irrigation water during irrigation;
- from runoff during field irrigation;
- under-utilization of irrigation water, such as a farmer requesting irrigation water and then not irrigating, allowing too much water down a channel when it is not required;
- allowing water to flow during the night and not utilizing it.

Table 5.29 outlines some measures to reduce losses and increase the efficiency of irrigation at the tertiary unit level.

Rotation of Irrigation Water Supplies

Irrigation water supplies are regularly rotated at the on-farm level as the available supply is shared between different users. It is particularly used when irrigation water is in short supply in order to maintain high flows in canals and reduce the distribution losses. Some rotation plans are carefully calculated and formalized as with the Warabandi method described in Chapter 4; others are more informally calculated and planned each day by the

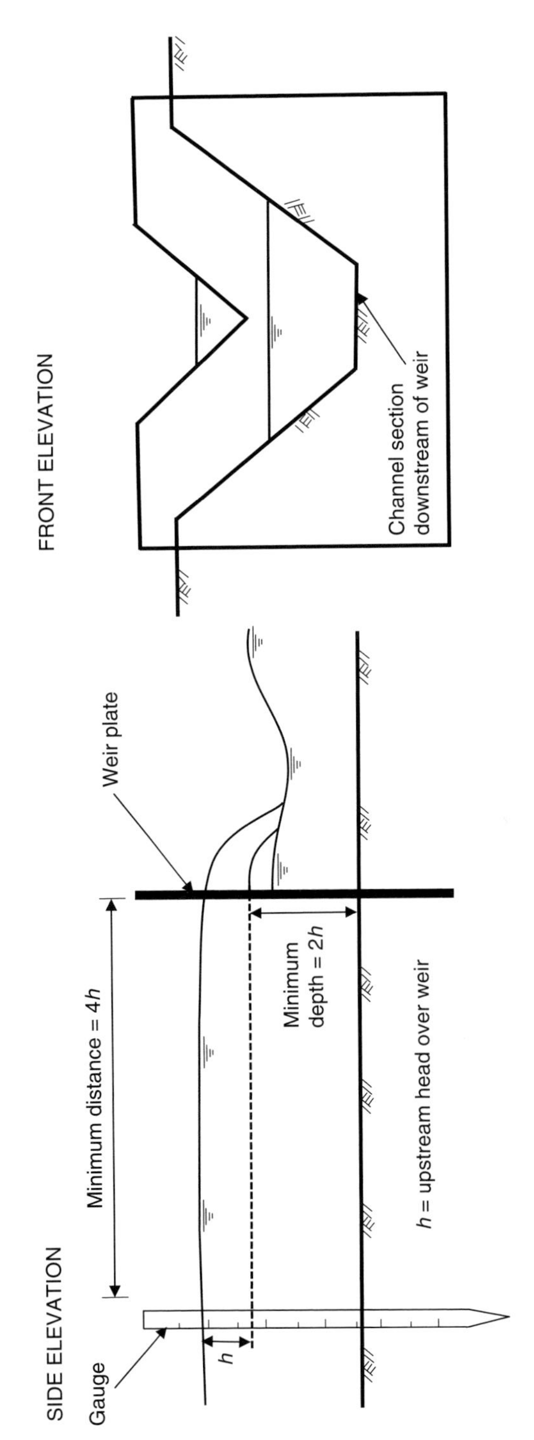

Fig. 5.37. V-notch weir for discharge measurement.

Table 5.28. Flow rates for a 90° V-notch weir.

Head (cm)	Flow rate (l/s)	Head (cm)	Flow rate (l/s)	Head (cm)	Flow rate (l/s)
–		11	5.5	21	27.9
–		12	6.9	22	31.3
–		13	8.4	23	35.1
4	0.4	14	10.2	24	38.9
5	0.7	15	12.0	25	43.1
6	1.2	16	14.1	26	47.6
7	1.8	17	16.4	27	52.3
8	2.5	18	18.9	28	57.3
9	3.3	19	21.7	29	62.5
10	4.4	20	24.7	30	68.0

Table 5.29. Measures to reduce losses within the tertiary unit and increase irrigation efficiency.

No.	Measure	Description	Impact
1	Schedule irrigation to match crop water needs	One of the main measures to reduce irrigation water losses is to schedule irrigation water to match crop needs. This requires knowing when and how much irrigation water the crops require	High. With proper scheduling, wastage and loss of water can be significantly reduced. Reducing on-farm losses has beneficial knock-on effects such as reduction in waterlogging and salinization
2	Maintain tertiary unit canals	High levels of seepage, leakage and over-topping are a feature of poorly maintained on-farm canals. Flow capacity is considerably increased, and losses reduced, if weeds are cut back	High. Well-maintained canals pass higher volumes of water more quickly, reducing contact time and reducing losses. Spillage and over-topping are also reduced
3	Line tertiary unit canals	Lining of canals reduces seepage losses and travel times	Moderate to high. Lining of canals can have a significant impact in locations with light soils and can significantly reduce travel times. Lining is not a substitute for proper maintenance, however
4	Charge farmers for water consumed	In some schemes charging farmers for the time that they irrigate, or the volume they use, has significantly reduced water wastage	High, if the price charged is significant in relation to the value of the product
5	Monitor water delivery and use	Monitoring of the water delivered to each field/farmer can show which farmers are using water efficiently and which are not. Training can be carried out to improve the capabilities of the less well performing farmers	Moderate to high. Identifying poor performers and improving their capability can have a significant impact, releasing water for others to use
6	Schedule filling and emptying of canals	Water is lost in filling and emptying canals. The quantity lost can be reduced if the irrigation schedule is organized in order to optimize filling and emptying. This can be a significant factor in large on-farm systems, or in long sections of canals	Moderate to high. Can be significant with some soil types. Keeping soils moist can prevent cracking and the associated increased losses due to wetting and drying effects
7	Limiting the irrigation duration	Restricting farmers in the amount of time they are allocated water can help to reduce losses. Water users are more willing to maintain tertiary, quaternary and field channels to increase flow rates when supplies are restricted	Moderate to high. Keeps water users on their toes and keen to make the most use of the time available

water master. The sections below outline the basic principles for organizing rotations at the on-farm level.

Approach

In organizing a rotation of irrigation water supplies the application depth and irrigation interval are not fixed solely from crop, soil and climatic factors, but must also take into account practical constraints related to

- the irrigation frequency (or interval);
- the rate of supply;
- the duration of supply.

Irrigation frequency or interval

With the demand mode of irrigation water distribution, the irrigation interval may vary throughout the season to meet moisture requirements as closely as possible. With a rotation system however, planning and organization is much simpler if irrigations are scheduled on a basic interval (or a multiple of this). Common irrigation intervals are 5 days, twice weekly, 7 days, 10 days, twice monthly and monthly.

Rate of supply

The rate of supply is governed by:

- the capacity of the canal;
- the supply available;
- the capacity of the farmer to manage the supply.

The capacity of the canal is governed by the design, which should have taken into account the need to rotate irrigation flows. The closer to the field one gets, the greater the need to increase the capacity of the canal to cater for rotation of irrigation supplies.

Losses are greater (as a percentage of the flow) when the flow in the canal is low relative to its design capacity. As a rule-of-thumb, canals should not flow with discharges less than 50% of the design capacity.

The maximum discharge that a farmer can control in earth canals and surface irrigation is generally in the range 30–50 l/s. This is commonly adopted as the design discharge for quaternary canals (or multiples of this figure; with two or more farmers irrigating at the same time).

Duration of supply

The duration of supplies to individual fields will vary with their size, the streamflow size and the level of demand. In some systems farmers irrigate round the clock (24 h/day); in others the irrigation day extends from early morning to the evening, with no irrigation at night. The irrigation day might then range from 8 to 16 h/day.

Example of rotation calculations

The parameters which need to be considered in determining the rotation flow rate and duration are:

- available water supply;
- canal design capacity;
- crop area supplied;
- canal filling time;
- losses in canals;
- total hours available in a given period.

Table 5.30 shows the irrigation water demand for five fields in the period 16–21 July (the data are taken from the example given in Appendix 1). As can be seen the discharge required varies from 10 l/s on 19 July to 75 l/s on 20 July. In order to reduce the variation in the daily flow rate a rotation plan can be worked out with a uniform flow rate for the rotation period.

The process for determining the rotation flows and durations is outlined below and in Table 5.31.

Step 1: Determine the total irrigation demand at the field intake in m^3

- Convert the demand from mm depth into m^3/ha.
- Multiply mm depth by 10; thus for F2-2 irrigation depth of 25 mm=250 m^3/ha.
- Divide by application efficiency to get demand at the field intake (250/0.6=417 m^3/ha).

Table 5.30. Irrigation water demands in the field.

Date	F2-1 Cotton 1.6 ha	F2-2 Cabbage 1.4 ha	F2-3 Beans 1.7 ha	F2-4 Maize 1.4 ha	F2-5 Maize 1.6 ha	Discharge required (l/s)[a]
	Irrigation water demand (mm depth)					
16 Jul						
17 Jul						
18 Jul						
19 Jul		25.0				10.1
20 Jul			50.0	125.0		75.2
21 Jul						

[a]With application efficiency=0.6.

Table 5.31. Volume share, time share and rotation timetable; irrigation available for 16 h/day.

Field	F2-1	F2-2	F2-3	F2-4	F2-5	
Crop	Cotton	Cabbage	Beans	Maize	Maize	
Area	1.6 ha	1.4 ha	1.7 ha	1.4 ha	1.6 ha	
(a) Share of flow						Daily volume (m^3)
19 Jul		0.12				583
20 Jul			0.29	0.59		4333
21 Jul						0
	Total volume=4917 m^3					
(b) Time share (h)						Hours
19 Jul		5.7				6
20 Jul			13.8	28.5		42
21 Jul						0
	Total hours=48					
(c) Adjusted flow allocation (h)						Hours
19 Jul		5.7	10.3			16
20 Jul			3.5	12.5		16
21 Jul				16.0		16
	Discharge=28 l/s; total hours=48					

- Multiply by the area of the field (417×1.4=583 m^3).
- Calculate the demand for each field and also the total demand (F2-2=583 m^3, F2-3=1417 m^3, F2-4=2917 m^3, total=4917 m^3).

Step 2: Determine the number of days to supply water

- By inspection it seems sensible to irrigate over 3 days, from 19 to 21 July.

Step 3: Calculate the proportion of the total volume to be given to each field

- Divide the demand for each field by the total demand: for F2-2=583/4917=0.12; for F2-3=0.29; and for F2-4=0.59.

Step 4: Calculate the time share hours based on the volume proportions

- Determine the total hours available. In this example the maximum hours of irrigation per day is 16. The total hours for three days is thus 48.
- Time share hours: F2-2=0.12×48=5.7 h; F2-3=13.8 h; F2-4=28.5 h.

Step 5: Calculate the flow rate required during the rotation period

- Flow rate, Q=volume/time; thus $Q=4917\times1000/(48\times3600)=28.5$ l/s.

Step 6: Adjust the time share over the 3 days, with a maximum of 16 h/day

- There is no set formula for this; it is a question of looking at the relative delivery volumes and time shares and deciding how to split the flow.
- In this case: on day 1 irrigate F2-2 for 5.7 h and F2-3 for 10.3 h; on day 2 irrigate F2-3 for 3.5 h and F2-4 for 12.5 h; on day 3 irrigate F2-4 for 16 h.

Step 7: Check that the volumes delivered match those required

- Multiply the flow rate by the flow duration; thus for F2-2 = $28.5\times5.7\times3600/1000=583\ m^3$. Correct value.

Endnote

[1] The *potential evapotranspiration rate* is the rate at which the crop will consume water when sufficient water is available in the crop's root zone.

6

Maintenance

This chapter details processes and procedures for identifying, planning, costing, prioritizing, implementing and recording maintenance work on irrigation and drainage systems (I&D systems). Maintenance needs are discussed and the maintenance cycle detailed with its component parts. Typical maintenance machinery and equipment are identified. The increasingly important topic of asset management is introduced and a detailed account given of the processes and procedures involved.

Introduction

The need for maintenance

An I&D system which is inadequately maintained will fall into disrepair. Gates will become inoperable, measuring structures will drown out, canals and drains will silt up, vegetation will block canals and drains, canals will overtop and breach. As a result irrigation water supplies will become irregular, unreliable, untimely, inadequate and uncontrolled. Drainage water removal will be hindered, leading to a rise in the groundwater table and salinization. The ultimate consequences of a lack of maintenance are a reduction in crop yields and overall crop production, leading to a reduction in farmers' incomes and the ability to pay the service fees (Fig. 6.1).

Unless preventative action is taken an I&D system will deteriorate over time as a result of natural forces, as well as from human and animal activities. The forces acting on the physical infrastructure include: rainfall; wind; erosion by surface runoff, flow of water in canals and drains; transportation and deposition of silt in rivers, canals and drains; vegetative growth in and around canals, drains and structures; rodents and burrowing animals (in embankments); human and animal traffic across canals and drains; corrosion and rusting of gates; biological degradation of organic matter (e.g. wooden gates); thermal expansion and contraction.

The main reason why this natural process of deterioration is allowed to occur unchecked is often the lack of adequate funds for maintenance. It is not the only cause, however. Other factors include: a lack of understanding of the need and priorities for maintenance; poorly defined maintenance procedures; lack of staff training in the identification, reporting and processing of maintenance requirements; poor allocation of available resources, incorrect or undefined maintenance priorities; poor supervision and monitoring of maintenance work; poor design and construction of the system, or parts thereof, in the initial instance; and poor operation practices.

 ©Martin Burton 2010. *Irrigation Management: Principles and Practices* (Martin Burton)

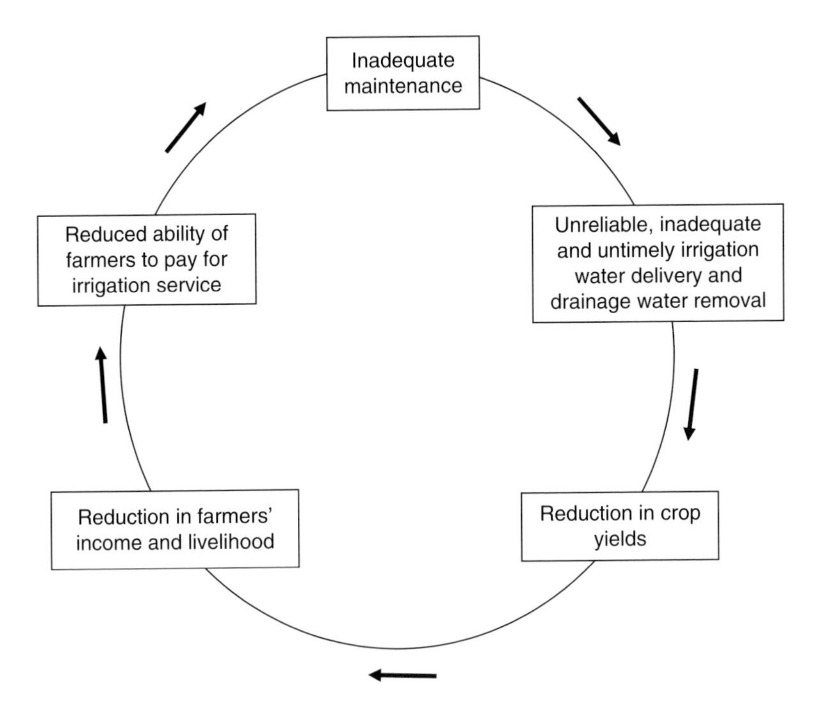

Fig. 6.1. The vicious circle of inadequate maintenance.

Poor operation of the I&D system can play a major role in the speed at which the system deteriorates. Incorrect operation of the gates at the river intake, for example, can result in unnecessarily large quantities of silt entering the canal system, while filling and emptying canals too rapidly can cause embankments to slip, collapse or breach. Incorrect operation of cross regulator gates can result in overtopping and breach of canals, and a failure to close down the irrigation system during periods of heavy rainfall can lead to overloading of the drainage system as unused irrigation water is added to surface water runoff.

When looking at why an I&D system has deteriorated it is important to look at the global picture and consider all influencing factors. It is not sufficient to conclude that the problem stems solely from inadequate funding, and that little can therefore be done. If lack of funding is the key issue then it is important to quantify the level of funding required, the scale of the shortfall and the consequences. The cost to individual farmers and to the local and national economy in lost agricultural production of failing or failed I&D systems will almost always be more than the costs associated with providing adequate maintenance. Figure 6.2 shows a possible scenario for an irrigation and drainage scheme (I&D scheme), with a rapid growth in agricultural production following construction of the irrigation and drainage infrastructure. With a period of stability, good operation and adequate levels of maintenance the productivity may increase over time as a result of new crop varieties, improved seeds, etc. However, if there are inadequate funds for maintenance and the standard of operation declines, the condition of the system will deteriorate and agricultural production will decline, resulting in lost productive potential. This productive potential may then be returned if the system is rehabilitated. The longer the rehabilitation is delayed the greater its cost, and the greater the level of lost production. If the system is not rehabilitated it will be abandoned or return to rainfed agriculture.

Rehabilitation of I&D systems will always cost more than a programme of regular maintenance, on three fronts. First in terms of the lost production as the system

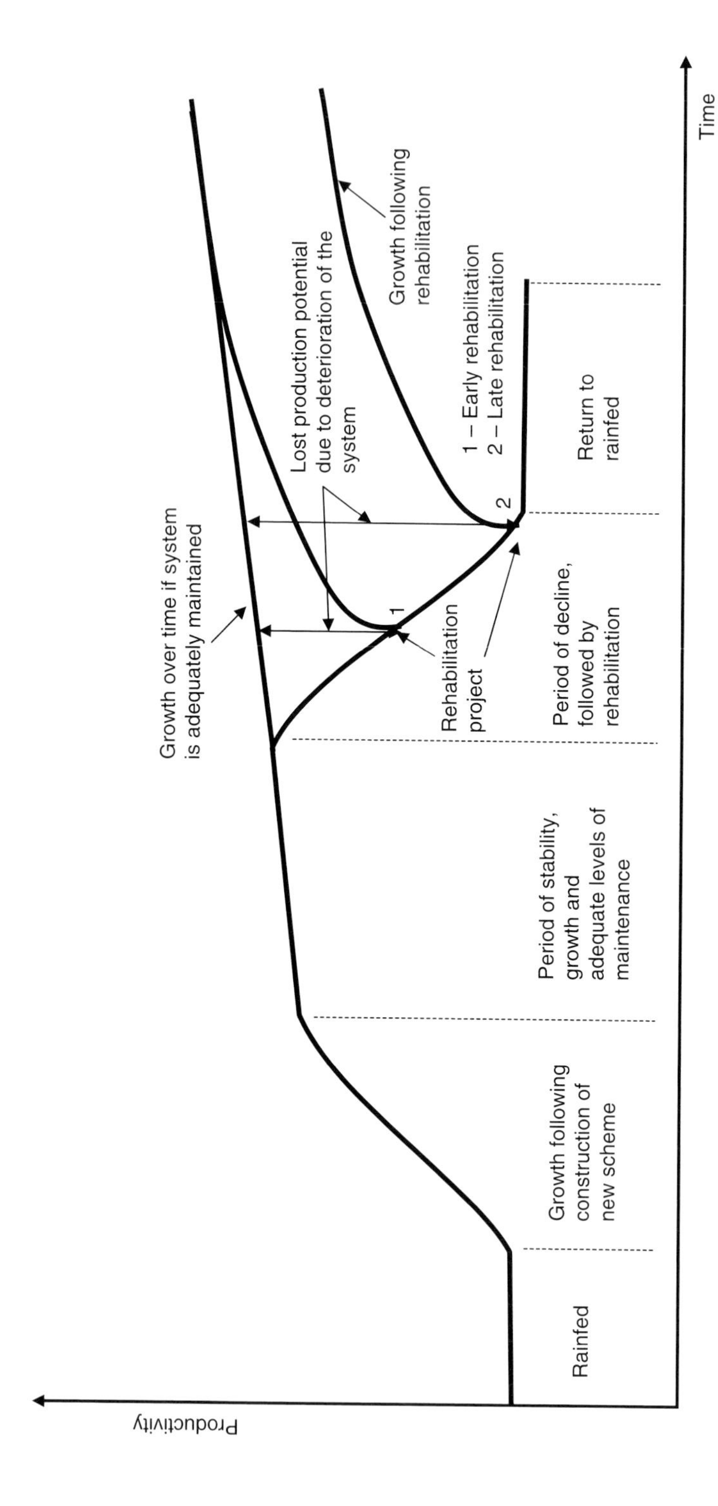

Fig. 6.2. Possible stages of growth and deterioration of irrigation and drainage systems with and without adequate levels of maintenance.

deteriorates over time, second in terms of the increasing rate and extent of deterioration as irrigation and drainage components are allowed to deteriorate (the adage 'a stitch in time saves nine' is applicable here), and third in the actual costs of rehabilitation itself, where consultants and contractors are employed to carry out surveys, studies and construction.

As shown in Fig. 6.3, the longer a system goes without adequate maintenance, the higher the costs for maintaining, rehabilitating or rebuilding the system. Rehabilitation and maintenance costs vary from country to country and system to system depending on a number of factors including the type of irrigation system, the costs of materials, fuel and labour and the specific features of the system (large-scale, small-scale, topographic conditions, etc.). In several Central Asian countries total costs for rehabilitation of a gravity-fed surface irrigation system, including all overhead costs, are typically in the range US$200–600/ha, while costs for a new scheme might be of the order of US$3000–5000/ha. At the same time costs for maintenance of the system will be in the range US$15–50/ha. On a simple analysis rehabilitation costs are sufficient to cover maintenance for 12–15 years, with no lost production resulting from the poor condition of the system.[1]

Increased levels of funding may result in improved maintenance of the system, increased staffing and increased maintenance facilities (equipment, stores, etc.). In conjunction with such funding it is important to increase the efficiency with which the maintenance funding is used. This requires attention to strengthening maintenance processes and procedures, staff training, and studies/research to establish maintenance needs, costs and returns to expenditure, including levels of service delivery.

Objectives for maintenance

The objectives for maintenance of an I&D system can be stated as:

1. Keeping the system in good operational order at all times;
2. Obtaining the longest life and greatest use of the system's facilities;
3. Achieving the above two conditions at the least possible cost.

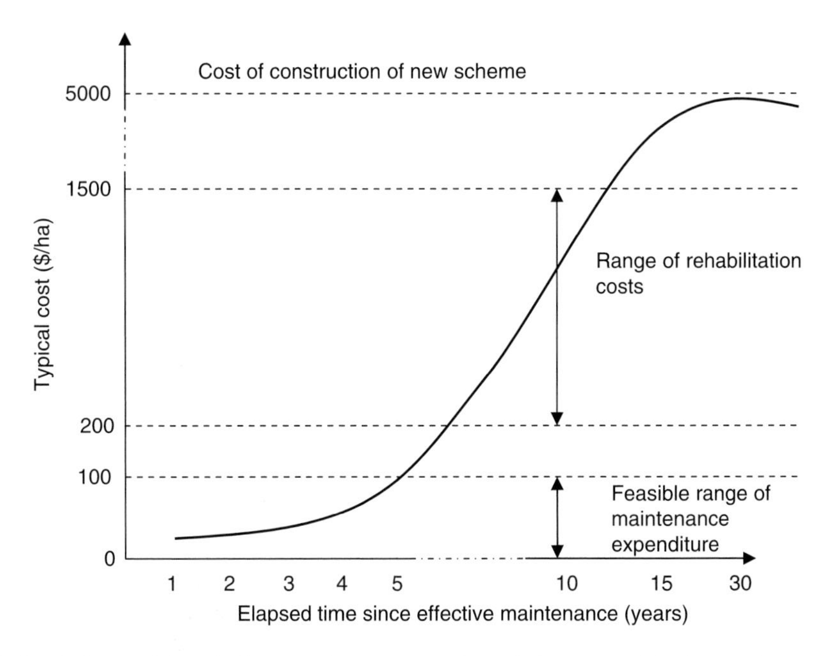

Fig. 6.3. Cost of remedial works if no maintenance work is carried out on an irrigation system.

Maintenance Categories

Maintenance can be classified into six main categories:

- routine;
- periodic;
- annual;
- emergency;
- deferred;
- preventative.

Maintenance work can be carried out under these categories by one, or a combination, of the following:

- direct labour, either as individual labourers responsible for certain sections or components of the I&D system or as maintenance gangs – this labour may be employed full- or part-time;
- contractors;
- local communities.

Routine maintenance

Routine or day-to-day maintenance is small maintenance work that is carried out on a regular basis. It is usually carried out by manual labour. Such work includes, but is not limited to, the following:

- minor repairs to earth embankments – small gullies from rainfall runoff, animal damage, machinery damage, cracks and small seepage holes;
- clearance of silt in canals and drains near structures, especially near gates, measuring structures and siphons;
- clearance of floating rubbish from canals and structures, rubbish screens and gate wells;
- removal and cutting back of vegetation from within canals and drains, from embankments (trees and bushes) and from around structures;
- greasing and oiling of gates.

Routine maintenance work is usually done by a gatekeeper, maintenance labourer or by farmers working individually or in groups.

Periodic maintenance

Periodic maintenance is small-scale, often preventative, maintenance work that does not pose any immediate threat to the functioning of the system. Such work may require skilled labour or machinery and should be carried out at intervals during the irrigation season, as required. This work includes but is not limited to the following:

- repairs to concrete canal lining and structures;
- repairs and maintenance to wood and metal works, in particular gates;
- repairs to measuring structures, and installation of gauges;
- repairs to canal embankments if there is leakage or overtopping;
- painting of metal and woodwork;
- repairs to machinery such as pumps and engines;
- access road upkeep.

Some of this work could be carried out though small contracts but can also be done by an in-house maintenance team. This team might comprise a foreman, concrete/masonry artisans, carpenters, fitters/mechanics, maintenance plant operators and labourers. The maintenance team would be mobile and have a pick-up truck and possibly some maintenance plant such as an excavator.

Annual maintenance

Annual maintenance is work that is planned as a result of maintenance inspections, which is too large or on too wide a scale for periodic maintenance work. It could also include work related to the improvement of the system rather than maintenance. Contractors are generally hired to carry out this work.

The maintenance work is carried out when the canals or drains are not in use, either at the end or the beginning of the irrigation season. Such work includes but is not limited to the following:

- major desilting work in main canals and drains;
- repair of canal lining;

- repair of headworks and canal/drain structures;
- maintenance of canal embankments, service roads and flood bunds;
- repair or replacement of equipment, gates, pumps, motors, etc.

Emergency maintenance

Emergency maintenance is work that cannot be planned for and is carried out as the need arises. The uncertainty of what and where the problems are going to be makes coping with the problems difficult. Flexibility of working practices throughout the system is required as a result. Work in this category may include:

- temporary repairs to river, canal or flood bund embankments in the event of a breach or possible breach;
- preventative work to avoid structure failure, or temporary repair as a result of a structure failure;
- work to alleviate flooding, landslides or mud flows.

The nature of the work requires it to be carried out quickly. Prompt action minimizes the extent of any damage and of the repair work required. Good communication systems are extremely useful in these circumstances, for example with a canal breach to communicate with the headworks to close down or reduce the intake discharge.

Carrying out a risk assessment for the scheme to identify areas where emergencies might occur can save time, resources and expense when these events actually occur. The risk assessment will review historical emergency events, inspect the site and talk with scheme staff and water users to identify areas of risk and measures to prevent, mitigate or deal with them if they occur. This might, for example, take the form of storing sand and sandbags in villages near areas of river prone to overtopping during extreme river flow periods, maintaining a list and contact details of village headmen, and organizing a practice emergency call-out with the villagers.

Deferred maintenance

Deferred maintenance is work that has been identified following inspection of the infrastructure but which is either of low priority or cannot be carried out due to lack of sufficient funds. The work is recorded in the maintenance register and periodically reviewed. Some of this work may be related to system improvements such as:

- improved footbridge crossings, road culverts;
- improvements to access along canal embankments.

The phrase 'deferred maintenance' is also sometimes used to refer to work carried out under rehabilitation projects, where maintenance work has been 'deferred' and carried out under the rehabilitation project.

Preventative maintenance

Preventative maintenance is work that, if carried out, will result in preventing more expensive maintenance or repair work at a later date. A classic example of preventative maintenance is the prevention of seepage around or under hydraulic structures; if seepage is identified and remedial action taken in good time, the collapse of the structure can be prevented, saving considerable expense.

Priority areas for preventative maintenance include:

- checking for seepage around or under structures, especially if there is a high pressure head across the structure;
- grading of embankments and canal/drain inspection/access roads to avoid ponding of water and gullying;
- closing river intake gates before high flood levels in the river, both to avoid excessive discharges in the canal and intake of water with high sediment loads;
- painting of metal and wood components, particularly gates and gate frames.

Maintenance Cycle

The maintenance cycle is shown in Fig. 6.4 and discussed in the sections below.

Maintenance inspections and reporting

Inspection of irrigation and drainage works for maintenance can be carried out by engineers, operation and maintenance (O&M) staff or field staff. There are two forms of maintenance inspection:

1. Inspections as part of the day-to-day work;
2. Annual or seasonal inspections.

Standard procedures for inspection and reporting of maintenance are an obvious prerequisite for effective maintenance. Unfortunately such procedures are not always properly developed. The following are required:

- a set of clearly defined instructions and procedures detailing when inspections should be carried out, by whom and how often;
- clearly defined reporting procedures, comprising a set of reporting forms and a maintenance register – the maintenance register should have a record of all the maintenance work required, and its current status and categorization (required, periodic, annual, deferred, etc.).

Field staff should have field books in which identified maintenance work can be written down and then reported to the office. Daily routine maintenance, such as greasing of gates, need not be reported and booked, though the annual and periodic inspections should check that this work is being carried out by field staff.

Inspections as part of the day-to-day work

Inspection and monitoring of maintenance needs is part of the field staff's work, and should be part of their daily routine. Gatekeepers and pump operators will also be responsible for identifying and reporting any maintenance requirements.

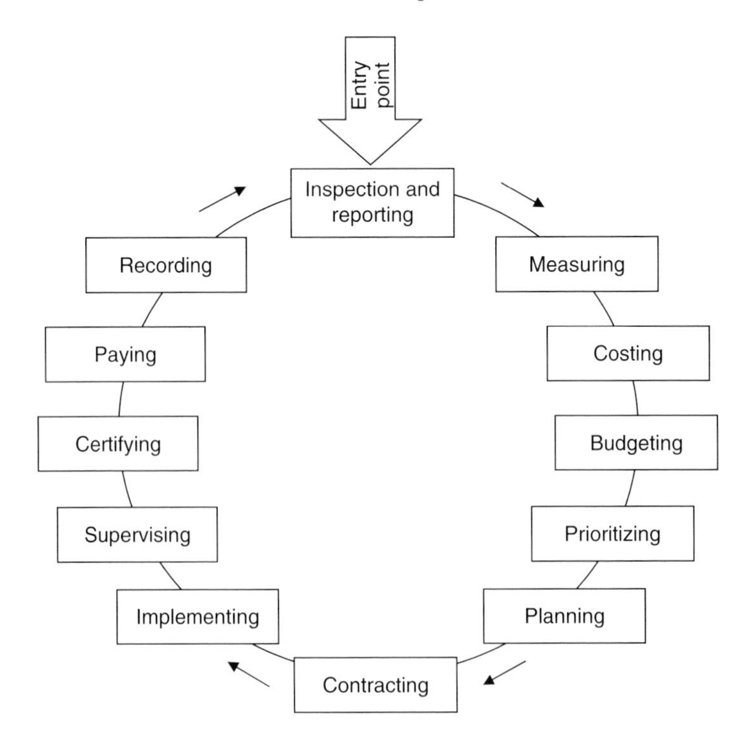

Fig. 6.4. Maintenance cycle.

Any maintenance requirements observed by these staff that they cannot carry out should be reported and recorded in the maintenance register. In the case of emergency maintenance the field staff should take action immediately, and do what they can to get help in dealing with the emergency.

The sort of maintenance needs that should be looked for during the irrigation season is listed in Table 6.1.

Annual or seasonal inspections

Annual or seasonal maintenance inspections should be carried out by experienced engineers. There should be one pre-season inspection to identify work that has to be carried out before the irrigation season starts, and one inspection at the end of the season that identifies work that may need to be contracted out and completed before the following irrigation season commences.

Ideally the annual or seasonal inspections should take place under two conditions: (i) when the canals are empty of water; and (ii) when the canals are flowing at design capacity. Inspection when the canals are empty enables inspection of infrastructure below the normal water line, while inspection when the canals are full and flowing at design capacity allows assessment of the carrying capacity of the canals, and the functioning of conveyance, control and measuring structures. For drains similar practices apply, with inspections when the drains are relatively dry and when they are flowing full. Efforts should be made to inspect the drainage system during or immediately after periods of heavy rainfall and runoff.

Points to look for during the annual/seasonal inspection are presented in Table 6.2.

Maintenance reporting

A key part of both the day-to-day and annual inspections is the recording of the maintenance work required and the details of when, how and by whom it was carried out when it has been dealt with. For this purpose a Maintenance Register is useful in order to:

- help in processing the data collected on maintenance requirements;
- assist in prioritizing and allocating maintenance work;
- record the maintenance work carried out in a transparent and accountable format.

As discussed above, the maintenance work is identified in the field and the work required measured and quantified. To assist in the measurement and quantification a Maintenance Work Sheet can be used (Fig. 6.5), or alternatively the data can be recorded in a notebook.

The data collected from the field (measurements and quantities) can be recorded in the maintenance register and data entered on the unit costs of the work items to determine the total estimated cost of the work (Fig. 6.6). The work can then be prioritized and a decision made as to who will do the work (in-house maintenance team, contracted labour, contractor, local community voluntary labour, etc.).

Once the work has been completed details of the work done will be recorded, including the sum paid, the name of the contractor and the date completed.

A maintenance coding system is required for creating a computer database of maintenance work and can simplify the maintenance work identification and recording process. The coding system can be divided into the main categories of work, such as:

A Control structures
B Measuring structures
C Canals
D Drains
E Conveyance structures
F Access/inspection roads
G Buildings

with further sub-division providing more detail for main categories, such as for control structures:

	Component	a	b	c
A1	Gate spindle	Grease	Straighten	Replace
A2	Gate nut	Repair	Replace	–
A3	Gate plate	Paint	Repair	Replace
A4	Gate frame	Paint	Repair	Replace
A5	Masonry	Repair	Replace	–
A6	Concrete	Repair	Replace	–

Table 6.1. Points to look for during in-season maintenance inspections.

Where to look	Typical problem and maintenance need	Consequence	Possible solution
Canal section	Vegetation obstructing flow	Capacity of canal is reduced	Cut or remove vegetation
	Rubbish obstructing flow at siphons, aqueducts, culverts, etc.	Capacity of canal is reduced. In severe cases may cause overtopping of the canal embankment resulting in a breach in the canal	Remove rubbish
	Undersized culverts or structures	Pipe culverts placed in the canal will obstruct and may reduce the maximum flow capacity of the canal	Do not allow construction of undersized pipe culverts in canals, insist on bridges. Remove and replace culverts that are obstructing flow
	Siltation	Canal capacity reduced	Remove sediment
Canal embankments	Seepage through embankments	Loss of water, but in the longer term the embankment may collapse. Large breaches in canals often start with small leakages	If severe close the canal, excavate damaged section and refill with compacted material
	Erosion	Overtopping and eventually breaching of canal	Early identification of problem and cause. If due to human or animal traffic put protection in places (steps, stones, etc.), if from rainfall grade embankment top
Structures	Seepage through structures (through concrete or masonry)	Loss of water, but in the long term the seepage through the structure may lead to piping undermining of the structure. Seepage through reinforced concrete rusts the reinforcement and leads to spalling of the concrete	Need to break out the poor concrete or masonry section and replace with sound concrete or masonry, as well as replacing and compacting any eroded backfill
	Seepage or piping around structures	Loss of water from the canal, but very likely hazard that the seepage will erode the soil material around the structure and it will collapse. This form of structural failure is one of the most common, and the most expensive to repair	As soon as possible close the canal and repair by excavating eroded backfill material and replacing with well-compacted backfill. If necessary extend wingwalls or cut-offs to increase the seepage path
Gates	Leakage through closed gates	Loss of water	Some leakage is unavoidable. If excessive then replace the gate plate or the whole gate. For gates with rubber seals, replace seals as they wear out
	Unable to operate gate properly	Inability to control water, resulting in wastage of water and inability to deliver water according to demands. Serious consequences for downstream users	Replace broken or damaged portion of gate (spindle, nut, plate, frame) or whole gate
	Corrosion	Leakage through or around gate plate. Inability to move gate plate	Preventative maintenance a priority through regular painting with protective paint. Very cost-effective
Measuring structures	Drowned out or damaged measuring structure	Cannot measure flow. Inability to match supply and demand for water leading to inefficient operation and either shortage or wastage of water	Repair damaged section. If drowned out look for cause of drowning and either raise measuring structure crest level (if head available) or remove vegetation/ obstructions downstream, or calibrate canal section

Table 6.2. Points to look for during annual or seasonal maintenance inspections.[a]

What to look at	What to look for (typical problem and maintenance need)	Consequence	Possible solution
Canal or drain section	Vegetation in canal or drain section	Capacity of canal or drain is reduced	Cut or remove vegetation
	inadequate functioning of weep holes to relieve pore pressures	Pressure will build up behind the canal lining and the lining will collapse	Clean out weep holes or install new ones
	Sediment	Reduced capacity of canal/drain	Survey and remove
Embankments	Vegetation along canal or drain embankments	Roots of large vegetation such as trees and bushes can damage canal or drain embankment	Cut down, also remove roots
	Vegetation obstructing access	Cannot fully inspect or move along the canal embankment or drain, operation and maintenance will be impaired	Remove vegetation
	Low spots in embankments	Possibility of overtopping of embankment and breach of canal	Raise section of embankment to design level with compacted fill material
Structures	Cavities beneath concrete or masonry floors or side walls (test by banging with a stout pole – a hollow sound indicates a cavity). Test for cavities in lined canals, particularly with masonry lining	Indicates seepage or piping behind the concrete or masonry. If not dealt with the structure or lining may collapse, requiring costly repairs	Locate and repair the cause of the loss of backfill (e.g. piping, seepage, etc.). Break out the concrete or masonry and backfill the affected areas. Alternatively, excavate behind the concrete or masonry, place compacted backfill
	Cracks in masonry or concrete. Check depth and extent of cracking. Check if reinforcement exposedw	Water is lost through the cracks. This can result in undermining of the backfill material and eventual collapse of the lining or structure. If reinforcement is exposed, or water leaks through reinforced concrete, the reinforcing steel will rust and the concrete will spall	Cut out affected area and replace with well-compacted concrete
	Scour hole downstream of structures, such as cross regulators or drop structures. Plumb holes with plumb line, or drain with a pump to inspect fully	The structure may be at risk of collapse	Check if the situation is stable or not. If scour is continuing then a full engineering inspection may be required
	Partial blocking of culverts, siphons, etc.	Impeded flow, possible backing up and overtopping of canal/drain section upstream	Remove sediment, rubbish and vegetation causing blockage
	Condition of metal and woodwork, like gates and stop logs	Deterioration of wood or rusting of metal can lead to failure of the component	Protect wood and metal parts with creosote, varnish or paint
Gates	Inoperable gates (test if gate can be fully opened/closed)	Inability to control water, resulting in wastage of water and inability to deliver water according to demands. Serious consequences for downstream users	Replace broken or damaged portion of gate (spindle, nut, plate, frame) or whole gate

[a]In addition to items listed in Table 6.1.

MAINTENANCE WORK SHEET

CANAL/DRAIN: ____________________

LOCATION: ____________________

Decription of work required:	Units	Quantity

SKETCH (If required, showing location and/or damage details)

Fig. 6.5. Maintenance work sheet for in-the-field recording of maintenance work required.

MAINTENANCE REGISTER

LOCATION		WORK REQUIRED	BILL OF QUANTITIES					ACTION TAKEN		
Canal/structure type	Chainage (m)	Description and drawing of the work required	Unit	Quantity	Rate	Amount	Priority	Date completed	Cost	Name of contractor

Fig. 6.6. Example sheet from a maintenance register.

For canals and drains the degree of work required for vegetation removal can usefully be codified using diagrams or photographs (Fig. 6.7).

Maintenance measurement and costing

Measurement of the maintenance work is needed to quantify the work to provide a basis for estimating the time required to do the work, and the cost.

Typical work items and measurement units for different types of work are presented in Table 6.3, while Table 6.4 gives an example of a costed summary table of a maintenance inspection and measurement.

Maintenance budgeting, prioritization and planning

It is often not possible to carry out all the required maintenance work, generally due to financial, resource (labour, machinery, etc.) or time constraints. In some cases it is not efficient to carry out the maintenance work each year, for example in the case of sedimentation of canals or drains where it is more efficient and cost-effective to remove sediment once every 3–5 years rather than on an annual basis.

Once the required maintenance work has been identified it can be prioritized and planned to fit within the available budget and resources. An example of priorities for maintenance work is presented in Table 6.5, emphasizing the importance of considering the

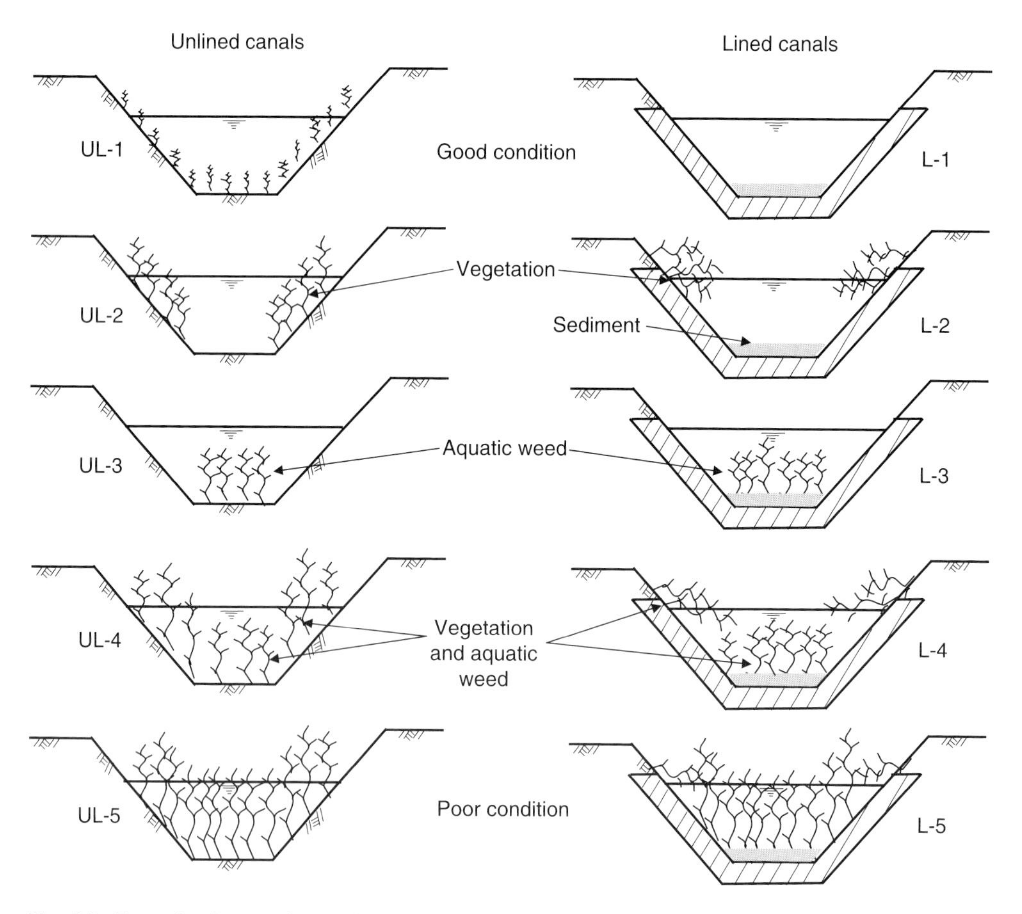

Fig. 6.7. Example of vegetation and sedimentation condition coding system for canals and drains.

Table 6.3. Example of maintenance items, measurement units and maintenance costing.

Item	Measurement unit	Annual quantity	Unit rate ($)	Amount ($)
Earthworks				
Compacted fill for embankment construction	m^3			
Removal of sediment from canal	m^3			
Repair of access road	m^3			
Removal of sediment from drains	m^3			
Canal lining				
Excavation of unsuitable material	m^3			
Placement of compacted backfill	m^3			
Concrete for lining	m^3			
Repair canal lining joints	m			
Structures (and associated earthworks)				
Excavation of soil	m^3			
Placement of compacted backfill	m^3			
Concrete (including shuttering)	m^3			
Masonry	m^3			
Stone rubble protection (rip-rap)	m^2			
Steel reinforcement	kg			
Concrete pipe 40 cm dia.	m			
Concrete pipe 60 cm dia.	m			
Concrete pipe 80 cm dia.	m			
Steel pipe <60 cm dia.	m			
Control and measurement				
Greasing/oiling of gates	no.			
Painting of gates	no.			
Repair to gates – small, <60 cm wide	no.			
Repair to gates – medium, 60–120 cm wide	no.			
Repair to gates – large, >120 cm wide	no.			
Replacement of gates <60 cm wide	no.			
Replacement of gates 60–120 cm wide	no.			
Replacement of gates >120 cm wide	no.			
Repair to measuring structure	m^3			
Replacing/painting of depth gauge	no.			
Miscellaneous				
Removal of floating vegetation	h			
Removal of vegetation from canal section	m			
Removal of vegetation from canal embankments	m			
Removal of vegetation from drain section	m			
Other items				
			Sub-total	

location, nature of the work, and the potential problems if the required work is not carried out. For any given I&D system such a list of priorities should be drawn up by experienced personnel to act as a guide for the selection and prioritization of maintenance work.

It is difficult to set a generic set of rules for prioritization of work for I&D systems; for some systems with heavy sediment loads in the river the priority is sediment removal, in a system with low sediment loads the priority might be vegetation removal (as weeds grow

Table 6.4. Example of identified maintenance work within a secondary command area.

Location/chainage (measured from secondary canal intake)	Structure	Problem	Description of work required	Priority	Quantity[a]						Cost ($)
					No.	*L* (m)	*B*/*W* (m)	*H*/*D* (m)	*A* (m^2)	*V* (m^3)	
U3-5 0+000	U3-5 Secondary canal measuring structure	Discharge measurement difficult as no gauge	Paint new gauge upstream of measuring structure	High	1						40
U3-5 0+040 to 0+100	Canal lining, both banks	Canal lining is crumbling and holed. Loss of function and rapid deterioration	Repair canal lining	Medium		200	0.10	1	200	20	5,000
U3-5 0+200 to 0+400	Both canal banks	Left bank covered in reeds, impairing canal flow. Right bank needs vegetation cutting back	Left bank – cut down reeds and remove roots. Right bank – cut back vegetation (brambles) at start of next season	High		100	–	–	–	–	500
U3-5 0+500 to 0+550 (approx.)	Left canal banks	Left bank top is low, resulting in high risk of overtopping	Raise bank top level by 30 cm	High		80	1.5	0.30		43.5	600
U3-5 0+600	Tertiary gate	Whole gate frame and plate badly rusted	Install new gate	High	1						400
U3-5 0+800	Tertiary gate	Gate not operable, no spindle or plate	New gate plate and spindle	High	1						300
U3-5 0+950	Left bank just upstream of B-18	Canal at risk of overtopping	Raise embankments with compacted fill	High		30	1.0	0.20		6	400

U3-5 1+000	Tertiary gate	One gate inoperable, no spindle	New spindle required	High	1						200
		One gate partially operable, no spindle nut	New nut required		1						
U3-5 1+178 to 1+300	Both canal side slopes	Canal lining in poor condition, crumbling and holed.	Repair and replace canal lining	Medium		20	0.10	0.50	10	1	1,000
U3-5 1+401	Secondary canal	Very badly damaged secondary canal left bank. Severe overtopping of canal and loss of water to drain	Reform canal bank, placing compacted backfill behind canal lining and compacted fill to level of 20 cm above top of lining	Very high		50	1.5	0.30		22.5	3,000
U3-5 1+405	Aqueduct over drain	Some leakage from aqueduct	Monitor situation. If leakage gets worse seal leakage	Low	–		–	–	–	–	0
U3-5/2 0+190 to 0+390	Tertiary canal	Heavily weeded	Remove weeds and kill roots	High		100	0.30	0.60	18		100
U3-5/4 0+050 to 0+400	Left and right banks of tertiary canal	Heavy vegetation growing over canal banks	Cut back and remove vegetation	Medium		200	0.30	0.60	36		200
				Total cost							11,740

[a]No.=number, *L*=length, *B*=breadth, *W*=width, *H*=height, *D*=depth, *A*=area, *V*=volume.

Table 6.5. Example of priorities for maintenance work.

Priority	Type	Comment
1	Diversion weir and intake	Failure of this structure would have serious consequences for the operation of the system. Therefore it has to have top priority for maintenance, particularly the gates
2	Leakage, unauthorized offtakes and overtopping	Leakage of water through canal banks, unauthorized offtakes and overtopping of the canal embankments can lead to failure of the embankments with serious cost consequences
3	Gates and control structures	Without gates control of water is difficult. The system cannot be operated efficiently without control structures in good condition
4	Masonry repair	Repairing of cracks in masonry is necessary before water gets in behind the masonry and causes cavities and piping around structures, leading to collapse
5	Embankment protection	Protection of canal embankments takes several forms, i.e. from: • erosion by canal water • gullying caused by low spots and crab and rodent holes • removal by farmers cultivating close to or even on top of embankments • erosion by human and animal traffic across the canal • growth of trees and deep-rooted shrubs on or near embankments
6	Measuring structures	Inefficient and incorrect water management will result from having measuring structures in poor condition
7	Silt removal[a]	Silt removal upstream of measuring structures has higher priority, general silt removal has a lower priority, except where excessive silt build-up has reduced canal capacity or caused the water level in the canal or drain to rise leaving inadequate operating freeboard
8	Vegetation removal	Removal of weeds and vegetation from the canal or drain section is important to maintain the carrying capacity. Such work can have a very high priority in locations where vegetation growth is rapid, such as in Guyana where some types of grass grow at 2 cm/day. Removal of vegetation from canal embankments and besides drains is important to maintain access. Removing grass from cracks in masonry and removing strong-rooting shrubs and trees from canal banks or drains and in the vicinity of structures is also important

[a]The systems in this example generally have low silt loads, hence its low priority. Where the silt loads are higher this item will move up the priority list, in some cases to the top of the list.

more quickly in clear water). Factors influencing the setting of priorities are the following.

- *How sophisticated is the system?* In simple systems measurement may not be as important as conveyance, while in more sophisticated systems measurement has a high priority as it is the basis for charging for service delivery. In systems with automatic downstream control or automated gate control timely maintenance is essential.
- *What are the consequences of not doing the (maintenance) work?* What is the risk of failure, and what is the cost of such failure on crop yields, agricultural production and repair work?
- *Will water be lost or used inefficiently?* If the system is water-short then conserving water will be a priority, if there is sufficient water then the loss of water may be less important, but waterlogging and salinization may be issues.
- *Will control be lost or impaired?* An inability to control the flow at division points can mean that some downstream users get too much water while others do not get sufficient, leading to wastage on the

one hand and possible crop yield reduction on the other.

- *What command area is affected by the maintenance work?* Is the work in an upstream location (large area downstream affected) or a downstream location (smaller area affected)?
- *How cost-effective is the maintenance work?* A classic example here is masonry lining of canals, which has little effect on seepage losses, versus repair of damaged control gates. Repairs to gates are often cost-effective relative to canal lining; leakage and wastage are reduced as flows can be stopped to locations where water is not required and distributed to where it is required.
- *Can it wait until next year?* In some cases work can be deferred, in other cases there is a high risk of failure and increased costs if the work is not carried out and preventative maintenance is not done.

Figure 6.8 presents an example of a flowchart for the prioritization and selection of maintenance work. Such flowcharts are useful in structuring the approach to selection and allocation of maintenance work.

A key part of maintenance planning is to schedule maintenance work to come within the annual maintenance budget. The maintenance budget should be set at a level such that all the required maintenance work can be carried out over a period of 5–20 years. Peaks and troughs in expenditure should be avoided, with the work required spread out over time such that the expenditure is smoothed and an annual budget can be set. Thus, for example, the main drains in a system might need desilting every 5 years; to avoid a peak each 5 years this work will be scheduled so that different reaches of the drainage network are cleaned each year, with a return to a given reach in 5 years time. Table 6.6 shows an example of such a schedule used to estimate the budget required for maintenance of a rehabilitated system. Note that the rehabilitation costs are US$1731/ha, annualized to US$115/ha per

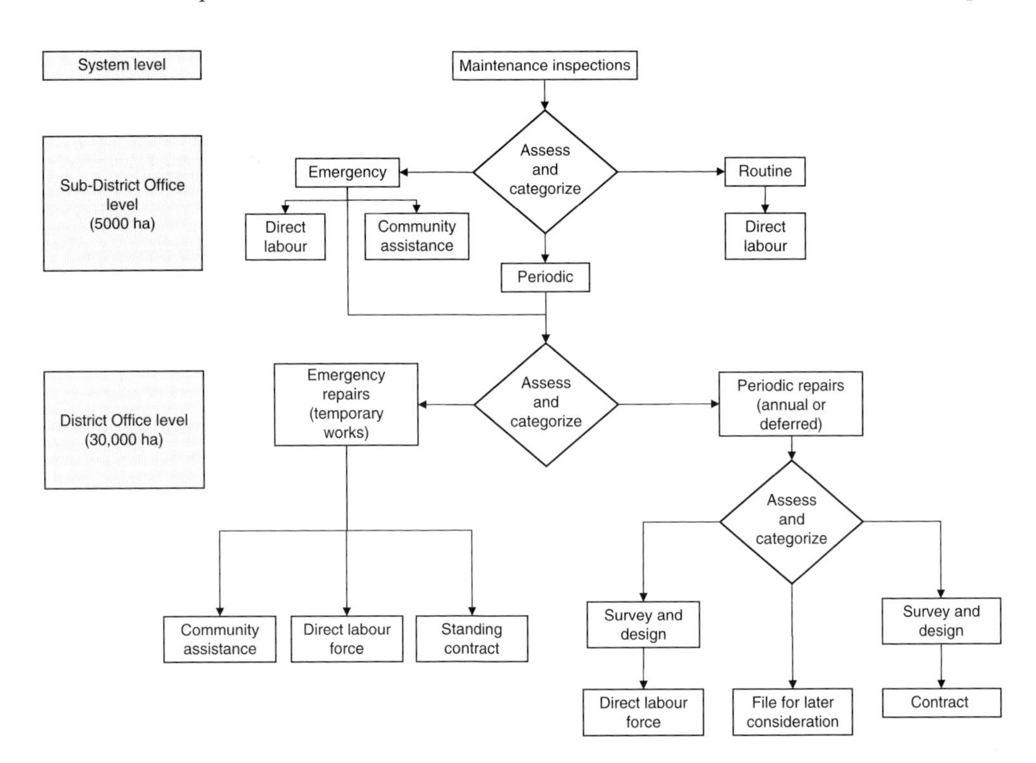

Fig. 6.8. Example of a flowchart used to categorize and plan maintenance work.

Table 6.6. Example of estimating the maintenance costs based on annual and multi-year replacement and maintenance work.

Rehabilitation design service area (ha)	29,000	
Total rehabilitation costs (000s US$)	50,188	
Rehabilitation costs per unit area (US$/ha)	1,731	Spread over 15 years=US$115/ha

Item	Total annual cost (000s US$)	US$/ha	Maintenance cost build-up[a] (000s US$)				% total MOM costs
			Annual	5 years	15 years	20 years	
Annualized maintenance costs							**83**
Main canal and collector drains	461	15.9	193	367	2674	316	25
Secondary canals, tertiary canals and field drains	1081	37.3	691	655	3602	367	58
Management costs							**8**
Main canal	87	3.0					5
On-farm	70	2.4					4
Operation costs							**8**
Main canal	87	3.0					5
On-farm	70	2.4					4
Total main canal MOM costs	635	21.9					**34**
Total on-farm MOM costs	1221	42.1					**66**
Total MOM costs	1856	64.0					**100**
Property tax	175	6.0					
VAT on main canal MOM income (ISF)	114	3.9					
Total MOM costs, including taxes	**2145**	**74.0**					

MOM, maintenance, operation and management; ISF, irrigation service fee.
[a]These columns are the summation of cost items recurring each year, 5 years, 15 years and 20 years. They are then annualized to give the total annual cost.

annum over 15 years, whereas estimated maintenance costs are only US$74/ha per annum. The rehabilitation cost of US$1731/ha is just the physical works and does not include the project overhead costs, or the costs of other items such as the feasibility study costs. In this example the annual budget allocation to cover maintenance costs and repay the rehabilitation costs amounts to US$189/ha per annum (US$115 + US$74/ha per annum). Note also in this case the relative management, operation and maintenance costs of the main system (US$21.9/ha per annum) and the on-farm systems (US$42.1/ha per annum).

Maintenance contracting

Once the maintenance work has been drawn up, costed and prioritized, contracts can be let for the work (unless it is to be carried out by direct labour or by the water users themselves). Tender documents with bills of quantities, specifications and the contract terms are drawn up and contractors invited to bid. There are different processes for engaging contractors for this work; in some cases tendering is open to any contractor, in other cases contractors have to pre-qualify and a short list is prepared of those with relevant experience and credentials.

It will be important to include guidance in the contract and penalty clauses to ensure that the contractor takes due account of the constraints that it will be working under. This may include ensuring that irrigation water supplies are maintained to water users during the maintenance period, and that the maintenance work is completed before the start of the irrigation season. Delays in re-opening canals can have serious financial consequences for farmers, and must be avoided. Ideally the contract should allow for compensation payments to be made to farmers where water supplies are delayed or impaired, or allowance made in the contract for compensation payments to farmers where diversion channels are built on their land to bypass construction work.

In some cases a long-term framework contract might be let, which will state the general type of work and invite the contractor to submit rates for stated types of work. The contractor will then be given specific work to be carried out each year, based on these rates. Such contracts are usually let for several years at a time, and enable the client to budget for the cost of the work and call on the contractor in case emergency work is required.

Contracting out maintenance work is increasingly being used in many countries as the private-sector contracting industry strengthens; formerly it was only government agencies that had the financial resources to purchase construction machinery and equipment. Contracting out maintenance work can have financial benefits over direct labour maintenance work provided that the tendering process is open and transparent, and there is a vibrant contracting sector where competitive bidding exists.

Implementation and supervision of maintenance work

Once the maintenance work is underway it is important that it is properly supervised, whether the work is carried out by direct labour or by a contractor. All relevant persons should be involved in the supervision process – if the work is being carried out at the on-farm level then farmers should be informed of the nature of the work so that they can keep an eye on the work, as well as the formal supervising body, which might be the water users association management team. At the main system level the field staff should be kept informed of the work that is being carried out, and should take a keen interest in ensuring that the work is done properly and to an adequate standard.[2] For some of the smaller work the field staff may be delegated to carry out the day-to-day supervision, for more major works the engineer will be responsible for day-to-day supervision. For large maintenance works a full-time supervisor may be placed on site.

A key role of authorized supervision personnel will be authorizing interim (often monthly) payments for maintenance work carried out by contractors. The procedures for measurement and authorization of these interim payments need to be clearly specified in the contract.

The timing of carrying out the maintenance work is important. Considerations to be taken into account include the following.

- *The cropping season*: If possible, maintenance and construction work should be avoided during the irrigation season. Where this is not possible (e.g. due to a short interval between cropping seasons) agreement must be reached with water users on procedures to minimize the disruption caused by maintenance work.
- *The climate*: It is advisable to avoid adverse climatic seasons, such as rainy seasons when access is difficult, flood periods or winter when fresh laid concrete can be damaged by frost and freezing conditions.
- *The availability of labour*: If it is intended that work is to be carried out with community assistance, then the work has to be timed to avoid peak agricultural labour demands.

Certification and payment for maintenance work

Following completion of the work it must be certified as having been completed, and completed to the specified standard. Such certification will usually involve a final inspection of the completed works, following which payment will be made. It is increasingly the case that water users are involved in the identification of maintenance work required on the main system, and also in the certification of the work done. This involvement increases transparency and accountability and enables the water users to see how, where and for what the service fees are being used.

Recording maintenance work done

It is important to record that maintenance work has been carried out, and to document the time that the work has taken, where it was located, who carried it out and how much it cost. These data can then be used to build up a database of the type and cost of work carried out; this will be of considerable assistance in the planning and costing of future maintenance work. The maintenance register should be completed to show these data.

In practice it is often disappointing to see how little recording of maintenance work is carried out. If better use is to be made of available funds then proper recording of maintenance work is a fundamental component of improved maintenance management systems.

The maintenance 'bicycle'

Figure 6.9 shows the 'maintenance bicycle' linking the various elements related to maintenance management. The key processes and categories of maintenance are linked by the organizational framework and its processes and procedures, with the direction set by the organization (the irrigation and drainage service provider – a government agency, private entity or water users association). Social and political will and finance are important factors in the 'pedal power' driving maintenance, while the vision and direction can be identified through studies and research, avoiding or mitigating where possible natural hazards.

Maintenance Plant and Equipment

The main items of maintenance plant and equipment are listed in Table 6.7 together with a summary of their main uses.

Asset Management

Overview

The term *asset management* originates from the business and finance sector. Formerly it applied to the assets of a business or trading company, but has been adapted to apply to the physical assets of engineering infrastructure, such as roads, railways, bridges, water supply pipelines, canals and drains. Asset management is in wide usage by utility companies in Europe, North America, Australia and New Zealand. A fundamental principle behind asset management is that assets such as canals and drains serve a function from which benefits can be derived. Maintaining or enhancing this function results in sustained or enhanced benefits, either financial or social. Asset management can be defined as:

> A structured and auditable process for planning, implementing and monitoring investment in the maintenance of built infrastructure to provide users with a sustainable and defined level of service.

Asset management planning identifies asset stock (canals, drains, structures, roads, buildings, etc.) and quantifies its condition and performance. From the assessment of the asset condition and level of performance estimates can be made of the investment required to:

- maintain the existing asset condition and system performance;
- enhance or extend asset condition and system performance.

Asset management planning is at the core of planning for long-term investment and expenditure in irrigation and drainage infrastructure (Burton *et al.*, 1996). Asset management planning seeks to relate investment and expenditure to specified, user-defined levels of service. The process (Fig. 6.10) involves

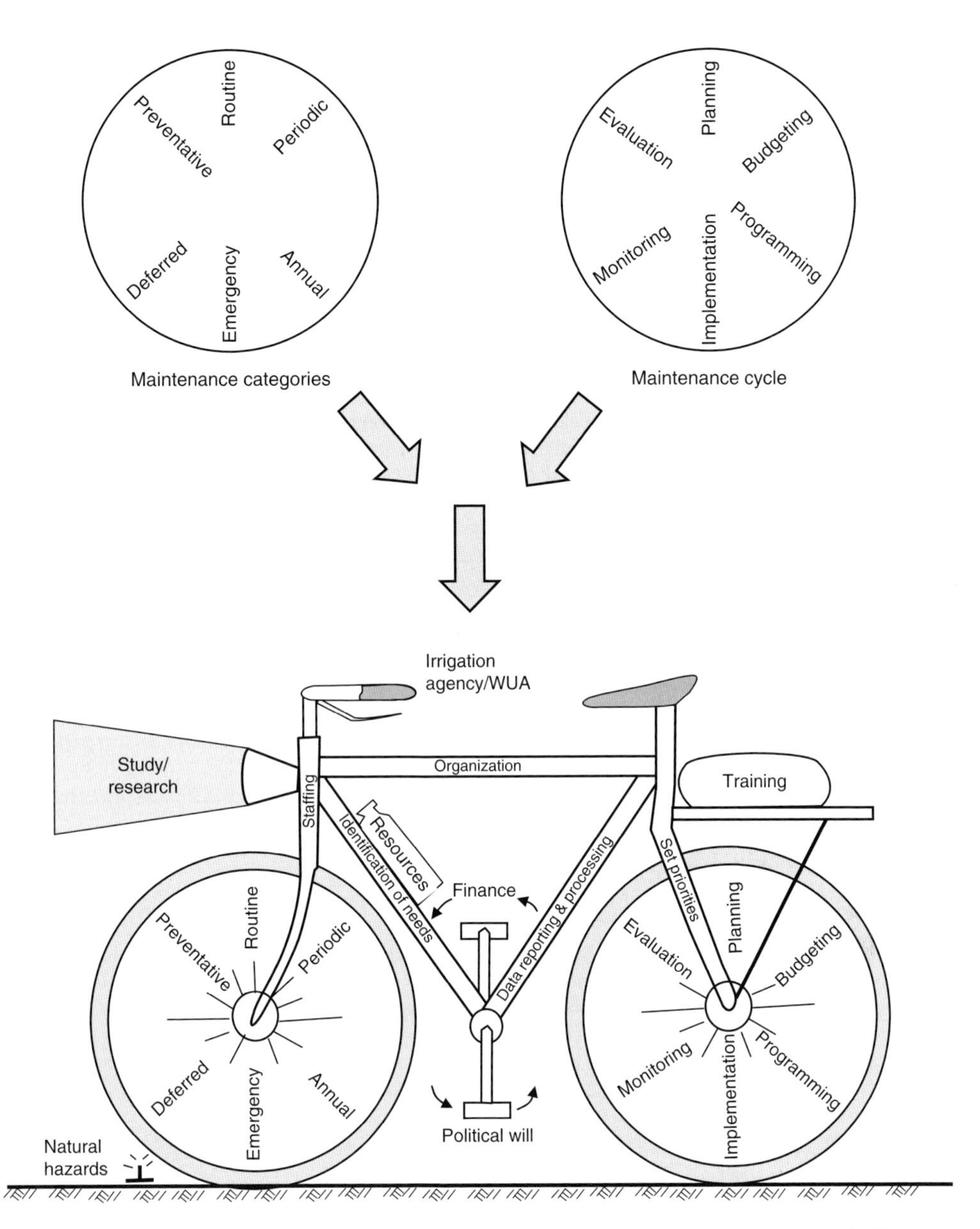

Fig. 6.9. The 'maintenance bicycle' framework for maintenance planning and implementation (WUA, water users association).

defining the level of service to be provided, quantifying the ability of the water users to pay for the specified service, identifying the condition and performance of the assets (canal, drains, structures, roads, etc.) and quantifying the investment and expenditure required to maintain, improve or extend the assets in order to satisfy the specified levels of service.

An explanation in terms of the asset management of a group of houses owned by a housing association helps to explain asset management (Fig. 6.11). In the group of 30 houses there are, say, ten houses which are Grade A (four bedrooms), ten which are Grade B (three bedrooms) and ten which are Grade C (two bedrooms). The monthly rental

Table 6.7. Summary of key maintenance plant and equipment and their uses.

Maintenance plant and equipment	Application
Dragline excavators	Used to clean sediment and weeds from large channel sections. Also used to clean sediment upstream of river pump stations where sediment exclusion works have not been provided. Efficiency and effectiveness declines rapidly if used on small channel sections where a hydraulic excavator would be better suited
Hydraulic excavators	Most commonly used piece of maintenance plant. Either tracked or wheeled. Used to remove sediment and vegetation from canals and drains
Hydraulic backhoes	Popular item of maintenance plant for water users associations wanting to clear out on-farm irrigation and drainage channels. Mobility is a key advantage
Dredgers	Used to remove sediment from canal and drain sections and sediment/settling basins. Occasionally used in river section upstream of pumped or gravity intakes. Suitable where sediment load is high and round-the-year removal is required
Bulldozer	Commonly used item of maintenance plant. Used to flatten spoil heaps left by excavator following sediment and vegetation removal, also used in river beds following flood season to form temporary diversion structure for low-flow season
Scraper	Used to move large volumes of earth over relatively short distances (generally up to 1 km). Can be motorized or pushed by a bulldozer. Good for rapid rebuilding of embankments
Tipper truck	Used in conjunction with an excavator to remove sediment from the vicinity of the canal or drain and dispose of at some distance away. Also used to bring in soil for rebuilding of embankments or for compacted backfill
Tyre or sheep's foot vibrating roller	Used to compact soil following placement in embankments. Used in conjunction with a water bowser to maintain optimum soil moisture content for compaction
Water bowser	Used to wet soil prior to compaction and maintain optimum soil moisture content for compaction
Grader	Used to grade roads and embankments to maintain uniform surface and avoid formation of ruts and gullies
Front-end loader/ shovel	Used to lift soil into tipper trucks, or to move materials (gravel) for concrete construction
Tractor-powered attachments	There is a wide range of tractor-powered attachments, including weed and vegetation cutters, water pumps, etc. Can be very versatile and can make good use of limited resources by using tractors for both farming and maintenance activities
Tractors and chain (vegetation removal)	Can be used in the absence of an excavator or dragline. Two tractors, one on each bank, pull a heavy chain along the canal or drain to tear out vegetation. Disposal of the vegetation can be a problem though
Flat-bed loader	Required to transport tracked maintenance plant such as bulldozers, draglines and excavators from one location to another
Flat-bed ditchers (Briscoe type)	V-shaped ditchers for cleaning out small on-farm channels. Attach to the back of a heavy-duty tractor and pull along the length of the channel
Concrete mixer	Essential for mixing concrete. Portable so that it can be moved from site to site
Concrete vibrator	Required to compact concrete. Essential item
Hand-moved soil compactor	Required to compact soil around structures. An essential item of equipment to avoid piping and undermining of structures

value of Grade A, B and C houses is $500, $400 and $250, respectively. The houses will require different levels of maintenance at different intervals, possibly painting of the exterior woodwork every 3 years, painting of the interior woodwork and walls every 6 years, etc. In addition there will be major capital expenditure at generally longer intervals: rewiring of the electricity circuit every, say, 20 years. A fundamental principle in this

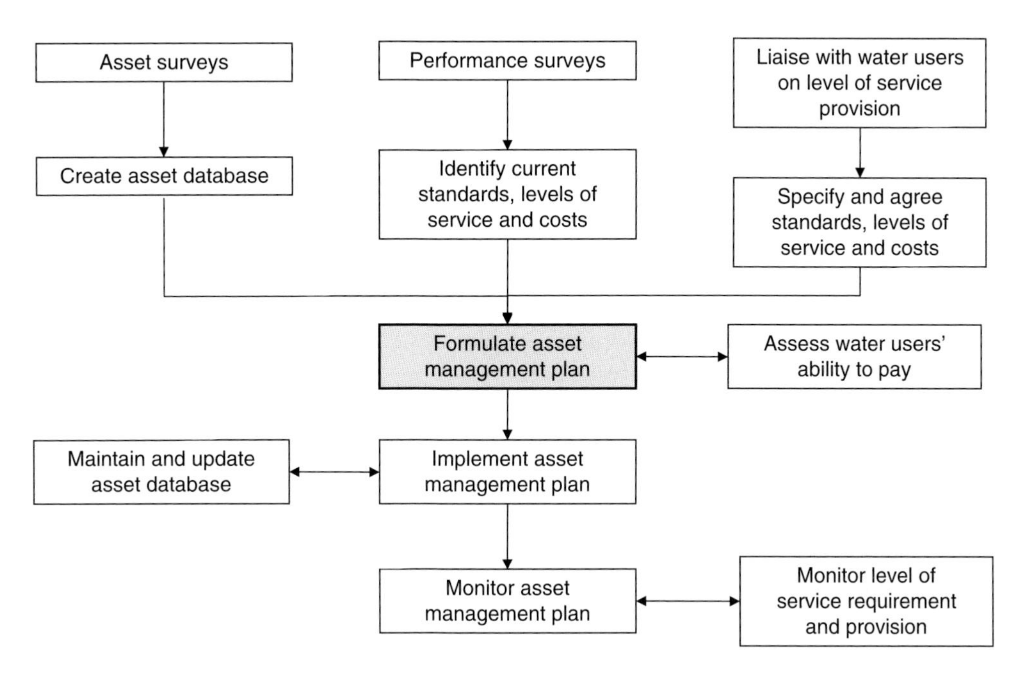

Fig. 6.10. Framework for asset management and strategic investment planning for irrigation and drainage infrastructure.

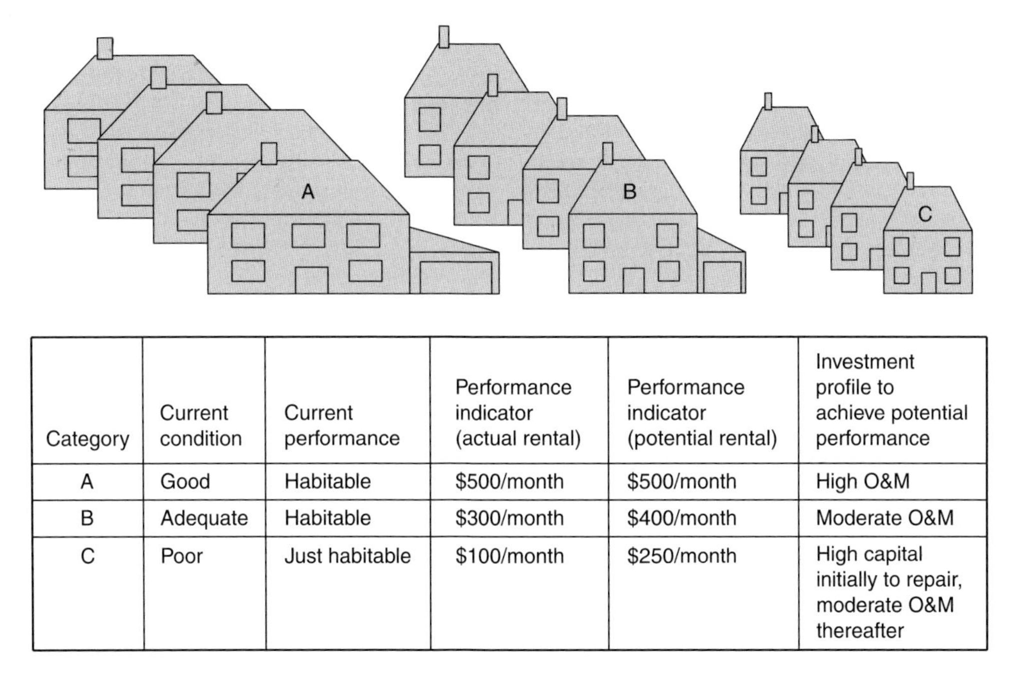

Category	Current condition	Current performance	Performance indicator (actual rental)	Performance indicator (potential rental)	Investment profile to achieve potential performance
A	Good	Habitable	$500/month	$500/month	High O&M
B	Adequate	Habitable	$300/month	$400/month	Moderate O&M
C	Poor	Just habitable	$100/month	$250/month	High capital initially to repair, moderate O&M thereafter

Fig. 6.11. Example of asset management planning for a group of houses (O&M, operation and management).

process is that the income from rental is able to cover these costs, including an allowance for management overheads. It may also be that the housing association at some stage decides to modernize the houses by providing new kitchens. This modernization will enhance the level of service provided to the tenants for which an increased rental may then be charged.

A similar process can be applied to irrigation and drainage infrastructure. The function and value of the infrastructure can be assessed and the infrastructure categorized according to the potential level of service that it can provide (ability to deliver water to match crop demands).[3] The level of expenditure required to keep the system operational over time at a specified level can be ascertained and the fee level to be charged to water users determined. If further investment is made in the irrigation or drainage system and the system is modernized, then the fee level can be changed to reflect the increased level of service provision. For example, the conversion of a system with manually operated gates to a system with automatic level control gates will increase the level of service by facilitating water distribution on-demand, thereby better matching supply and demand and facilitating enhanced agricultural production. There will be capital expenditure to remove and replace the control structures while the day-to-day operation costs may be reduced due to the saving of labour costs. The balance of the costs and savings will need to be determined by discounting over a 10–20 year timeframe to ascertain if the irrigation service fee level needs to be increased or decreased to pay for the changes made. Table 6.8 shows conceptual relationships between level of investment, canal control systems, level of service, O&M costs, and potential income levels. The level of service potential outlined in Table 6.8 assumes a close relationship between the control infrastructure and the management capability.

The interacting factors of asset condition and performance, current and desired levels of service are incorporated into the asset management plan and the investment over time calculated. The resultant expenditure profile (Fig. 6.12) is compared with the ability of the water users to pay, and in some cases the standard of the desired level of service may need to be reduced to match the users' ability to cover the planned expenditure.

The asset management plan is then implemented through shorter-term implementation plans, often of 5 years duration. The asset database will be upgraded as work is carried out, and the implementation of the plan and the level of service provision will be monitored.

Asset management can be used by the owners and managers of infrastructure as part of the process of assessing, monitoring and maintaining the value and utility of the assets. It can also be used by regulatory authorities where publicly owned infrastructure has been sold, franchised or transferred to non-governmental bodies. Such infrastructure often serves a monopoly function (delivery of irrigation water, potable water supply and sanitation, etc.), and the government has a duty of care to ensure that the infrastructure is properly managed and sustained over time. Failure on the part of government in this respect may mean that the management entity 'mines' the value of the assets by failing to invest sufficiently in the infrastructure over time, leading to failure of the system in the longer term.

An important current application of asset management is in the process of transferring the management, operation and maintenance of the I&D system to water users associations. Applying asset management procedures at the transfer stage can have important benefits, including: identification and audit of all infrastructural assets; identification of water users' desired level of service; identification of the cost of maintaining the system over time commensurate with the agreed level of service provision; understanding by the water users of the relationship between infrastructure condition and system performance; and development and ownership by water users and irrigation service provider of the relationship between fee payment and service provision.

A word of caution is required. Asset management is a management tool; how it is used, and how effective it is, depends entirely

Table 6.8. Indicative relationship between level of investment, canal control, level of service, and operation and management requirements and costs.

Type	Canal control system	Water delivery system	Level of service potential	O&M requirements	O&M costs	Capital investment level	Indicative O&M cost level ($/ha)	Possible potential income level
1	Fully automated downstream level canal control, fully adjustable and responsive to farmer demands	Demand	Very high, fully responsive to farmers' demands for water. Highly efficient in water use	Low staffing levels due to automation, but work force needs to be highly skilled	Low on day-to-day basis but may be high on occasion as control equipment is expensive. High capital cost, moderate O&M cost	High	35	High
2	Manual control with some automation at key locations. Discharge measurement at flow division and delivery points	Arranged-demand	High, responsive to farmers' demands for water though farmers need to order water in advance. High interaction between service provider and farmer	High staffing levels due to manual operation and need for measurement to match supply to demand	High due to cost of O&M staffing and associated facilities (offices, motorbikes, etc.). Maintenance costs high to maintain and replace gates over time	Moderately high	40	Good
3	Manual control throughout the system. Discharge measurement at flow division and delivery points	Supply-demand	Moderate. Supply-driven with irrigation service provider controlling/allocating available water taking into account farmers' cropping patterns. Relatively low interaction between service provider and farmer	Moderate staffing levels due to manual operation and need for some measurement to match supply to demand	Moderate due to O&M staffing and need for some O&M facilities. Maintenance costs high due to need to maintain control gates	Moderate	25	Moderate
4	Manual control at main control points, ungated and/or proportional distribution at lower locations. Limited measurement	Supply	Moderate, not responsive to farmers' demands, limited control over water distribution to match demands	Moderate to low staffing levels due to manual operation, though little measurement	Moderate to low due to O&M staffing and need for some facilities. Maintenance costs moderate due to need to maintain main control gates, kept lower by low-cost control at delivery points	Low	10	Low
5	Fixed proportional control system, supply-controlled, not responsive to demand. Measurement at water source intake only	Supply	Moderate to low, not responsive to farmers' demands for water but farmers can plan ahead and adjust cropping pattern to suit supply. Inefficient in water use	Low level of staffing, only low skill levels required	Low due to low O&M staffing levels and to low-cost proportional division structures	Very low	5	Subsistence

O&M, operation and management.

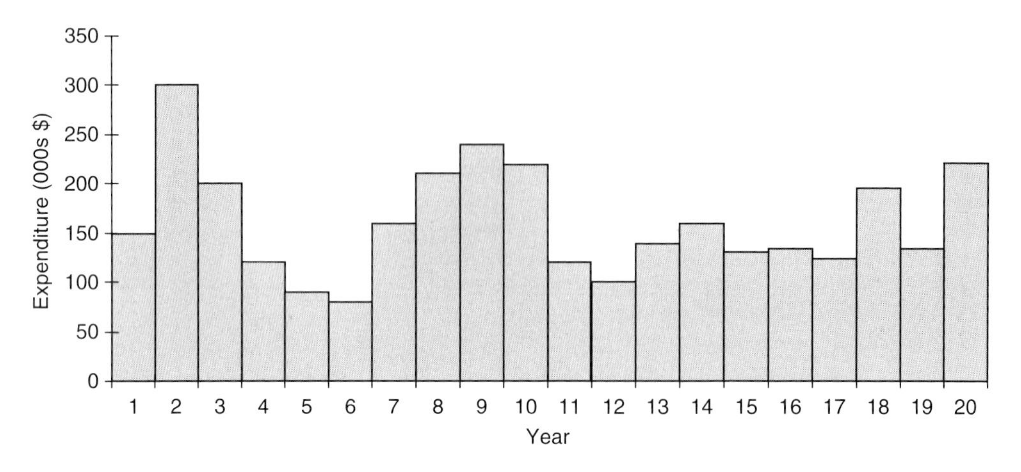

Fig. 6.12. Example of a 20-year investment plan profile.

on who uses it, and in what context. In the wrong context, where management is weak or lacks control over finances and budgeting, asset management will not work. What asset management can do, if used correctly, is identify infrastructural constraints to performance, and formulate plans to address them within the context of the ability and willingness of the users to pay for a specified level of service.

Asset management processes

Asset surveys

Asset surveys are the starting point for asset management planning. The asset survey determines the following.

- The *category* of components of the system (canal, head regulator, etc.).
- The *extent* of the assets that exist (how many and in what categories).
- The *size* of the asset (these can be grouped into size bands to facilitate costing).
- The *'importance'* of the asset. This relates to the impact that malfunction of the asset might have on the system as a whole. The head regulator at the river intake is more 'important' than a secondary canal head regulator lower down the system.
- The *value* of the assets in each size band. The value is based on the modern equivalent asset (MEA), which is the cost of replacing the structure at today's costs.
- The *components/facets* of each asset (e.g. gates and masonry in a head regulator structure). Different asset components/facets may deteriorate at different rates.
- The *condition* of the asset and its components/facets. The condition will affect the level of investment required. Condition grades are used to categorize condition.
- The *serviceability* of the asset; that is, how well it performs its function. An asset may be in a poor condition (masonry damaged) but performing its function satisfactorily (gates operating and passing design discharge). For irrigation, serviceability of structures can be divided into *hydraulic function* (ability to pass design discharge) and *operations function* (ability to control flow across a specified range, ability to provide command level, etc.). Serviceability grades are used to categorize performance (Table 6.9).

The assets can be grouped into categories based on their function (water capture, conveyance, control and measurement, ancillary, etc.) and can be grouped within these categories in terms of their size (Table 6.10). The size can be based on one or two leading variables (such as crest length and height for a river weir, or design capacity for a canal). Grouping

Table 6.9. Examples of condition and serviceability grades for canal cross regulators. (From IIS, 1995a, b.)

(a) Condition grade

	Condition grade (implying cost)			
	Grade 1	Grade 2	Grade 3	Grade 4
Component	Good	Fair	Poor	Bad
Structure Upstream wingwalls Downstream wingwalls Superstructure Notice board Control section (note type)	Structurally sound with no deformation of dimensions or profile. Well-maintained with little or no signs of deterioration. Upstream and downstream bed having only minor, or no, silt deposition and clear of debris	Generally sound but with some deterioration of structure and/or dimensional deformation. Needing maintenance attention with a review of condition in the medium term OR Structural and dimensional condition as (1) but with silt and/or debris significantly affecting functionality	Significant deterioration of structure and/or dimensional deformation, requiring urgent corrective work OR Structural and dimensional condition worse than (1) with silt and/or debris significantly affecting functionality	Serious structural problems causing actual or imminent collapse and requiring partial or complete reconstruction
Gauge(s)	Gauges securely fixed and readable	Gauges generally satisfactory but may be difficult to read under some flow conditions	No proper readable gauge but level mark present from which to measure	No gauge or level mark available OR unreadable OR unreliable
Bench mark	Bench mark secure, apparently undamaged and readable	Bench mark condition generally as (1) but difficult to read	Bench mark present but of uncertain reliability	Bench mark missing, damaged or unreadable

(b) Serviceability grade

	Serviceability grade (implying priority)			
	Grade 1	Grade 2	Grade 3	Grade 4
Function	Fully functional	Minor functional shortcomings	Seriously reduced functionality	Ceased to function
Hydraulic To pass the design flow safely Operations To control 'command' (water level) across the required range (except for a fixed crest) AND to allow measurement of flow	Apparently properly designed and constructed with capacity to pass the design flow safely AND fully capable of being operated to control command across the desired range AND allowing measurement of flow by means of its own components or an adjacent measuring structure. Performance unaffected by silt or debris	Normally able to pass the required flows AND capable of being operated to control command in a measured manner BUT performance likely to be unsatisfactory under extreme conditions of demand or climate. Deficiencies may be due to design or construction inadequacies, insufficient maintenance, measuring devices that are difficult to read or due to the presence of silt and/or debris	One or more of the three defined functions seriously impaired through deficiencies in design, construction or maintenance, or due to the presence of silt and/or debris. (Likely to have a significant detrimental effect on system performance)	Complete loss of one or more of the three functions or serious reduction of all three for whatever reason

Table 6.10. Examples of asset types, function, components and estimated lifespan. (From IIS, 1995a, b.)

Asset type	Size measures to be recorded	Functions to be assessed	Components to check	Depreciation life (estimate)
River weir	crest length crest height	Hydraulic: • provide level • pass offtake design flow • pass design flood Operations: • gates • gauges	weir wall dividing walls abutments crest apron sluice gate offtake gate stilling basin superstructure	Civil: 50 years M&E: 10 years
Head regulator	total gate width design flow	Hydraulic: • pass design flow Operations: • control flow • gauges	gate(s) structure notice board shelter	Civil: 25 years
Cross regulator	total gate width design flow	Hydraulic: • pass design flow Operations: • control command (level) • gauges	control section[a] structure notice board upstream wingwalls downstream wingwalls gauge(s) shelter	Civil: 25 years M&E: 10 years
Measuring structure	total crest width design flow	Hydraulic: • pass design flow Operations: • measure flow	control section gauges structure upstream wingwalls downstream wingwalls stilling box	25 years
Canal Linings earth masonry concrete tile continuous concrete	design flow length	Hydraulic: • pass design flow Operations: • not applicable	embankment side slopes (note type) bed	Civil: 25 years
Drain Linings earth masonry concrete tile continuous concrete	design flow length	Hydraulic: • pass design flow Operations: • not applicable	embankment side slopes (note type) bed	Civil: 25 years
Hydraulic structure aqueduct culvert drop structure escape structure (note type)	(depends on structure) design flow length fall	Hydraulic: • pass design flow Operations: • not applicable	conveyance support structure upstream wingwalls downstream wingwalls stilling basin	Civil: 25 years M&E: 10 years
Supplementary structure Examples bridge cattle dip	(depends on structure) design flow length	Hydraulic: • pass design flow Operations: • not applicable	structure safety other features	Civil: 25 years M&E: 10 years
Access roads	width length	Operations: • access to system	structure surface drains	Civil: 25 years

M&E, mechanical and electrical.
[a]Options fixed crest gate(s) stop logs flume.

in this way means that average costs can be determined for categories and size bands of assets for maintenance and for assessing the MEA value. The MEA value represents the cost, in today's prices, of replacing the asset, and as such builds to give a complete valuation for the asset base.

To carry out the survey the asset surveyor first gathers available data (maps, design drawings, structure inventories, etc.) before starting on the field work. For the field work the surveyor generally commences at the top of the primary canal system and works down to the tail, then returning to survey each secondary canal in turn. The distance along the canal is measured using a tape or measuring wheel, and condition and performance assessments made of each stretch of canal, and at each structure. The level of detail collected depends on the resources available, in some cases full profiles of the canal are measured each 100 m, in other cases, only observations are taken. For structures key measurements are taken (gate widths, height, etc.) and in some surveys full measurements are taken for all components/facets of each structure. Standard forms are used to record the survey data (Fig. 6.13). The survey may need to be carried out first with the canals flowing and then with them dry to capture all the data required.

With the advent of digital cameras photographs are increasing being used as an integral part of the asset survey, both to record the condition of the asset and to use a part of the process of grading the condition and serviceability of the asset.

Asset database

Data collected from the asset survey need to be recorded in a systematic manner in the asset database. The asset database can be created using a spreadsheet, a standard database or a tailor-made database. A possible structure for such a relational database is presented in Fig. 6.14, with data on different aspects being entered and stored in separate, but related files. Photographs will form an important part of the database, and will act as a record to be able to assess the performance of the asset management programme by being able to visually review asset condition over time.

Performance surveys

Performance surveys are required to assess the current and potential performance of both the I&D system and the I&D scheme. Strictly speaking those responsible for the assets are interested in the performance of the system (the network of canals, drains, structures) rather than that of the scheme (the physical system plus the land plots and crops) as they have control over the performance of the system, but not, generally, over the performance of the scheme as a whole. The exception would be where the system and scheme are under the control of one organization, such as on a privately run sugar estate.

For asset management planning, assessment of the performance of the system relates mainly to the delivery of, and removal of excess, water in a reliable, adequate, timely, equitable and cost-effective manner. For assessment of scheme performance additional indicators such as crop production, crop yields and crop income are important. In some cases, such as in Australia where irrigation systems are equipped with accurate measuring structures, separation of the performance of the system and the scheme is feasible. In other irrigation systems where measurement of water at the system delivery point (the head of the secondary or tertiary canal) is not possible, or not well done, then the performance of the scheme, and sections within the scheme, is required as a proxy for assessing system performance.

Performance assessment is discussed in greater detail in Chapter 9. In the context of asset management planning it is important to distinguish between the performance constraints arising from the *condition* of the infrastructure and those arising from the *operation and use* of the infrastructure. Asset management seeks to minimize infrastructural performance constraints in order that system performance is not constrained; it does not deal directly with operational issues.

Defining and agreeing on standards and levels of service provision

A key feature of the asset management planning process is to specify the desired

ASSET SURVEY

Form CR for Cross Regulator

CR

System Details

Section ____________________

Sub-Section ____________________

Irrigation Command ____________________

IC Reference No. ____________________

Data Collected

By (Name) ________________

On (Date) ___◊___◊___

Asset Details

Area served (ha) __________ Asset Ref. No. ______________

Location (km) __________ Canal Name ____________________________

Type of canal: Primary ☐ Secondary ☐ Supplementary ☐

Reported age (years): 0–5 ☐ 5–10 ☐ 10–20 ☐ 20+ ☐

Control section width (m) ______ Design flow (l/s) ______

Control section type: Gate(s) ☐ Fixed crest ☐ Stop logs ☐ Flume ☐

Component Condition

	General Condition Grade				Worst Case Local			
	1	2	3	4	1	2	3	4
Structure	☐	☐	☐	☐	☐	☐	☐	☐
Control section	☐	☐	☐	☐	☐	☐	☐	☐
Upstream wingwalls	☐	☐	☐	☐	☐	☐	☐	☐
Downstream wingwalls	☐	☐	☐	☐	☐	☐	☐	☐
Gauges	☐	☐	☐	☐	☐	☐	☐	☐
Bench mark	☐	☐	☐	☐	☐	☐	☐	☐
Superstructure	☐	☐	☐	☐	☐	☐	☐	☐
Notice board	☐	☐	☐	☐	☐	☐	☐	☐

Asset serviceability

Overall Serviceability Grade

1	2	3	4
☐	☐	☐	☐

Notes:

Fig. 6.13. Example of asset survey form for cross regulators. (From Davies, 1993.)

level of service and then to determine the performance shortfall by measuring the current levels that are being provided by the assets (assuming there are no management constraints).

The ability to deliver the desired level of service will primarily depend on:

1. The type of irrigation infrastructure provided;
2. The condition and serviceability of the infrastructure;
3. The capability of the O&M management.

Assessment of the desired level of service can be made prior to the preparation of the

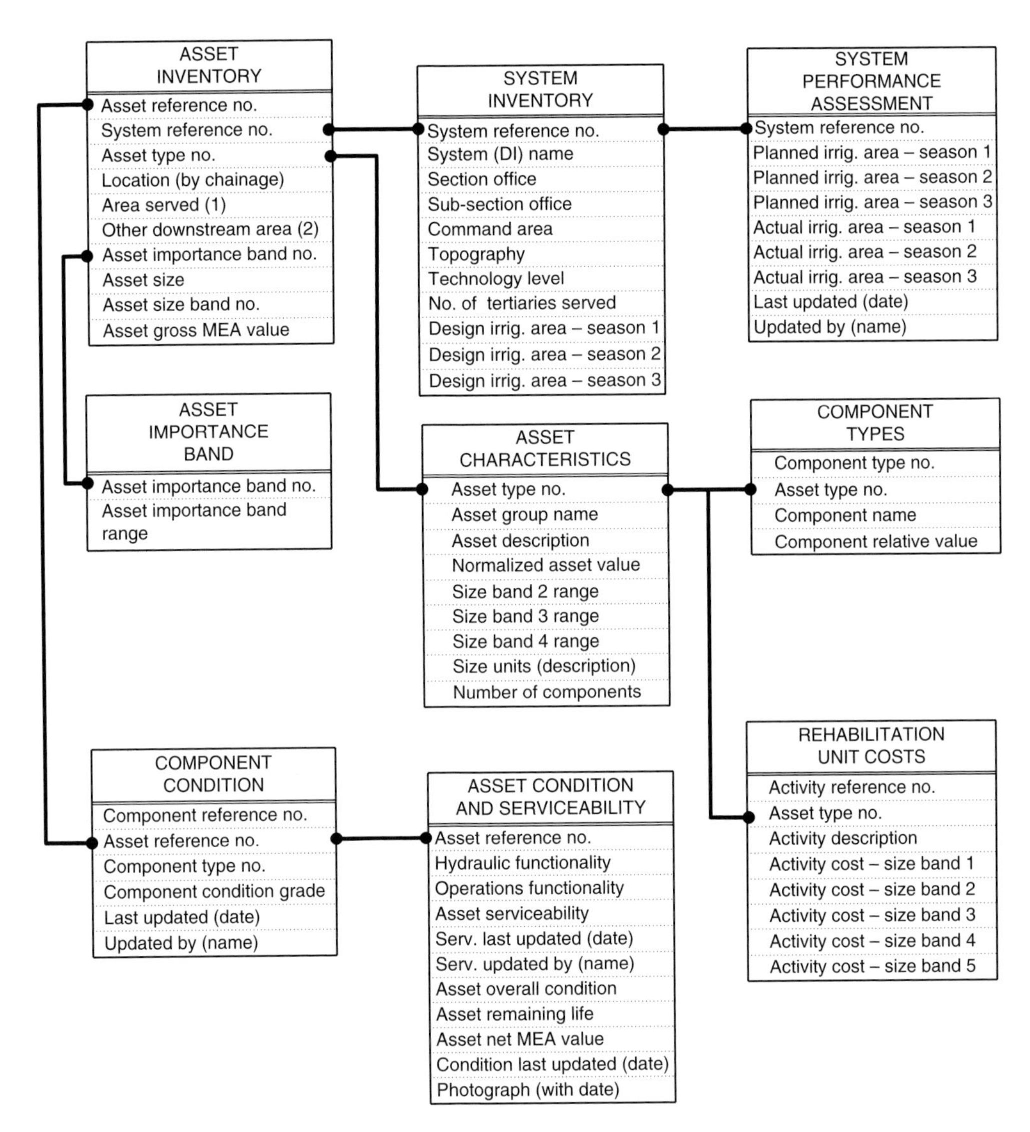

Fig. 6.14. Possible structure of a relational asset database (MEA, modern equivalent asset). (From IIS, 1995a, b.)

asset management plan through interviews and discussions with water users, though the cost of providing a given level of service will not be known until the asset survey has been completed and the asset management plan prepared. Establishing the desired level of service will not be easy, as in many schemes such a concept has often not been communicated explicitly to water users. The Warabandi system used in northern India and Pakistan is an exception. In this instance farmers are well aware of the stated level of service provision, with time shares, and times and duration of water turns, being set out well in advance of each irrigation. One of the benefits of the asset management process is that it requires the stipulation of the standards by which performance will be measured, and that it also requires the stipulation of the desired level of service. Making these explicit facilitates communication between the irrigation service provider and the water user.

The relationship between the potential level of service provision, investment levels, and management, operation and maintenance requirements are outlined in Table 6.11 for schemes with different types of control systems. The O&M costs given in this table are indicative only; they are used to emphasize the point that different types of system will have different O&M costs composed of the asset value (influencing replacement costs), the maintenance costs and the operating costs (including staffing costs).

From the engineering studies (discussed below) an understanding will be gained of the anticipated improvements in performance benefits arising from different levels of investment. These improvements need to be assessed against the investment costs. The benefits will accrue to the irrigation (investing) service provider from the revenue generated from the water users, who will, in turn, derive their income from agricultural production generated as a consequence of the (improved) water delivery service provided by the irrigation service provider. The link between level of service provision and fee payment is central to the process of asset management.

Engineering studies and costs

Engineering studies are required to study generic issues such as:

- the deterioration rate of different types of assets and asset components (facets);
- the development of cost models (costs for rebuilding/upgrading/rehabilitating assets);
- the relationships between individual asset performance and system performance.

Through engineering studies the cost database for maintaining or enhancing the condition/performance of each *type* of asset (river weir, canal head regulator, aqueduct, culvert, etc.) can be ascertained and applied to the asset condition/performance of each asset. In this way the cost of maintaining or enhancing the condition/performance of the I&D system is determined. The deterioration rate of individual components, such as rubber gate seals, or pumps and motors, is estimated and standard profiles drawn up for each type of asset.

The *importance* of the asset will influence the priority given to investment in it. An asset's importance relates primarily to the asset's function, position in the irrigation or drainage network, and its replacement value. A river diversion weir is more important than a secondary canal head regulator, for example, because of its central function in diverting and controlling inflow to the scheme, its position at the head of the system, and its (usually) significant replacement cost.

An additional feature of the engineering studies is to look at alternatives, for example

Table 6.11. Relationship between investment and service delivery for different types of irrigation control systems.

Control system	Water delivery	Potential level of service	O&M required	O&M costs[a]	Capital investment	Income potential
Fully automated	Demand	Very high	Low but skilled	High over time, $35/ha	High	High
Full manual control & measurement	Arranged-demand	High	High staffing	High due to staff, $40/ha	Moderately high	Good
Partial manual control & measurement	Supply/arranged-demand	Moderate	Moderate to low	Moderate to low, $25/ha	Moderate	Low
Fixed proportional distribution	Supply	Moderate to low	Low	Low, $5/ha	Very low	Subsistence

O&M, operation and maintenance.
[a]These are illustrative estimates of O&M costs, actual comparative cost estimates are difficult to find for these different types of system.

replacing manually operated gates with automated gates to save operating (OPEX) costs, or replacing a structure that is at the end of its useful life with a new structure, possibly of a different design, or with different features. Replacing a structure may cost more in terms of capital invested (CAPEX) but less in terms of OPEX costs.

Historic records of capital and O&M expenditure provide a valuable basis for assessing the future capital and O&M expenditure. Past expenditure figures can be brought up to date using standard cost index tables. Records of maintenance work done and costs can inform on cost items and recurrence intervals (e.g. how often the main canal is desilted, what volume and at what cost, etc.). Figure 6.15 shows an analysis of a pumped irrigation scheme where the funding for OPEX costs has declined significantly in real terms. As a consequence the physical condition of the assets had declined markedly, requiring (expensive) rehabilitation in 2003. In the meantime the productivity of the scheme declined markedly, in part due to due to poor water delivery caused by improperly functioning infrastructure, especially the pumps.

Preparing the asset management plan

Utilizing information developed from the asset surveys, the performance surveys and the engineering studies, the investment requirement in the assets over time is determined. This calculation leads to the formulation of the long-term investment profile as presented earlier in Fig. 6.12. This long-term plan needs to be broken down into a schedule of planned activities, and a short-term budget prepared for a 2–5 year period.

Financial modelling is an integral part of the preparation of the asset management plan, as adjustment may be needed to the initial plan to match the investment required with the finances available. Alternative strategies may be need to be looked at; for example reducing the specification for the desired level of service in order to save investment costs, or accelerating or delaying investment. These strategies will take account of the source and profile of funding available (such as capital loans or grants from government, irrigation service fees, etc.). Figure 6.16 shows examples of different investment profiles that can be generated depending on the level of service required. In the first case the level of service required is high, resulting in high initial investment and high operational expenditure (Fig. 6.16a). In the second case the level of service is lower, with deferred investment and lower operational expenditure (Fig. 6.16b). From these calculations the average annual budget can be prepared and linked to the irrigation service fee.

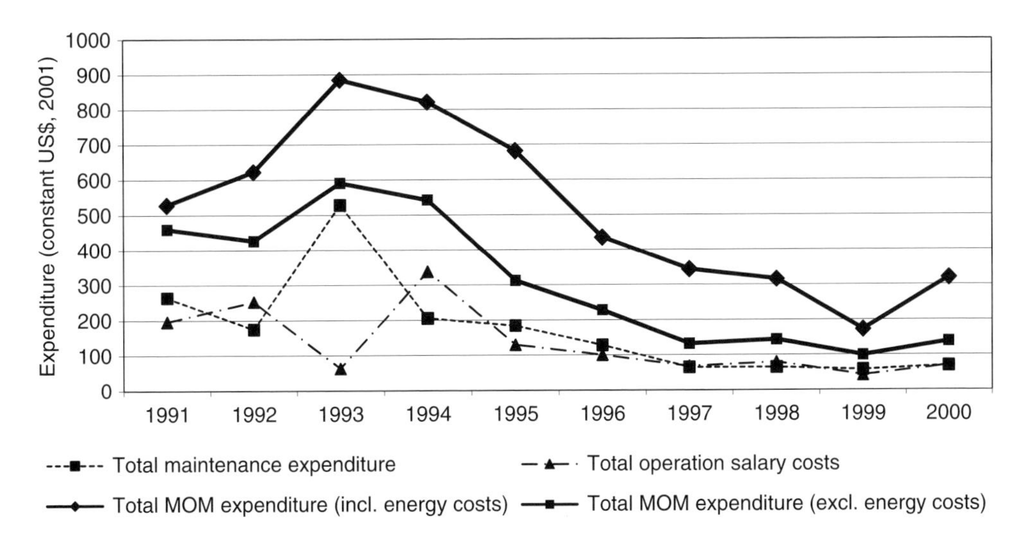

Fig. 6.15. Analysis of historical operating costs using constant US dollar prices, 2001 (MOM, management, operation and maintenance).

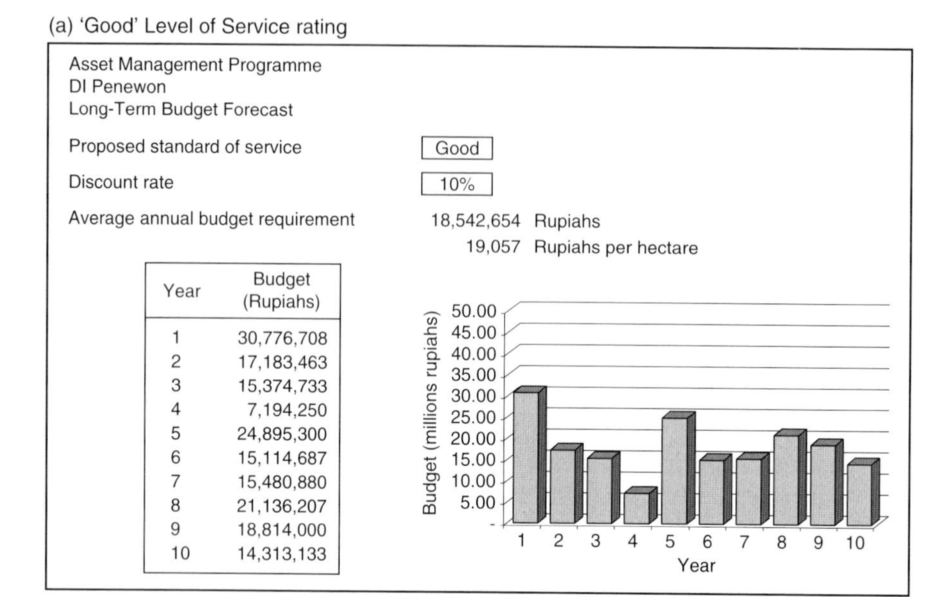

Year	Budget (Rupiahs)
1	30,776,708
2	17,183,463
3	15,374,733
4	7,194,250
5	24,895,300
6	15,114,687
7	15,480,880
8	21,136,207
9	18,814,000
10	14,313,133

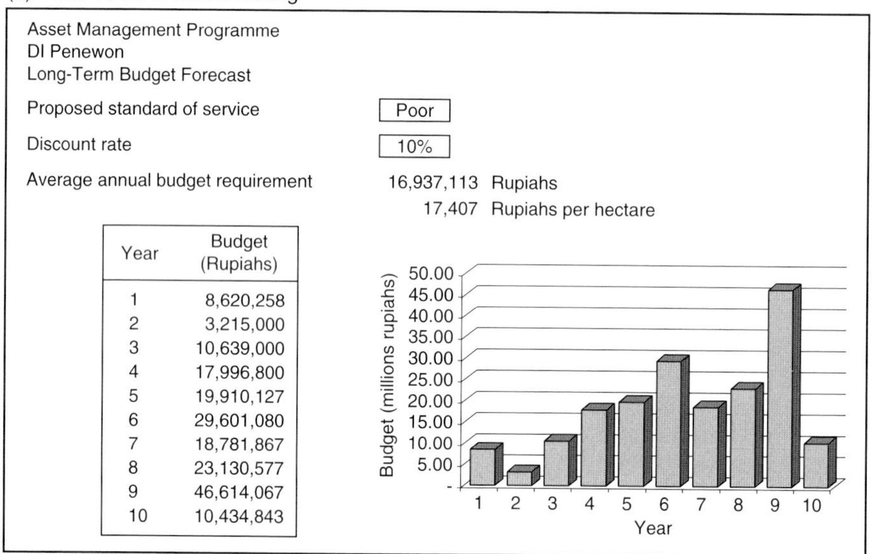

Year	Budget (Rupiahs)
1	8,620,258
2	3,215,000
3	10,639,000
4	17,996,800
5	19,910,127
6	29,601,080
7	18,781,867
8	23,130,577
9	46,614,067
10	10,434,843

Fig. 6.16. Examples of investment profile for different levels of service provision. (Modified from Davies, 1993.)

The final asset management plan comprises the information outlined in Table 6.12.

An indication needs to be given in the asset management plan on the accuracy and reliability of the data used in preparation of the plan. Tables 6.13 and 6.14 present guidelines used by the UK Office of Water Services for confidence grades.

There are a number of sources of variance in the data – cost variations for physical works, differences in asset survey assessments, engineering judgement on lifespans of assets, etc.

Table 6.12. Summary of information contained in the asset management plan.

Report	Content
Asset stock, condition and serviceability profile	A statement of all the assets, divided by category and size. Total value of the assets is quoted as gross MEA and net (depreciated) value. Condition and serviceability profiles provided for all assets, together with an importance profile
Unit costs report for MEA value and capital investment activities	Presentation of the information contained in the cost model; provides the build-up of costs for work required on each type of asset
Investment programme	Report on the total investment estimates for CAPEX and OPEX as programmed by each 5-year period for the next 20 years.[4] Investment is presented in different formats: by each asset category; by each importance band; by each purpose category
Activity report	Complements the investment programme by detailing the timing of the activities to be carried out. Details how many kilometres of canal to be relined, desilted, etc., each year
Benefits report	Provides details of the historical trends and the anticipated future benefits of the investment programme, based on the identified performance indicators. Maintaining or improving the asset condition profile will be an important output performance measure
Asset depreciation categories	A summary report on the assumptions made in the AMP about asset depreciation rates and lifespan

MEA, modern equivalent asset; CAPEX, capital expenditure; OPEX, operational expenditure; AMP, asset management plan.

Table 6.13. Data accuracy bands. (Data from OFWAT, 1992.)

Band	Definition
1	Better than or equal to ±1%
2	Not band 1, but better than or equal to ±5%
3	Not bands 1 or 2, but better than or equal to ±10%
4	Not bands 1, 2 or 3, but better than or equal to ±25%
5	Not bands 1, 2, 3 or 4, but better than or equal to ±50%
6	Not bands 1, 2, 3, 4 or 5, but better than or equal to ±100%

Assessing water users' willingness and ability to pay

The investment plan may need to be revised to match the ability of the water users to pay for the service. If this occurs, the potential level of service provision arising from the condition and performance of the infrastructure may be reduced. A reduced level of service may result in a reduction in crop yield and a diminished ability to pay for water. There is obviously a balance to be struck between these two factors.[5]

It is important to note that there is a difference between the water users' *ability* to pay and their *willingness* to pay. For this reason it is important that the asset management process is clear, transparent and auditable, and that the water users are active participants in the process.

Implementing the asset management plan

Though asset management plans generally look at a longer-term timeframe (15–20 years), they are implemented in short-term time segments. The asset management plan will have given a profile of the investment needed in the infrastructure over time, and will have been used to establish the financial plan to sustain the assets over time. This plan may incorporate contributions from different sources, including the irrigation service fees and government subsidies. The short-term budgeting and expenditure sets out to manage the investment such that necessary

Table 6.14. Data reliability bands. (Data from OFWAT, 1992.)

Band	Description	Definition: Actual	Definition: Forecast
A	Highly reliable	Data based on sound records, procedures, investigations or analysis, which is properly documented and recognized as the best method of assessment	Based on extrapolations of high-quality records covering or applicable to more than 100% of the study area, kept and updated for a minimum of 5 years. The forecast will have been reviewed during the current year
B	Reliable	Generally as A but with some minor shortcomings; for example the assessment is old, or some documentation is missing, or some reliance on unconfirmed reports, or some extrapolation	Based on extrapolations of records covering or applicable to more than 50% of the study area, kept and updated for a minimum of 5 years. The forecast will have been reviewed during the previous 2 years
C	Unreliable	Data based on extrapolation from a limited sample for which grade A or B data is available	Based on extrapolations of records covering more than 30% of the study area. The forecast will have been reviewed in the previous 5 years
D	Highly unreliable	Data based on unconfirmed verbal reports and/or cursory inspections or analysis	Based on forecasts not complying with bands A, B or C

maintenance and replacement work is carried out to sustain the agreed level of service. Cost control and performance monitoring are key parts of this process, as are making sure that the expenditure is made transparent and accountable to users.

Maintaining the asset database

The asset database will undergo continuous revision. Maintenance work will be recorded, and periodic updates made to asset condition and performance gradings through further asset surveys. With experience adjustments will be made to the information available on deterioration rates, cost models, CAPEX and OPEX costs, etc. and the asset management plan refined.

Monitoring service provision and the implementation of the asset management plan

Monitoring and evaluation (M&E) are important parts of the asset management process, allowing for the monitoring of the levels of investment, and its impact on the service delivery. M&E systems need to be set in place which are transparent and accountable, so that those paying for the investment (water users, and/or government) can be satisfied that their money is being efficiently and effectively used. Feedback mechanisms are an important part of the M&E process.

Asset surveys will monitor the condition and performance of the infrastructure, while monitoring of key indicators (such as water delivery versus water demand) coupled with user surveys will assess the level of service provision.

Endnotes

[1] In practice, a more detailed analysis would be required using discounted cash flows and net present values to make a true comparison of the alternatives.
[2] It is the field staff who will bear the brunt of criticism from water users if the maintenance work is not carried out properly.
[3] It is important to note that there are at least two aspects here: the condition and performance of the physical infrastructure, and the performance of the people and organizations that operate the infrastructure. While asset management primarily focuses on the infrastructure, an assessment of the ability of management to use and operate the infrastructure is also required.
[4] The selected short- and long-term timeframes may vary depending on the situation.
[5] In practice this is not a direct one-to-one linkage, it has to be moderated by other factors.

7

Training

This chapter details the steps required in setting up and running training courses in the irrigation and drainage sector. Training is required for a wide range of personnel, from system managers to water users; and in a wide range of disciplines, from general management to specific and detailed technical procedures. Training principles and procedures are outlined together with more detailed information on key aspects such as carrying out a training needs analysis and organizing and running courses. Finally a detailed checklist is provided for setting up, running and costing training courses.

Overview

Training is an integral part of the effective functioning of any organization. Training is the process by which staff are taught the skills necessary to perform their job functions, such that they can carry out those functions effectively and efficiently. Poor or inadequate training of staff can be a major cause of poor performance within an organization.

Training is a difficult task, which requires specialist expertise to carry out effectively. It requires a thorough understanding of the subject matter and of the training needs of the participants. Preparation of targeted training material and the use of appropriate training methods are fundamental to successful training outcomes.

There is a wide range of training needs in the irrigation and drainage sector, ranging from training for management and administration staff through to technical training of engineers and field staff. A feature of irrigation and drainage is also the degree of human interaction that exists between the service provider and the water user; often irrigation and drainage training programmes will include training in liaising and working with people.

Training is often a major component of irrigation and drainage projects. In recent years there has been a trend towards incorporating training into irrigation and drainage rehabilitation projects, with the aim that the training will enable management to adequately manage, operate and maintain irrigation and drainage systems such that rehabilitation is not required again at some future date. Institutional development is increasingly being allied to physical rehabilitation projects, a key component of which is the formation and support of water users associations. In some countries reform is now also focusing on the main irrigation and drainage service provider, with restructuring and reform of government-run irrigation and drainage agencies to provide a more effective and efficient service provi-

 ©Martin Burton 2010. *Irrigation Management: Principles and Practices* (Martin Burton)

sion to water users (see Chapter 8). In both cases, the formation and support of water users associations and the restructuring of irrigation and drainage agencies, training has a central role to play.

The forms of training required in the irrigation and drainage sector are summarized in Table 7.1.

Training Principles

What is training?

In training, generally speaking, we are interested in effecting *change* in a person's behaviour. We may want gauge readers to be more accurate with their gauge reading, we may want farmers to use less water and be more efficient with application of irrigation water to the fields. In order for that person to change their behaviour they must go through certain processes:

1. They must be aware that their performance, or their situation, could be better;
2. They must want to learn how to improve the situation;
3. They must do some learning;
4. They must implement what they have learnt.

The outcome of the training must be a *change in performance* of the activity for which the person has been trained (i.e. the gauges must be read more accurately, or irrigation water used more productively). Training is not *education*; in education the objective is improving a person's general level of knowledge, it is not always specific to a defined task or activity. Even if a person does not use their education we still feel that they, and society, have benefited. If a person has been trained and does not apply it, we consider that to be a loss, we could have spent the time and money on training someone who would apply it.

Training could thus be defined as:

> The process of bringing about change in behaviour of an individual or group which results in improved performance in their work or situation.

Another important distinction can be drawn between education and training. In an educational system, if a student does not achieve the required pass marks in the examinations it is the student who has failed, not the teacher (though the teaching may well have been poor!). However, if a trainee fails to achieve the desired level of performance at the end of a training programme then it is the trainer who has failed, not the trainee. For training, the content of the training programme must be such as to raise the trainee from a certain level of capability to another, higher level. The training programme must be designed and implemented to effect this change, and the responsibility for this rests with the trainer.

This leads on to the issue of training needs assessment. Training needs assessment can be summarized as:

Required knowledge and skills minus
Existing knowledge and skills equals
Training need

Before the training course the trainee's level of knowledge and ability should be ascertained and compared with the required level of knowledge and ability. The difference is the training need. Knowing the start and end points the most effective path to effect the desired change can be plotted. Different situations will require different routes. For example, for senior management a series of lectures is effective in getting information across in a short time, with reinforcement through discussion sessions; for gatekeepers very short lectures (on basic principles) followed by demonstrations and field practicals are appropriate.

Training can effect changes in three broad, interrelated domains:

- cognitive learning (*knowledge*) – improvement in mental skills;
- psychomotor learning (*skills*) – improvement in manual or physical skills;
- affective learning (*attitude*) – growth in feelings or emotional capabilities.

These domains are more commonly referred to by trainers as KSA (Knowledge, Skills and Attitude), and effecting change in one or more of these domains is the goal of the training process.

Table 7.1. Forms of training required in the irrigation and drainage sector.

Focus	Target audience	Typical training elements
Design	Designers	Design principles; design criteria; modern design techniques; CAD; GIS; bills of quantities; specifications; tendering procedures
Construction	Construction supervisors Contractors	Construction principles and procedures; construction supervision; quality control
Management	Management personnel – I&D service provider	Management principles and practices; management information systems; resource management (including human resource management)
Operation and maintenance	O&M managers and staff	O&M principles and practices; planning; budgeting; determining water requirements; scheduling; liaising with water users; maintenance management; asset management
Formation and support of WUAs	Water users WUA Councils WUA staff Water users	Rationale for WUA formation; WUA and water users' roles and responsibilities; principles and practices of service provision; communication and liaison with water users; WUA staff duties and responsibilities; fee setting, collection and financial management; irrigation planning and scheduling; maintenance planning and management
Finance	Financial personnel – I&D service provider WUA accountants	Financial management; accounting and book keeping; cash flow; service fee setting; maintenance costing based on asset management principles
Administration	Administrative personnel – I&D service provider WUA management	Administrative processes and procedures; record keeping; support service provision; human resource development
Performance assessment	I&D service provider management WUA management	Purpose of performance assessment; identifying suitable performance indicators; data collection, processing and analysis; diagnostic analysis; implementing results of performance assessment
Monitoring and evaluation	Project management unit M&E personnel	Purpose and value of M&E; M&E principles; establishing M&E programmes; data collection, processing and analysis; M&E reporting/feedback

I&D, irrigation and drainage; O&M, operation and maintenance; WUA, water users association; M&E, monitoring and evaluation; CAD, computer-aided design; GIS, Geographic Information Systems.

The knowledge or cognitive domain relates to the development of thinking or intellectual skills, ability and knowledge. It can be broken down into six categories, often referred to as Bloom's taxonomy of educational objectives (Table 7.2).

An understanding of these different levels of application of knowledge and understanding is important in training, as it guides us in the development of the training programme and training material. For instance, for design engineers it is essential that they can make evaluations based on their knowledge, as each design situation is different. For farmers the knowledge that fertilizer increases yields might be sufficient, but the farmers' comprehension and application might be improved if they are taught how fertilizer works.

The skills or psychomotor domain relates to physical manipulation and use of the body's motor skills. Training related to development of these skills usually involves performing the task, initially under guidance and later independently. This domain requires abilities of perception – being able to see, feel, hear, smell or taste – and judgement – being able to manipulate the body or other items to the required degree to achieve a desirable outcome.

The attitude or affective domain relates to the way in which people address things emotionally, and covers values, feelings, opinions; motivation, attitudes and understanding. It deals with people's ability to receive and respond to information and situations arising from living and working together, based on their own and other people's values,

Table 7.2. Bloom's taxonomy of educational objectives. (Modified from Bloom, 1956.)

Category	Meaning	Example	Keywords
Knowledge	Ability to recognize and recall information	Can quote the rules and regulations	Know; recall; recognize; reproduce
Comprehension	Ability to interpret, translate, summarize or paraphrase given information. Knowledge is required to demonstrate comprehension	Can explain the rules and regulations to a client or new employee	Comprehend; explain; interpret; summarize
Application	Ability to apply information in a new situation. Applies classroom learning to situations in the work-place. Comprehension is required to apply information to a new situation	Can use an O&M manual to schedule irrigation water	Apply; modify; solve; use
Analysis	Ability to dissect and analyse information into component parts and meanings. Ability to apply information is required in order to analyse	Can interpret and interrogate data in a computer spreadsheet	Analyse; compare; explain; show
Synthesis	Ability to pull together diverse sets of information to formulate a structured solution to a problem or situation. Analysis is required in order to synthesize	Can write an O&M manual to suit the identified audience	Synthesize; summarize; précis; compose; create; modify; formulate
Evaluation	Ability to form an opinion or judgement based on criteria and rationale. Synthesis is required in order to be able to evaluate	Can make choices between different options, e.g. selecting new staff	Compare; contrast; evaluate; interpret; see through; discriminate

O&M, operation and maintenance.

understandings, aspirations and beliefs. Training in this area often relates to helping people understand other people's views and situations. An example is training aimed at sensitizing operation and maintenance (O&M) staff to the importance to farmers of delivering irrigation water at the right time, in the right place and in the right quantity to suit their needs. Changing people's attitudes is perhaps the most complex task in training; it requires an understanding of psychology to be able to set up a learning environment that effectively alters people's attitudes or levels of motivation. The Irrigation Management Game (Box 7.1) is an example of a training exercise designed to change understanding, attitudes and behaviour in relation to irrigation water delivery in the main system. In addition, attitudes are strongly affected by the day-to-day environment within which people live. Thus, while a training programme may improve trainees' level of motivation, such motivation may not endure in their working environment if there are too many constraints.

In summary, these three categories are all interacting within a training programme; for instance, individuals who are being taught to drive will gain in confidence, skill and knowledge as they learn and their ability develops. Their attitude to other road users will also change in this process, though not always for the better!

How people learn

People learn by observation, listening to others and doing. In training we use a variety

Box 7.1. The Irrigation Management Game: A Simulation and Role-playing Exercise for Training in Irrigation Management (Burton, 1989a, 1994)

The Irrigation Management Game places participants in the position of either irrigation agency staff responsible for managing the main canal system or farmers responsible for managing irrigated landholdings within the main canal command area. Usually one or two people take on the role of the main system service provider and eight to 16 people take on the role of farmers managing landholdings within the eight tertiary units (with one or two participants per tertiary unit). The exercise is run by two trainers, one as the Game Controller, the other as the Trader. The game usually takes a full day to play, including a debriefing and discussion session at the end.

In the game the tables and chairs in the training room are set out following the layout of the main canal and eight tertiary units. The available water (represented by blue counters) at the river intake is distributed by the main system management staff to the eight tertiary units within the system, working down the system from top to bottom. The farmers take their allocation of water from the main system managers and distribute it among their four fields.

The farmers have to decide on the crops to be grown on each of their four fields (based on data provided on crop costs, yield response to water and prices), and then use yield response to water graphs to decide how to allocate the available water among the four fields. Water is generally in short supply, so the final crop yield is dependent on water allocation decisions made in each of the three crop growth stages.

The main system management staff have to make decisions on the water allocations to each tertiary unit based on different water allocation procedures for each allocation round. In the first round allocation is in proportion to tertiary unit command area, in the second round in proportion to irrigation water demand, and in the third round based on demands and actions at the tertiary unit gate by the farmers. In the third round farmers can override the allocation by the main system managers by 'breaking' padlocks on the gates and adjusting the gate settings to suit their needs. These actions tend to benefit the upstream farmers, and lead to (simulated) conflict between head and tail-end farmers.

The exercise serves to demonstrate the interactions between the main system management staff and the farmers, and the impact that their decisions and actions have on farmers and agricultural output from individual tertiary units within the system. It also raises issues of system maintenance, corruption, water trading, value of irrigation water, yield response to water, performance assessment and inter-personal relations, both between the main system managers and farmers and between the farmers themselves.

of methods for communicating with our trainees, the least effective method of which is acknowledged to be through lecturing, the most commonly used medium! Research has shown that:

We learn	We remember
3% through taste	10% of what we read
3% through smell	20% of what we hear
6% through touch	30% of what we see
13% through hearing	50% of what we see and hear
75% through sight	80% of what we say
	90% of what we say while doing

The following Chinese proverb summarizes this well:

If I hear, I forget
If I see, I remember
If I do, I understand

From the above it is clear that training must actively involve the trainees in the training process as much as possible. Lectures are the most widely used medium for conveying information; they can be significantly enhanced with the use of visual images and can be reinforced through practical exercises that put into practice the taught word.

Establishing an effective learning environment

The following points are useful in setting up an effective learning environment for trainees.

1. Trainees learn more effectively when there are clear and explicit objectives. Setting clear objectives for a training programme helps both the trainees and the trainers. Trainees are more relaxed if they have been given a sense of the end point and how they are going to get there.
2. Trainees are more comfortable and prepared to accept the training when the training builds on their existing understandings, knowledge and skills.
3. Learning is most effective where the trainee can see an application for the knowledge and skills involved.
4. Variety in the training methods used helps sustain the trainees' level of interest and concentration. It has been found that after a period of about 20 min listening to a lecture a person's concentration starts to fall off. The trainees' attention can be regained by breaking up the lecture with a slide show or a brief discussion session. There should be variety within a training programme, mixing lectures with exercises, demonstrations, video films, exercises, field visits and the like.
5. Exposing the trainees to role models can be beneficial. Extension programmes often make use of progressive or lead farmers to encourage other farmers to adopt new techniques.
6. Learning is enhanced in supportive, safe and non-threatening environments. Management games, role plays and simulation models are useful in this context.
7. Trainees should be encouraged to work together during the training course and to share knowledge. Learning should not be competitive.
8. An autocratic, top-down approach by trainers alienates trainees and is counterproductive to learning. In contrast trainees are encouraged to learn, participate and share knowledge in a participative environment. Participation starts with the trainer, in the design of the course and his/her behaviour on the training course.
9. Trainees learn more readily when they can see the impacts arising from the application of their learning. Support and feedback from the trainer are important in this context.
10. Learning is reinforced through practice.
11. Repetition enhances learning and retention of taught material. Summaries should always be given at the end of a lecture, an exercise or a field visit.

Communication and learning

Communication is a means of exchanging information; it should be a two-way process (MacDonald and Hearle, 1984).

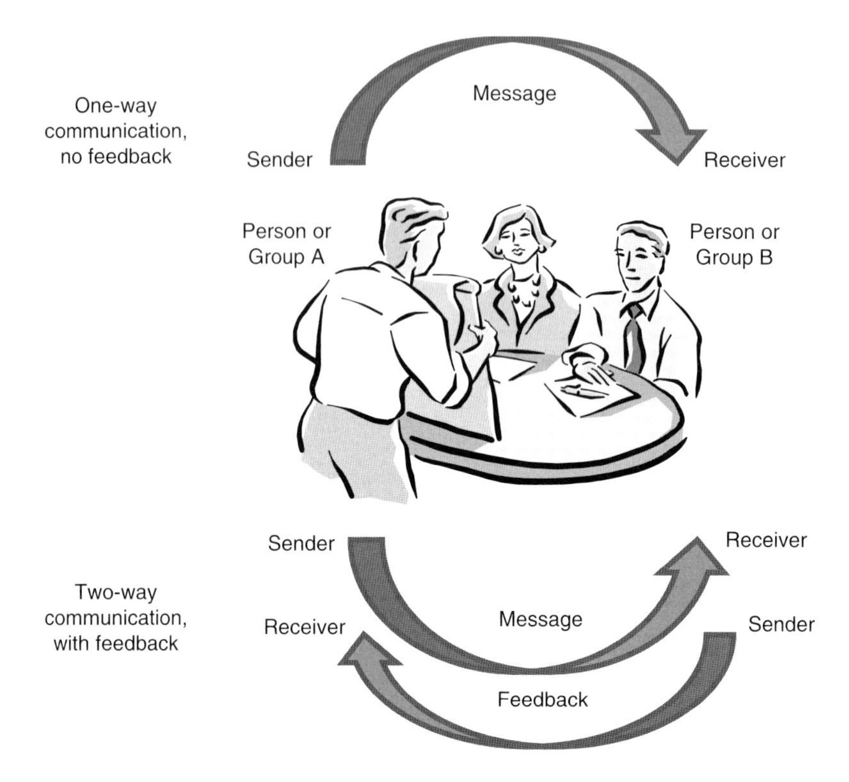

A person or a group should send as well as receive messages in order to communicate. One-way messages represent poor communication, and can appear to the receiver as being commands rather than messages. Communication means sending messages and listening to, and in many cases acting upon, the response.

In wanting to communicate with water users we need to consider the following:

- it is important to get water users to express their needs and wishes and to understand the context within which these are set;
- water users have a considerable amount of experience, which should be respected;
- water users will take on board new concepts and ideas that are useful to them;
- training and extension messages should be tailored to meet the needs of the water users, and the context within which they live and work;
- care should be taken not to confuse government information needs with those of the water users;
- there is a variety of ways of imparting ideas and information, some of which are more useful than others in certain contexts;
- people accept most easily ideas that they think are their own, or which are based on their own understanding of reality;
- water users are not a homogeneous group, they are made up of many different groups with different needs for information and different motivations.

Similar considerations apply when communicating with system O&M staff, more particularly:

- communication between senior and junior levels of staff can significantly improve job satisfaction and levels of motivation at the junior level;
- field staff have a wealth of knowledge about actual field conditions and the issues affecting farmers, and they can pass this information on to senior management.

A useful exercise for demonstrating the importance of communication is outlined in Box 7.2.

Box 7.2. Communication Exercise

This simple yet effective communication exercise can be used to show the value of two-way communication. It is applicable for all levels of trainee, but especially for extension staff and O&M field staff working with farmers.

Equipment and materials

The exercise requires a table and chair for each pair of trainees, and two sets of five identical objects such as a pencil, a biro, a ruler, a piece of paper or card and a rubber.

The exercise

The trainees are divided into pairs or small groups of five or six people and assigned to pairs of tables and chairs in different parts of the training room. The chairs are placed back-to-back with the tables in front of them and the five objects placed on each table. Two trainees are asked to sit down, back-to-back.

One of the trainees is designated as the Sender of the messages, the other as the Receiver. On the Sender's table the five objects are set out in some positional order by the Trainer.

The Sender is then asked to describe to the Receiver the layout of the five objects on his/her table. The Receiver must then endeavour to replicate this layout on his/her table using the same objects following the instructions received. *Neither party is permitted to turn round and look at the other's table, nor is the Receiver allowed to say anything other than 'Message received'. If other members of the group are watching they should not interfere or give advice.*

The first part of the exercise ends when the Sender has passed on all the information he/she wishes. The Sender and Receiver can then turn round and see how closely the layout on the Receiver's table resembles that on the Sender's table. Often it bears no resemblance at all!

In the second part of the exercise two new trainees sit down back-to-back and the Trainer rearranges the objects on the Sender's table into a new layout. The exercise is repeated, except that this time the Receiver can ask questions to clarify the messages received. At the end of the exercise the trainees can turn round and view the outcome; generally it is an improvement on the first effort.

The exercise demonstrates:

- the value of receiving feedback in achieving the desired outcome (namely, the same layout on the Receiver's table as on the Sender's);
- how different the Receiver's image of the message can be from that of the Sender if the Sender does not ask for and receive feedback.

Steps in Establishing Training Programmes

The following steps and associated activities can be identified in setting up a training programme for staff on an irrigation and drainage scheme (Burton, 1986, 1988):

1. Establish training needs;
2. Identify suitable trainers;
3. Plan training programme and course structures in detail;
4. Prepare and test training material;
5. Implement training programme;
6. Monitor and evaluate the training given.

Step 1: Establish training needs

The first step of preparing a training programme is to identify who needs to be trained, and what they need to be trained in. If this is not specified within the terms of reference for the training specialist, then he/she must carry out a detailed Training Needs Assessment (TNA's are covered later in this chapter).

The TNA involves carrying out a survey of all staff within each organization under review (be it the irrigation service provider, water users association (WUA), extension service or related organization). This survey will ascertain the structure of the organization, the numbers of staff at each level, age, educational background, current and required capabilities, etc. The process will require collecting secondary data (such the names, age, educational background, etc.), which will be held by the human resource department, and primary data through interviews with individuals or groups at different levels within the organization. For example additional information required might include procedures and operations manuals that stipulate the duties and responsibilities for each position within the organization.

The output of the TNA will be a Training Plan, which will detail who will be trained, and how that training will be accomplished. A useful structure of a Training Plan is provided by answering the following questions.

- Why?
  - Why is the training required?
- Who?
  - Who will be trained?
  - Who will carry out the training?
- What?
  - What are the objectives of the training programme/course(s)?
  - What are the desired outcomes?
  - What are the training needs?
  - What are the key features of the training programme/course(s)?
  - What are the training topics?
  - What will it cost?
- How?
  - How will the training be imparted?
  - How will the course be structured?
- Where?
  - Where will the training be carried out?
- When?
  - When will the training be carried out?

The Training Plan will outline the scale of the training to be carried out and the resources required. The variables which influence the scale and extent of the training proposed include:

- the total time available for the training programme;
- the time required for training each trainee;
- the number of trainees;
- the availability of suitable trainers;
- the costs of training;
- the budget available.

Step 2: Identify suitable trainers

If the training is to be carried out in-house it is essential at the start of the training process to identify suitable trainers. Great care should be taken in the selection of these trainers, as the success of the training programme will rest mainly on their shoulders. Criteria that may help in identifying suitable trainers include:

- experienced at the level at which the training is being carried out;
- having an interest in training and education of others;
- an ability to communicate and empathize with people;
- having an interest and willingness to share information and knowledge with others;
- organized;
- energetic, but also patient!

As with other disciplines training is not something that everyone is capable of, some people are good at it, some are not. It is worth noting that there are also different roles on a training team, some people are interested in training but prefer a supporting, rather than lead role. Such people can be invaluable in managing the organization and running of the training programme, while the front-line trainers get on with the job of imparting knowledge, skills and understanding.

Step 3: Plan training programme and course structures in detail

The Training Plan should provide an outline of the training required by answering the questions set out in the section above. It will provide sufficient detail for management to decide on the training priorities and to allocate a training budget. Following this, individual training courses can be prepared in detail. For this the set of questions outlined above can be elaborated upon (Table 7.3).

Training methods

The trainees will have been identified in the Training Plan, together with an outline of the training course(s) required. In the detailed planning of each training course decisions will need to be made on the training methods to be used. These can include:

- lectures/presentations;
- case studies;
- practical exercises;
- site visits or field trips;
- discussion groups;

Table 7.3. Typical questions to ask when establishing a training programme or course.

Why?	Why is the training required?	Is it to provide a more efficient service? Is it for induction of new staff?
Who?	Who will be trained?	Where is the training most required? What will be the scale of the beneficial impact of the training for the selected group?
	Who will carry out the training?	Will the training be carried out in-house or contracted out? How capable/experienced are the trainers? Will training of the trainers be needed?
How?	How will the training be imparted?	What are the options – lectures, exercises, field practicals, case studies, management games, role plays, etc.?
	How will the training course(s) be structured?	Will it take place all in one session or in several sessions over a period of time? Will it be all classroom-based, or a mix of classroom and field visits, or classroom and on-the-job?
What?	What are the objectives of the training programme/course(s)?	What is the overall objective? What are the specific objectives?
	What are the desired outcomes?	What will the trainees be able to do following the course that they could not do before?
	What are the training needs?	What change is required? Where are the gaps in performance that training can assist in closing?
	What are the key features of the training programme/course(s)?	Is the course to be theoretical or practical? Is it to be a one-off course, or multiple courses?
	What are the training topics?	What knowledge, skills and attitudes do the trainees need to have at the end of the course? What do they know now, and what do they need to know/be able to do?
	What will it cost?	What are the travel, accommodation and subsistence costs for the trainers and trainees? What resources are required – classrooms, vehicles, equipment – and how much does it cost?
Where?	Where will the training be carried out?	Is it to be held in a central training centre or is it to be mobile and travel round regional/field offices? Is the learning outcome likely to be higher if the trainers travel, rather than the trainees? What are the benefits/costs of different approaches to where the training is held?
When?	When will the training be carried out?	What are the seasons/periods to avoid for training? When are the trainees least busy and most likely to be able to attend the training? How urgent is the training requirement?

- individual guided reading;
- remote study.

It is necessary to combine the above methods into a balanced training course to suit a given training group. The correct balance will mainly depend on the trainees' prior level of education and their work function. A particular point to note in training for irrigation service staff and farmers is that the trainees are generally very knowledgeable about their own areas, often set in their ways, and not prepared to take risks. They will often only accept change and be motivated to adopt new procedures if they understand and accept the logic of it. Therefore a key of such training courses is to make them participative; involve the trainees in discussion, draw out their ideas and suggestions for improvements, and allow them to contribute to the training programme and in assisting others on the training course with the benefit of their experience.

Course and programme timetables

Individual training course timetables can be established once the training method and course content have been decided. A plan will need to be drawn up and made available to all those involved in the training programme prior to its implementation. A bar chart is the most effective form of plan, showing the dates of the course(s), the location and the number of trainees. Having decided on the dates it is important to notify the trainees in good time in order that they can make the necessary arrangements to attend.

A particular feature of training in irrigation and drainage is the timing of the training in relation to the irrigation season. Some training has to be carried out during the off-season when O&M staff, WUA management and water users are not busy irrigating, while other training is best carried out during the irrigation season. Theoretical training can be carried out in the off-season, while practical training such as that related to flow measurement or field irrigation is best carried out in-season. If carrying out the training in-season the training period can be divided into several sessions (e.g. 1 or 2 days a week) such that participants can get on with their main work activities on the other days.

As noted earlier it is essential to provide variety in the presentation of the training material. A full day of lectures to people used to doing field work (such as water masters) can prove very tedious, for both the trainees and the trainers. It is also important to provide short breaks with refreshments, and a good midday break with lunch provided.

More detailed information on planning, organizing and costing training courses is provided later in this chapter.

Step 4: Prepare and test training material

Preparation of training material can be a difficult and time-consuming process. Training material might be required for the following.

1. *The trainers*: The trainers will need handbooks to use as references for their lectures, practical work, field visits, etc.
2. *The trainees*: Trainees will require notes for each subject taught.
3. *Presentation*: Training material may be required in the form of audio-visual displays, overhead projection slides, or computer-and-projector based presentations.

The quality and clarity of the presentation material are obviously key features here, especially in the material used for lectures/presentations/exercises and the trainees' notes.

The value of good graphics cannot be overemphasized; a good technical illustrator is invaluable. Cartoons can be a powerful method of conveying a message.

Once the training material has been prepared it is as well to test it, and the format of the training course, on a small scale prior to the start of the training programme. The trainers must be careful to remain objective and to welcome constructive criticism. Changes are both easier and cheaper to make at this stage.

In some cases it can be useful to run through a training programme for middle or junior level staff with senior management in order that they can understand the training that will be carried out. This allows senior management to comment on the proposed training (the course structure, content, training methods, etc.) and to develop a sense of ownership for the training.

Step 5: Implement training programme

The beginning of a new training programme is always difficult, but provided the organization has been well thought out and planned the problems encountered can be minimized. The early days of a new training course can be stressful for the trainers, but as they gain experience so their confidence grows (another reason for having trial runs).

Time should be allowed at the start of each training course for registration and 'settling down', followed by a welcome and introduction to the training course. The introduction should detail the objectives of the course and the desired outcomes, and outline the structure and contents of the course. If the course is residential, details of accommodation, per diems, etc. should be discussed at this stage. It is important to allow sufficient time for the introductory session. In some cases a pre-course test may be given to establish the trainees' current knowledge related to the topics in the training course. The same, or similar test, is then given at the end of the course to measure the change in learning achieved.

Records should be kept of who has attended each training course; this is especially important for training programmes where there are a number of courses, such as for establishing water users associations.

An integral part of the training course will be to obtain feedback from the participants at the end of the course. This feedback can be obtained through a questionnaire and/or discussions on the training given. Feedback forms can be designed to cover a number of aspects, including training content, training methods, facilities, refreshments, accommodation, etc., together with requests for suggestions for improvement.

It is inevitable that the training material will require amendment in the light of experience using it. If the changes required are relatively minor it may be best to document proposed corrections and amendments over several courses and then implement the changes all in one go.

Step 6: Monitor and evaluate training given

The training programme will require monitoring once it has started, to check that it is being implemented as planned. The effect of the training given can be monitored and, if necessary, changes made to the training material or course structure if the desired levels of ability are not being achieved.

The monitoring exercise should be carried out by members of the training team in order that they see for themselves the effect that their work is having. Post-course monitoring also helps to reinforce the training given as the trainees perceive that training is a continuous exercise, not just restricted to a training course.

Evaluation of the training programme is a separate exercise from monitoring as its objective is to assess the success of the training programme in bringing about a positive change in the trainees' behaviour. Evaluation requires an objective and critical assessment of the training given.

In order to be able to evaluate the programme the amount of change brought about needs to be measured. For this reason it is advisable to prepare the evaluation procedures at the start of the training programme, and to test the situation before and after training using the same parameters.

Evaluation should not be carried out by the same person or team that planned, prepared or ran the training programme or course. It should be carried out by an independent evaluator (or team), preferably by someone who has experience of training for similar situations. An important part of the evaluation will be measuring the success of the training against the objectives and outcomes set by those who designed and implemented the training programme or course.

If the evaluation exercise finds that desired levels of change or outcome are not being achieved following the training course, then the course should be amended and improved. As noted above, training is different from education; it sets out to bring about change in a person's behaviour. If it fails to do that it is the training course that has failed, not the trainees, and the course must be amended accordingly.

Carrying Out a Training Needs Assessment

A TNA is a structured way to analyse the training needs of an organization. It provides the information required to enable planning of relevant and cost-effective training.

There are seven stages in carrying out a TNA (EDI, 1994):

1. Select assessors for the process;
2. Define performance shortcomings;
3. Identify how training can help;
4. Set performance and skills standards;
5. Determine current capabilities of personnel;
6. Determine gaps in capabilities;
7. Determine who to train and in what.

Stage 1: Select assessors to carry out the training needs assessment

The first stage in carrying out a TNA is to decide by whom it should be carried out. If it is related to a project the TNA is often carried out by an externally recruited Training Specialist. In a large organization with a well-established Human Resources and Training Department the TNA could be carried out by someone or a team from within the organization.

One benefit of employing an external person or team is that they may be more objective, and may be able to elicit more information from employees on issues that they face within the organization.

Stage 2: Define performance shortcomings

The next stage in carrying out a TNA is to look at the purpose and objectives of the organization and identify areas where there are performance shortcomings, or where training can have a significant impact on performance. It is important, for example, to have well-trained accountants in water users associations.

It is useful at this stage to list out the objectives of the organization and to then identify the role and contribution made to these objectives by categories of staff at each level within the organization (i.e. identifying who does what, and what impact their actions have on performance). It will also be important at this stage to identify what is meant by 'performance', and how it is measured.

Stage 3: Identify how training can help

Shortfalls in the performance of an organization can be caused by factors that are not amenable to improvement through training. Four possible management levels can be identified within an organization.

- *Policy level*: This level is least amenable to solution by training. It involves policy issues such as organizational structure, staffing, salaries, job functions, etc.
- *Institutional level*: This relates to the organizational structure, functions and rules and regulations at different levels within the organization. It may be amenable to improvement through training interventions, especially if associated with organizational restructuring.
- *Functional level*: This level (administration, technical or finance departments) is amenable to improvements through training interventions, particularly of senior and middle level management who may be responsible for reorganization/restructuring of their departments.
- *Individual level*: This level is very amenable to improvement through training interventions, though recruitment policy may affect the level from which the training must start. The TNA will look at individual and group capabilities and needs to assess skills, knowledge and attitude, level of motivation, salaries and conditions, etc. It will also need to take account of the quality of the personnel management processes and procedures within the organization.

In carrying out a TNA, account needs to be taken of the influence of these different management levels on performance, and to separate out where training can, and cannot, help. It also needs to identify where training will be adversely affected by factors at other levels; for example if training is required as part of a project-funded organizational restructuring programme but is not fully supported by senior management.

Stage 4: Set performance and skills standards

A key part of the assessment process will be to establish the required level of performance for each individual or group. Having detailed job descriptions is essential at this stage; from these the required knowledge, skills and attitudes required to perform the defined job functions can be identified.

Relevant levels of management should be involved in the determination of the required standards for their respective teams, and care should be taken to ensure that these standards are realistic and achievable.

Stage 5: Determine current capabilities of personnel

This stage of the process can be very time-consuming, and a policy decision needs to be made at the outset as to whether or not all staff that are to be included in the training programme need to be interviewed and their current capabilities assessed. For key positions where there are a small number of staff then individual interviews are required, for positions with large numbers of staff (such as water masters or gatekeepers) then interviews can be held with a sample number of staff.

The assessment of each individual's (or group's) capabilities needs to be made against the defined job description.

Stage 6: Determine gaps in capabilities

The 'training gap' is the difference between the required capability and the current capability. Correct identification and assessment of this gap is obviously central to the design and outcome of the training programme. Failure to correctly identify this gap in capabilities can be costly in terms of time, money and the organization's credibility.

Stage 7: Determine who to train, and in what

An equally important process to identifying the training gap is the identification of appropriate measures to close the gap. Different approaches will be required for different positions and levels of staff (Table 7.4). Where training is to be carried out for a number of levels and disciplines within the organization (technical, financial, administrative) it will be important to select priority levels and subjects, and to prepare an integrated strategy for all training.

As mentioned previously, the TNA culminates in the Training Plan, where the required training, resources required, timing, etc. are defined. Where several courses are required they can usefully be summarized in summary sheets, as shown in Fig. 7.1.

Organizing and Running Training Programmes and Courses

Training methods

The type of training methods used has a significant impact on the effectiveness of the training given. The theoretical effectiveness of different training methods is usually stated as:

Most effective	Least effective
Practical work	Lecture

- demonstrations
- management games/role plays
- practical exercises
- television or slides
- still pictures
- books and print

There are no hard-and-fast rules for which training method will work best for a given group of trainees. Different methods have to be adopted to suit each particular group. Some useful methods are presented in Table 7.5 and Fig. 7.2.

It is important to vary the method used within the training programme. Adoption of a variety of training methods within a training course breaks up the monotony of lecturing, maintains the level of interest of trainees, and increases the degree of learning.

Small group discussions

If they are structured and organized well, small group discussions (Fig. 7.21) can be a powerful training tool for adults. The following is a proven technique for small group discussions

Table 7.4. Examples of irrigation and drainage staff categories and possible training needs.

	Category	Personnel	Possible training needs
Management	Senior Management	Chief Executive Chief Finance Officer Chief Engineer	• Corporate planning • Strategic planning • Leadership • Financial planning and management • Communication • Human resource development • Media and public relations
	Personnel	HRD Manager Senior Trainer Trainers	• Human resource development • Personnel management • Interviewing skills and staff selection • Performance appraisal • Job descriptions • Staff motivation • Communication and liaison • Industrial relations and negotiation skills • Health and safety • Training needs assessment • Training methods and skills
	Accounting and Finance	Finance Director Chief Accountant Accountants Clerks Cashiers	• Financial planning and budget preparation • Financial management systems • Accounting systems • Auditing • Procurement
	Administration	Chief of Administration Clerks Secretaries Storekeeper	• Organizational procedures • Office procedures • Filing and record keeping • Maintaining inventories
	Information Technology (IT)	Senior IT Specialist IT Specialists Data processors	• Management information systems • Geographic Information Systems • Remote sensing • Network systems
	Legal	Legal Specialist	• Water legislation • Contract law
	Design Office	Chief Design Engineer Design engineers Quantity surveyors Draughtsmen Surveyors	• Design procedures • Surveying and costing work • Tendering and tender documentation • Computer-aided design • Contract management and supervision
Operation	Head of Operations	Operation engineers Headworks/structure operators Pump station operators Water masters Gatekeepers	• Determining irrigation water demands • Seasonal planning • Scheduling water supplies • Liaising with water users • Determination of service fees • Water saving measures • Rotation of water supplies • Canal regulation and operation
Maintenance	Head of Maintenance	Maintenance engineers Overseer Works inspector Machine operators Mechanics Artisans	• Identifying maintenance needs • Maintenance processes and procedures • Costing maintenance work • Tendering and tender documentation • Maintenance machinery • Workshop management

<table>
<tr><th colspan="2">MOM PROJECT – TRAINING COURSE INFORMATION SHEET</th></tr>
<tr><td>Course title: On-farm Water Management and Maintenance

Course ref: WU02

Venue:
/01: Pilot Area 1
/02: Pilot Area 2
/03: Pilot Area 3
/04: Pilot Area 4
/05: Pilot Area 5
/06: Pilot Area 6

Organizer: Senior Training Officer</td><td>Duration: 3 days

Date(s):
/01: 10–12 October
/02: 17–19 October
/03: 31 Oct –1 Nov
/04: 21–23 November
/05: 28–30 November
/06: 5–7 December</td></tr>
<tr><td colspan="2">Target group: WUA Water Managers and selected farmers in demonstration areas
No. of trainees: 20 per course   Fee: n/a ($/trainee)</td></tr>
<tr><td colspan="2">Objectives
The course is designed to enable participants to:
• calculate irrigation water requirements
• understand the principles of efficient water management
• use appropriate flow measurement techniques
• operate gates to control discharges to required values
• organize and monitor appropriate maintenance activities
• record and report on gate adjustments and discharges
• identify and report maintenance requirements and take appropriate follow-up actions</td></tr>
<tr><td>Content
• Crop water requirements
• Water management at village and farm level
• Rotation of water supply
• Reducing water losses
• Flow measurement
• Planning and scheduling supplies
• Data collection and reporting
• Maintenance
• Follow-up activities
• Action Plans</td><td>Trainers
• WUA Training Specialist
• O&M Training Specialist</td></tr>
<tr><td>Training methods
• Lectures
• Exercises
• Field demonstrations and visits
• Videos, slide sets, photographs</td><td>Timetable
Morning session: 08.45–12.15
Lunch: 12.15–13.15
Afternoon session: 13.15–16.45
Tea/coffee breaks in both sessions</td></tr>
<tr><td>Equipment/resources required
• Laptop computer and overhead projector
• Projector screen
• 35 mm slide projector
• Flip chart
• DVD/VCR and colour monitor/TV
• Flow measurement devices
• Transport for field visits</td><td>Evaluation methods
• Pre-course questionnaire
• Test at start of course
• Test at end of course
• Post-course questionnaire</td></tr>
</table>

Fig. 7.1 Example of a training course summary sheet.

Table 7.5. Possible training methods.

Method	Description/application
Lectures	Useful for conveying a lot of factual information in a short time
Case studies	Useful for management training where problem solving is part of the job. Useful for demonstrating the application of lectured material
Classroom exercises	Used to reinforce lectures and enhance understanding and retention of information. Useful for practical tasks such as how to perform calculations or fill in forms
Field exercises	Exercises carried out in the field which relate to required practice in the field, such as gauge reading or discharge measurement
Field trips	Visit to the field to reinforce taught material
Study tours	A series of visits over a period of several days to places or organizations to see and understand how the organization performs or how various functions are carried out
Exchange visits	Exchange visits between organizations to see how others are performing similar tasks. Useful for water users associations, with visits to best-practice associations
Demonstrations	Practical demonstration to reinforce taught material. Trainees learn from observation
Classroom practical	Practical work in laboratory or classroom to reinforce taught material. A typical example would be measurement of the water-holding capacity in a sample of soil
Discussion groups	Opportunity for exchange of views between trainees and trainers. Very useful with adults, encourages participation in training programme
Simulations	Develops skills in a non-critical environment. Useful for situations where risk is involved in the real-life situation. Can also be used to show complex relationships, such as the movement of water in soil using computer graphics
Role play	Places participants in a situation where they have to adopt a role. Useful to enable trainees to see a situation from another person's viewpoint
Models	Models can be a useful medium for explaining the workings of systems. Especially useful for technicians or farmers
Management games	Management games can be used to good effect in training. They have gained wide acceptance, and are commonly used in management training programmes. They allow trainees to learn for themselves in an enjoyable, non-threatening environment

which facilitates the exchange and sharing of views among a group of trainees. The key elements of the method are the following.

- A question, or questions, is (are) posed to the group on a key topic.
- Participants contribute through writing down their ideas on card during a period of 'quiet time' at the start of the exercise. Because the ideas are presented on card rather than spoken in discussion, all members of the group are able to contribute and the discussions are not dominated by a few members of the group.
- Ideas are displayed and retained for discussion (by being on cards).
- Further ideas can be added on new cards as the group discussion progresses.
- The cards can be rearranged into a suitable structure for presentation of the ideas to other groups.

The method comprises the following main steps.

1. The Trainer/Facilitator sets a subject for group discussion.
2. The participants are divided into groups of four or five persons.
3. Within each group each participant considers the task set and identifies the key elements for consideration and discussion by the group. A minimum time allocation of 5–10 min 'quiet period' is allotted for individual thought. No discussion is permitted in this period.
4. Each participant within the group writes down their individual ideas on card; keywords only, one or two keywords to one card, using large lettering. The Trainer/Facilitator may limit the number of ideas, say to five per person, in order to limit the time taken.

Fig. 7.2. Examples of different types of training methods. (a) Classroom-based practical exercise during a 6-day (2 days per week over 3 weeks) training course for irrigation agency water masters held in their office (Indonesia). (b) Irrigation agency water masters working in pairs on the practical exercise shown in (a) related to planning and monitoring water allocation on the main system (Indonesia). (c) Minivan used to transport the trainers and their equipment to the irrigation agency offices, also used for taking trainees on field trips (Indonesia). (d) Training water masters in the field about flow measurement and measuring structures (Indonesia). (e) Explanation for Water Users Association and Federation operations and management staff on the theory of discharge measurement prior to a practical exercise in a nearby canal (Albania). (f) Questions and discussions in the field between the trainer and trainees on discharge measurement (Albania). (g) Field exercise to determine the infiltration rate in a field of maize (Tanzania). (h) Models used to demonstrate different in-field layouts to show benefits of in-field channels for more efficient water distribution (Bangladesh). (i) The Trader in the Irrigation Management Game negotiating with a farmer for credit (Egypt). (j) Farmers negotiating and trading in water in the Irrigation Management Game (Egypt). (k) Two trainees in a role play about a conflict situation between two farmers over access to water through the upstream farmer's plot of land (Bangladesh). (l) Small discussion group using cards to capture the key points of their discussions.

Fig. 7.2. Continued.

5. Each participant then pins or sticks his/her cards to a notice board (or on to the wall). This does not have to be done in any order; each participant can put up his/her cards when he/she is ready.
6. The group then discusses the ideas raised, and adds cards with further ideas, if required.
7. The group may rearrange the cards into groups or clusters if wished. Different coloured or shaped card can be used to highlight different categories, if required. If the cards are stuck on to large sheets of plain paper, then annotations for groupings can be drawn on the paper around the cards.
8. The group may formulate a viewpoint for the whole group, or there may be divergences of opinion within the group. The group selects one member to make the group presentation.
9. Each group presents its findings to a plenary session of all the groups.

Requirements for the exercise

The materials detailed in Table 7.6 below are required for the exercise. Quantities are given for a typical session with 20 people divided into four groups.

Training material

Training material is an essential component of any training programme or course. Training material has to be prepared and requires differing degrees of planning, time and effort, and different levels of resources, both for preparation and use. The role and merits of several common types of training material are outlined below.

Film, video and television

Film, video and television are powerful media for imparting messages (Fig. 7.3). Television is particularly important nowadays for changing attitudes within society. With the advent of relatively inexpensive, good-quality video cameras, the in-house production of video films has become a feasible option for many training organizations. The approach of the video film can vary, depending on the

Table 7.6. Materials required for small group discussion exercise.

		Quantity (no.)				
No.	Item	Per participant	Per group	Number of participants	Groups	Total required
1	Coloured card (99 mm x 210 mm) – blue, pink, green, yellow, orange, etc. Cards can be made from A4 card (297 mm x 210 mm) cut into three equal portions (99 mm x 210 mm)	15	n/a	20	n/a	300
2	Marker pens – one broad-nib marker pen per person, with some spare	1	n/a	20	n/a	20+5 spare = 25
3	Brown paper – a large sheet of brown paper (approx. 1.5–2 m x 1.2 m) is fixed to the softboard or the wall first; this allows the participants to draw linkages on the paper, if they wish, between cards or groups of cards	n/a	4–5	n/a	4	16–20
4	Pins, Blutac or masking tape – the cards need to be fixed to the softboard or the wall. If a softboard is available then map pins (with a large plastic head) can be used; if a wall is used then Blutac or masking tape can be used	15	n/a	20	n/a	300
5	Softboard and stand – 2.5 m x 1.2 m (approx.) board to pin cards	n/a	2	n/a	4	8

n/a, not applicable.

(a)

(b)

Fig. 7.3. Use of films and cartoons for training. (a) Filming a sequence for a training video about forming water users associations (Indonesia); (b) cartoon used for training on maintenance of field channels (Indonesia).

requirement. It could, for instance, be a documentary showing how the water flows from the pump station to the fields. Alternatively it could show the same detail, but incorporate some human interest by telling the story as a drama of a farmer who wishes to find out where his water comes from and makes a journey to the pump station.

Overhead projection

Overhead projection can be a very valuable and versatile presentation medium. A major advantage is that text or diagrams can be used equally easily.

Two forms of overhead projection are available:

- using prepared acetate sheets and an overhead projector (OHP);
- using presentation software (such as Microsoft PowerPoint), a computer and projector.

In the first case a sequence of transparencies can be pre-prepared on acetate roll, or separate acetate sheets, and reused at different times. In the second case the training material can be prepared relatively easily using standard presentation software, and can incorporate colour photographs and video. The training material is also relatively easy to store (on disk or memory stick) and to disseminate or transport.

A drawback with overhead projection in some locations is the lack of a reliable and stable electricity supply, though this can be overcome by having a generator as backup. Care has to be taken with the (expensive) projector bulbs, which are susceptible to vibration damage.

Photographs

Photographs are an essential part of a training programme. It is said that a picture is worth a thousand words, so a well-organized library of colour photographs is an invaluable training resource. At the outset of the training programme a detailed list of the required photographs should be prepared. The library may take some time to build up, especially where the required photographs are of seasonal events such as planting and harvest.

Care has to be taken with taking photographs. There are several points to watch.

- Limit the amount of detail in the photograph.
- Take time to arrange the photograph and to limit the number of people in the photograph. Too many people in the photograph can be confusing.
- Get close up to the subject and take care to fill the full frame of the picture. The impact will be lost if the image is too small in the photograph.
- Bright and sunny weather gives a more vibrant picture, but take care to keep the sun behind the photographer, and to take

account of shadow caused by the sunlight. If necessary use a flash to highlight elements of the picture in shade.

- Take care to record and catalogue the photographs with date, location and subject matter.
- Take care to plan out the sequence of photographs that are needed. A well-organized sequence can convey a lot of information; a poorly organized sequence may well confuse the audience.

Real objects

Real objects are the best training aid. For items such as pests and diseased crops specimens can be collected and stored in specimen jars for use in the classroom. For items such as gates and structures, field visits should be made.

Models

Models can be useful to represent something that may be too large or complex to see in practice (such as the whole of the scheme area) or to simulate the operation or behaviour of the real object. For trainees with a limited education, physical models are most relevant; for more educated audiences computer models using colour graphics can be effective.

Simulation exercises or role plays

Simulation exercises or role plays such as the previously mentioned Irrigation Management Game can be a powerful medium for conveying otherwise difficult concepts and understandings. They can, however, be difficult to formulate and to develop into a functional training tool, and will often require many trials and iterations to get right.

Flip chart

The flip chart is a valuable training aid for making records of points raised in discussion. They can also be pre-prepared to convey a message or tell a story. The number of sheets should be limited to about ten, too many and the audience will become bored. They are valuable where an OHP cannot be used, such as in villages or the field where a reliable supply of electricity may not be available.

Posters

Posters are useful to convey and remind people of a specific message. They must attract people to read them; the use of diagrams and colour is helpful in this respect.

Presentation skills

Successful presentation of training material requires planning and adopting the right approach. Teaching adults differs from the teaching of children in several important respects.

1. Adults have to be motivated to learn. They are busy people with responsibilities, they cannot afford to 'waste their time' learning if it is not of benefit to them.
2. Adults have a wide range of knowledge and experience, in some respects they will know more than the trainer about certain issues. They must be treated as equals and with respect. Trainers who lecture 'down' to adults will not remain in the job long!
3. Adults have questions that they want answered, they expect to be able to discuss issues and obtain answers.
4. Adults can often learn as much from each other as from the trainer, if the right environment is established. This raises the issue of the role of the trainer; he/she should see themselves as a *facilitator of learning*.

Planning of presentations

Planning is the key to successful presentations. It involves the following processes:

- selecting the topic to be covered;
- establishing the objectives;
- writing a lesson plan based on the objectives and the time available (Box 7.3).

When preparing presentations it is often helpful to ask the following questions.

- Who are my audience?
- What am I going to teach them? What do they need to know?

Box 7.3. Elements of a Lesson Plan

- Preparation:
  - title;
  - lesson content;
  - objective;
  - location(s);
  - required seating arrangement;
  - equipment and materials required;
  - visual aid equipment required;
  - visual aids required;
  - training material;
  - estimate of times required for each part of the presentation.
- The presentation:
  - introduction;
  - set presentation in context within the training programme;
  - ascertain trainees' experience in the subject area;
  - core of presentation;
  - trainee participation or activity;
  - summary;
  - outline next related presentation and relationship to current presentation.

- What will they be better able to do at the end of my presentation?
- How am I going to present it?
- How much time have I got (or do I need)?
- Have I got enough material? Have I got too much material, can I cut back on extraneous material?

For the presentation itself there are three key stages.

1. *The introduction*: Tell them what you are going to tell them.
2. *The body of the presentation*: Tell them.
3. *Summary*: Tell them what you told them.

Repetition is an important element of training; always remember to draw your presentation to a close with a summary.

Organizing field visits

Field visits should be well-prepared and have specific aims. Preparation for a field visit includes:

- defining the requirements for the field visit;
- locating a suitable site;
- making necessary arrangements for the visit (notifying personnel, arranging transport, etc.).

The participants should be handed details of the field trip to include:

- the intended purpose;
- details of activities that the participants will be required to carry out;
- supporting material for these activities (discharge data collection forms, calibration forms, field observation forms, etc.).

Checklist for setting up, running and costing training courses

The key stages and resources for organizing and running a training course are:

- preparation of the training material (Table 7.7);
- pre-course organization (Table 7.8);
- running the course (Table 7.9);
- staffing, equipment and materials (Table 7.10);
- cost items (Table 7.11).

The activities and resources required in each of these stages are outlined in the form of checklists in Tables 7.7 to 7.11.

Table 7.7. Preparation of training material.

Main activity	Subsidiary activity	Status		
		Draft	Reviewed	Final
Course outline and content	Decide on course outline and syllabus			
	Prepare course timetable			
Course notes	Write training material for each session			
	Type training material			
	Prepare figures, diagrams, photographs, etc.			
	Review and edit training material			
	Prepare master copies			
	Translate training material (if required)			
	Type translated material			
	Prepare master copies			
	Print/photocopy training material for participants			
	File master copies			
	Purchase binders for course notes			
	Prepare front covers for course notes binders			
	Compile course notes			
Trainer's presentation notes	Prepare lesson plans for each session			
	Prepare trainer's presentation notes using course notes and supporting material			
	Type and print trainer's notes (if required)			
Trainer's presentation slides	Prepare draft material			
	Type/write final copy			
	Make final acetate copies (if using OHP)			
	Prepare presentation slides (if using computer-based projection)			
Administration material	Prepare application form			
	Prepare registration form			
	Prepare welcome note for participants			
	Prepare participants' (self) introduction note			
	Prepare questionnaire for participants on expected outcomes from the training course			
	Prepare questionnaire on course-related skills and experience (e.g. computing skills, field experience, etc.)			
Evaluation material	Prepare pre-course test			
	Prepare post-course test			
	Prepare course evaluation			
	Prepare post-training Action Plan note			

OHP, overhead projector.

Monitoring and Evaluation of Training

Monitoring and evaluation are important elements of any training programme. Monitoring is carried out during the training in order to assess whether the training is being carried out as intended and is to the required standard. Evaluation is carried out at the end of the training or some time after, in order to assess whether the objectives and required level of learning have been achieved.

Training can be monitored and evaluated by:

- monitoring classroom behaviour (whether people attended or dropped out, whether people asked questions, whether debate of issues was lively, whether participants felt involved and

Table 7.8. Pre-course organization.

Main activity	Subsidiary activity	Status		
		Plan	Reviewed	Final
Set up training office	Arrange staffing			
	Arrange office accommodation			
	Purchase/hire office equipment			
	Purchase office stationery			
Prepare training materials	(See Table 7.7)			
Organize participants	Send out letter of invitation			
	Screen applications and issue acceptance letters with joining instructions			
	Arrange board and lodging			
	Arrange travel			
Organize trainers	Prepare guidelines for trainers on house style for training material, training approach required, etc.			
	Delegate responsibilities among trainers for specified elements of the course			
Organize lecture room(s)	Furniture and fittings			
	Equipment			
Organize field visits	Define purpose, objectives and activities			
	Identify and visit potential locations for field visits			
	Arrange transport			
	Obtain permissions, notify local staff, farmers, etc.			
	Obtain necessary equipment			
	Prepare training material for field visits			
Organize opening and closing ceremonies	Invite guest speaker(s)			
	Arrange television, local press			
	Arrange banners, flowers, etc.			
Organize course stationery	Purchase course stationery for participants			

Table 7.9. Running the course.

Main activity	Subsidiary activity	Status		
		Plan	Reviewed	Final
Administration	Accommodation, subsistence and per diems for trainees			
	Honorarium for trainers			
Running the course	Opening ceremony			
	Administer pre-course test			
	Self-introduction by participants			
	Presentation of sessions			
	Field trip(s)			
	Monitor sessions			
	Administer post-course test			
	Administer course evaluation by participants			
	Closing ceremony and award of certificates			
	Course dinner			

Table 7.10. Staffing, equipment and materials.

Description	Unit	Rate ($)	Quantity	Amount ($)
Office staffing				
Secretary				
Office assistant				
Interpreters (if required)				
Computer/graphics assistant				
Office accommodation				
Tables				
Chairs				
Filing cabinets				
Office equipment				
Computers, complete with graphics, word processing and spreadsheet software				
Laser jet printer				
Colour printer				
Photocopier				
Spiral report binding machine				
Office materials				
Ring binders, file dividers, A4 plastic wallets				
Pens, pencils, erasers, rulers				
A4 pads				
Whiteboard marker pens, graphics pens				
Hole punches, staplers and staples, paper clips				
Envelopes – A4, A5 and letter				
Photocopy paper, cardboard wallets, OHP acetate sheets				
Corrections pens (white-out), glue sticks, highlighter markers				
Cutting board and cutting blades, scissors				
Sellotape, rubber bands, drawing pins				
Flip chart paper and marker pens				
Training room furniture and equipment				
Tables/desks				
Chairs				
Fixed blackboard/whiteboard				
Mobile whiteboard(s)				
Projector screen				
Flip chart stands				
Microphone and speaker system				
Softboard for card exercises				
Slide projector				
OHP				
Computer				
Computer graphics projector				

OHP, overhead projector.

felt they could contribute even if the material was new to them);

- measuring knowledge acquisition (testing recall of facts, procedures, techniques covered during the course);
- measuring any increase in skill (comparison of skill levels before and after training);
- measuring any changes in attitude towards the use of skills (for instance,

Table 7.11. Cost items summary.

Description	Unit	Rate ($)	Quantity	Amount ($)
Staffing costs				
Printing, photocopying and binding of training material				
Translation costs (if required)				
Office rental/furnishing				
Office stationery				
Course stationery for participants (bag, pen, pencil, ruler, eraser, calculator, notepaper, etc.)				
Purchase or hire of training equipment				
Training materials (OHP pens, whiteboard marker pens, flip chart paper and pens, OHP acetate sheets, spare bulbs for projector, computer disks, etc.)				
Travel costs for participants				
Accommodation costs for participants				
Per diems for participants				
Travel costs for trainers				
Accommodation costs for trainers				
Per diems for trainers				
Food and refreshments during training course for participants and trainers				
End-of-course dinner				
Printing of certificates				

OHP, overhead projector.

can the participants show changes in their appreciation of situations in which they might use particular skills);

- asking participants for their perceptions of the level and quality of learning they have achieved (asking whether the material in the course was new to them or not, whether they feel they can apply new knowledge confidently and can give examples of situations when they might, what they feel might stop them from using the new knowledge);
- asking participants to rate the quality of trainers (for instance asking them to say how satisfied they were with the length, style and content of training sessions, whether they felt the trainer was knowledgeable about the subject, well prepared in giving the training, etc.).

Table 7.12 outlines the assessment methods that can be used to evaluate the different types of learning. Questions or exercises to assess the degree of learning can be carried out at the end of a training course, usually through a questionnaire or test put to the participants. A further assessment may usefully be made 3–6 months later to determine the degree of retention of information, changes in attitudes and application of skills.

To evaluate the training carried out the following steps should be followed.

1. At the beginning and end of the course set pre- and post-course tests to ascertain the change in knowledge or skill of the participants. For knowledge and understanding the tests can be written, for skills the test may need to be more practical in nature (for instance, for someone trained to read river levels visit them in the field and ask them to take a reading on the gauge board). The tests should be marked and analysed to see if there has been an improvement in knowledge, skill or understanding.
2. At the end of the course set a course evaluation questionnaire to ascertain the participants' opinion of the course. Questions might include the following:

- Trainee's name and position, title of course, date of course.

Table 7.12. Evaluation of types of learning.

Types of learning	Assessment method
Knowledge and understanding	Test recall ability through written or verbal questions and discussions Set new tasks which require the application of new knowledge and understanding
Skills	Set new tasks which require the application of the taught skills
Attitudes and values	Through monitoring of the trainee's work, discussions with trainee, discussions with line manager, through relationships with others, especially clients/customers

- How effective has the course been in achieving the course objectives (scale: *very effective* to *not effective*)?
- What is your opinion of the course content (*very good* to *very poor*)?
- What is your opinion on the course material (handouts) (*very good* to *very poor*)?
- What is your opinion on the training methods used (*very effective* to *not effective*)?
- What is your opinion on the course duration (*far too long* to *far too short*)?
- Were you satisfied with the manner of presentation of the course materials (*very satisfied* to *not satisfied*)?
- Which parts of the course were *most* relevant to your work?
- Which parts of the course were *least* relevant to your work?
- How have the following been improved by the course (*list of key training needs*)?
- How do you think the course could be improved?
- Do you have any further comments about the course?

3. At the beginning of the course provide the participants with Action Plan sheets (Fig. 7.4) for them to write down what specific action they will take to apply their new understanding, knowledge or skills. Ask them to complete their Action Plan during the course and retain a copy on file.
4. At the end of the training course randomly select a sample of 10–15% of the participants with whom to carry out follow-up evaluation at a later date. Do not let these participants know that they have been selected! Retain copies of their pre- and post-course tests, their course evaluations and their Action Plans for later use.
5. Three to four months after the end of the course carry out a post-course evaluation by going to visit the randomly selected trainees. Using a standard questionnaire, record the objectives set for the course and the course contents. Then list several questions, or activities which will enable the effectiveness of the training to be evaluated. Look for the following points.

- What level of knowledge or skill has been retained? The post-course test could be set again to measure if there has been any change since the test taken at the end of the course.
- Has the trainee's attitude changed since attending the course? This is often difficult to ascertain, one way is to ask the trainee's supervisor or colleagues.
- What visible signs are there that the trainee has applied the training? Looking at the objectives set for the training, what should the trainee be doing as a result of the training and what should be the outcomes of his/her work? For example for WUA management training (see Box 7.4), have the WUAs had regular meetings? (ask to see the records of the meetings); has the WUA collected the water fees? (if so how much); has the service delivery been acceptable to water users? (ask the water users).
- Has the trainee carried out the activities specified in the course evaluation or their Action Plan? If not, why?
- Find out what constraints the trainee has experienced in applying the training and how they have overcome them.
- What is the trainee's opinion of the training now, has the training been of use in their work?

Concept/idea	Proposed actions to implement	Proposed timetable for implementing the action	Evidence of implementation
Example: Recording of maintenance work required and work completed	• Obtain A4 notebook and draw up tables • Record maintenance needs • Record maintenance work completed	Immediately following the course	Up-to-date Maintenance Register

Fig. 7.4 Action Plan proforma

Box 7.4. Key Evaluation Criteria for Assessing the Training of Water Users Associations

- Has the WUA been registered?
- How many of the total eligible water users are signed-up members?
- Have membership fees been collected, and if so how much? Have any irrigation service fees been collected? Ask to see the books.
- Are proper accounts being kept? Ask to see them and evaluate them.
- How is the WUA Management Board functioning – is it holding regular meetings, are there minutes of the meetings?
- Ascertain how many general meetings the WUA have had and how effective these have been – what topics were discussed, what actions agreed, etc. Ask to see the minutes.
- How are WUA staff performing? Meet with staff, interview them on the work they are doing and make a field trip around the WUA command area. Pose questions to the WUA staff to test if they know their job and are applying their learning – do they know the area, do they know who farms where, can they explain how they allocate and distribute water?
- Meet with farmers in the field and discuss how they order water, if they get reliable water supplies, what they do if they have a complaint or conflict with another farmer, etc.
- Find out if any maintenance work has been carried out and how it was done. Visit areas where maintenance work has been done, discuss with WUA field staff and assess if the work was worthwhile.

It will also be important to speak to third parties who work with the trainee to evaluate whether there has been any change brought about by the training. For a member of a WUA Management Board, for example, it could be ordinary members of the WUA or fellow board members.

A further important matter to ascertain is the level of follow-up and support given to the training by the organization. For example, has the trainee been given adequate support to implement the training, have there been follow-up meetings to reinforce the training?

To assist in the evaluation it is suggested that a checklist is prepared beforehand to use while carrying out the evaluation. This checklist would be different for each type of training course evaluated. In testing whether the WUA Management Board members know their job functions, for instance, the checklist would contain the full list of job functions. The evaluator could check off the replies and mark on the questionnaire form 'Knows 6 out of 7 of the duties' without having to write each one down.

6. It is important that the evaluation exercise, as far as possible, is carried out without prior warning. This means arriving unannounced, going out to the field and interviewing staff, water users, etc. at random.

7. When the full set of evaluations has been completed a final report should be prepared.

In any training evaluation report it is important to provide:

- the objectives of the training, course syllabus, duration, target audience, number of courses held, location where courses were held, names of trainers and their organization;
- how many people attended the course, where they were from and their designation, how many were expected, how many (if any) dropped out or failed to attend (and reasons);
- the results of pre- and post-course tests, if any;
- the data and findings from the evaluation, including examples of the evaluation questionnaires or evaluation methods used;
- conclusions on the effectiveness of the training and the training impact – whether the objectives set for the training were met, and to what degree.

8

Irrigation Management Transfer and Organizational Restructuring

This chapter deals with the topical issue of irrigation management transfer and the consequent restructuring of the way in which irrigation and drainage systems (I&D systems) are managed, operated and maintained. The origins of the move to transfer management from government to water users are discussed together with a summary of experience to date, not all of which has been successful. Processes and procedures for transfer are outlined and discussed and are then followed by a section on the emerging associated process of restructuring of government irrigation and drainage agencies (I&D agencies).

Irrigation Management Transfer

Background

The rapid increase of irrigated area in the 1970s and 1980s temporarily addressed the food crisis, but left governments with a heavy financial burden for the management, operation and maintenance (MOM) of irrigation schemes. Though money was available for capital works from international development funding agencies (such as the World Bank), many governments have had serious difficulties in providing adequate recurrent funds to sustain I&D systems. In addition operation of the system by government agencies has, in many cases, been poor, with operation and maintenance (O&M) staff poorly paid and poorly motivated. As a consequence of the failure to adequately operate and maintain them, the irrigation systems have fallen into disrepair, leaving many farmers with unreliable, inadequate and untimely supplies of irrigation water. Agricultural production and rural livelihoods have suffered, and the contribution to the national economy has declined.

Rehabilitation of existing schemes has been a feature of irrigation development since the late 1970s, with funding for capital works obtained from the international development agencies. Despite protocols between lending agencies and government requiring that government provide adequate funds for operation and maintenance, systems have continued to decline due to lack of recurrent funding. The lack of funds for O&M is such that in many countries rehabilitation is occurring of previously rehabilitated schemes.

To address this situation, and to improve the performance of the irrigation sector, a process of irrigation management transfer has been initiated. The top-down government-led technically driven developments of the 1970s and 1980s are giving way to bottom-up institutionally driven initiatives, which seek to fully involve the water users in the acquisition, management and use of water for irrigated agriculture. The transfer of irrigation

©Martin Burton 2010. *Irrigation Management: Principles and Practices* (Martin Burton)

management from well-established government agencies to groups of water users marks a significant change in the way in which irrigation and drainage is organized in many countries. The main objective is to shift ownership and operational responsibility of I&D systems from state to irrigators; the process is dynamic, progressing over time from lower to higher order infrastructure as mutual benefits are recognized. Institutional and legal reforms are essential for successful irrigation management transfer programmes.

The term 'irrigation management transfer' (IMT) is defined by FAO (1999) as:

> ... the relocation of responsibility and authority for irrigation management from government agencies to non-governmental organisations, such as water users associations. It may include all or partial transfer of management functions. It may include full or only partial authority. It may be implemented at sub-system levels, such as distributary canal commands, or for entire systems or tubewell commands.

Irrigation management transfer is distinguished from participatory irrigation management and decentralization where government still retains a significant role in the management of the irrigation system.

Irrigation management transfer involves changes in:

- public policy and legislation at national and local level;
- social attitudes, rights, roles and responsibilities;
- social and organizational arrangements at community level;
- financial arrangements for government irrigation agencies;
- financing of irrigation service provision;
- restructuring and reorientation of government agencies and redefinition of roles and responsibilities;
- nature of support services provided to farmers on irrigation and drainage schemes (I&D schemes);
- management, operation and maintenance procedures;
- relationships between government and water users.

A large number of countries are engaged in the IMT process (Table 8.1). Some countries such as the USA, Spain, France and Argentina have adopted irrigation management transfer

Table 8.1. Countries or states that have adopted irrigation management transfer in the past 30 years. (Modified from FAO, 1999.)

Latin America	South, South-east and East Asia	Africa and Near East	Europe and Central Asia
Brazil	Australia	Ethiopia	Albania
Chile	Bangladesh	Ghana	Armenia
Colombia	China	Jordan	Azerbaijan
Dominican Republic	India	Madagascar	Bulgaria
Ecuador	• Andhra Pradesh	Mali	Croatia
El Salvador	• Bengal	Mauritania	Cyprus
Guatemala	• Gujarat	Morocco	Georgia
Mexico	• Haryana	Niger	Kazakhstan
Peru	• Maharashtra	Nigeria	Kyrgyz Republic
	• Tamil Nadu	Senegal	Macedonia
	Indonesia	Somalia	Moldova
	Laos	South Africa	Romania
	Nepal	Sudan	Uzbekistan
	Pakistan	Turkey	
	Philippines	Zimbabwe	
	Sri Lanka		
	Viet Nam		

processes for over 30 years while others are just starting. Some countries such as Chile, Mexico and China are well advanced in the process, while others such as some states in India, Sri Lanka, the Philippines and Indonesia have started but have to some degree stalled. There have been success stories, such as Mexico and Turkey, and some failures, such as in Nepal.

The move to reduce the role of government in service provision has not been limited to the irrigation sector. Financial crises and poor progress with economic development in many developing countries has led to a rethinking of the role of government since the 1980s. At the heart of the debate on the role of the state in rural development have been the issues of *effectiveness*, *efficiency* and *accountability*[1] (Carney, 1998). Wider economic thinking in market-led economies has led to an evaluation of government's role in the provision of rural goods and services, ranging from seeds and fertilizer to veterinary and agricultural extension services.

In the 1980s there was increased research interest in traditional farmer-managed irrigation systems, with their comparative success often being used as justification for the transfer of government-run systems. Associated with these studies of farmer-managed irrigation were studies of the institutional and social issues related to how farmers organized themselves for irrigation. At this time Chambers (1988) identified three points of entry for action to improve performance:

- operational plans;
- rights, communications and farmers' participation;
- performance monitoring and computer analysis.

In practice, with the advent of irrigation management transfer, the rights, communications and farmers' participation has to some degree overtaken the other two points of entry. Associated with this way of thinking was seminal work by Elinor Ostrom (1992) who outlined measures for 'crafting' self-governing irrigation systems. Ostrom's work formed a sound basis for understanding the social interactions and institutions that govern successful water user organizations. A significant part of her work focused on rules (rules-in-use, psychological contracts, rules and culture, conflict resolution, etc.). Supporting this understanding was work by Uphoff (1990), which focused on the activities related to irrigation in terms of water (acquisition, allocation, distribution, use, disposal), infrastructure (design, construction, O&M) and organization (decision making, resource mobilization, communication, conflict resolution). Uphoff identified three 'ships' of water users associations as follows.

1. *Membership*: Definition of who should be members of the organization, and their roles and responsibilities.
2. *Leadership*: The calibre of the leadership mobilized from the farming community is the single most important factor in the effective and equitable functioning of water user associations (WUAs).
3. *Ownership*: Identification of the need for farmers to identify the irrigation system as 'theirs', and to take responsibility for it.

The experience gained during the 1980s and 1990s with irrigation management transfer resulted in the publication of guidelines for the transfer of irrigation management services (FAO, 1999). This comprehensive piece of work provides detailed guidelines for the IMT process, broken down into four phases:

- mobilization of support;
- strategic planning;
- resolution of key policy issues;
- planning and implementation.

The document outlines several phases of institutional development, with Phase 1 covering the need to identify the performance gap and to look at alternative options for bridging the gap. Phase 2 then discusses organizing the strategic change process, covering identification of stakeholders, identification of major issues, and identification and setting of objectives. Phase 3 involves investigation of the key policy issues related to financing, legal framework for transfer, extent of services/goods/infrastructure transferred, and ensuring accountability. Phase 4 outlines the development of the IMT plan, covering irrigation agency restructuring, forming and supporting water users associations, and measures to improve the condition of irrigation infrastructure.

One area that has not been adequately studied and understood is the varying ability of different communities and societies to adopt and make successful community-based management organizations. In some societies working together for the common good is an accepted norm; in others people are highly individualistic, tending to look after their own and their immediate family's own interests, sometimes at the expense of others. The factors that influence the ability and willingness of communities to work together on the management of I&D systems are a complex mix including local social history, social culture, value systems, the natural environment, access to water and other resources, and influence and interaction with external forces (including government).

Components of irrigation management transfer

There are three main components to irrigation management transfer:

1. Changes in the legislation related to water resources and irrigation and drainage management;
2. Formation of water users associations and transfer of management, operation and maintenance functions from government to water users;
3. Restructuring of government agencies responsible for water resources and irrigation and drainage management.

Changes to the legislation are an essential prerequisite for irrigation management transfer to allow formal recognition of water users associations as legitimate partners in the management of I&D schemes.

Formation of functional and effective water users associations is at the core of irrigation management transfer. The process of formation, establishment and support requires strong political will and support, in addition to significant resources during the formation and establishment phases. Though irrigation management transfer should result in savings to government in the long term, in the short term additional finance and resources will be required if the process is to be successful. Under-financed and poorly resourced transfer programmes have not been successful.

Finally, the government water resources and/or I&D agencies need to be restructured and reorganized to reflect the changes brought about by irrigation management transfer. These agencies need to adapt to withdrawal from the management of I&D systems to a greater regulatory and supervisory role. As part of the process there is often a downsizing of government staff, especially of lower-order O&M staff; this downsizing needs to be handled with care and consideration for the individuals involved. In addition there will be new functions and responsibilities within these organizations as they are modernized and their systems updated to cater for their changing role. Training is therefore a central part of the restructuring process.

Change management

Irrigation management transfer is one of the largest change management programmes many countries will have experienced, yet little is spoken about change management in the irrigation and drainage literature, and even less reference made to the wealth of knowledge and experience in this discipline. A possible reason for this might be that change management is part of general management theory and in the main refers to business and industrial organizations, though it has significant application in relation to irrigation management transfer.

In the business sector it is recognized that if an organization is to remain competitive then some organizational changes are required each year, and more major changes required each 4–5 years. Early warning signs that changes are needed within an organization include (Lorange and Nelson, 1987):

- overstaffing;
- restrictive and outdated processes and procedures;
- inflexible and time-consuming administrative procedures;
- poor levels of communication;
- lack of clear goals;
- lack of incentives to perform;
- tolerance of incompetence;
- outdated organizational structure.

Some or all of these signs are evident in government I&D agencies in many parts of the world, and signify a need for modernization and reform of these organizations. Key issues and factors that need to be addressed in modernizing and reforming these organizations are summarized in Table 8.2.

The process of change can be a daunting process, with key stakeholders experiencing a variety of emotions through the process. The process can be summarized in seven stages (Fig. 8.1).

- *Stage 1 – Immobilization*: A feeling of being overwhelmed, unable to make plans, reason or understand.
- *Stage 2 – Minimization*: A move to get out of the first stage by trivializing and denying the issues.
- *Stage 3 – Depression*: Realization that change is taking place, and its potential consequences on individuals. Individuals have difficulty knowing how to cope.
- *Stage 4 – Letting go*: An acceptance of change, letting go of the past and a sense of optimism for the future.
- *Stage 5 – Testing*: Individuals are energized and test out the new reality.
- *Stage 6 – Understanding*: Through testing and evaluation gradual understanding of how and why things are different, and an understanding of past reactions to the change.
- *Stage 7 – Internalization*: The change becomes a way of life. There is no longer a change, there is understanding and acceptance.

The above elements of change management were usefully summarized by Kotter when he outlined an eight-point strategy for implementing change (Kotter, 1995).

- Step 1 – Establish a sense of urgency;
- Step 2 – Form a powerful guiding coalition;
- Step 3 – Create a vision;
- Step 4 – Communicate the vision;
- Step 5 – Empower others to act on the vision;
- Step 6 – Plan for and create short-term wins;
- Step 7 – Consolidate improvements and produce yet more change;
- Step 8 – Institutionalise the new approches.

Kotter provided examples from his experience of over 100 companies of why change implementation worked or failed in relation to these factors. His overall summary was:

> The most general lesson to be learned from the more successful cases is that the change process goes through a series of phases that, in total, usually require a considerable length of time. Skipping some steps creates only the illusion of speed and never produces a satisfying result.

The first step is to create a sense of urgency and to communicate this information broadly and dramatically. Identifying the need (and causes) for change followed by frank discussion with key players is also required to galvanize action and participation. For the second step Kotter found that in cases of successful change management the leadership coalition grew over time. This process needs to be nurtured, with powerful players brought into this grouping at all levels, with senior managers always forming the core. Power might be in the form of titles, influence, access to information, expertise or relationships. Coalition building is a key part of this process, with off-site retreats a useful tool to build trust and communication. Kotter strongly emphasizes the need for the guiding coalition to build an easy-to-communicate and appealing vision for the future. Without this vision, in a form that can be grasped by all stakeholders, the transformation plan can dissolve into an unconnected set of projects and activities, leading nowhere. The need for communicating the vision and the change programme is another essential part of the change process. It is a process in which executives use all

Table 8.2. Key issues involved in change management. (Data from Kotter and Schelsinger, 1979; Plant, 1987; Hurst, 1995; Carnall, 1999.)

Issue	Explanation and actions to be taken
Resistance to change	• Self-interest and/or desire not to lose something of value • Misunderstanding and lack of trust of change and its implications • Different assessment and perception of proposed changes • Low tolerance of change • Fear of consequences of change, of the unknown
Overcoming resistance to change	• Education and communication • Participation and involvement leading to commitment, not just compliance • Provision of facilitation and support to combat fear and anxiety • Negotiation and provision of incentives for those likely to lose out as a result of change • Manipulation to influence key stakeholders • Use of both explicit and implicit coercion, if appropriate • Use a mix of approaches, not just one • Avoid moving too quickly • Avoid involving too few people
Change management processes	• Conduct an organizational analysis to identify the current situation • Conduct an analysis of the factors influencing the change process • Prepare a detailed change strategy • Monitor the change programme and address problems early
Components of change management	• Leadership, action planning, ability to cope with pressure and uncertainty • Ability to learn • Build an awareness of the need for change • Make a convincing and credible case for change • Change is a learning process – people don't get everything right initially • Change can feel chaotic and uncertain as people strive to come to terms with new understandings, tasks and functions, etc. • Attention must be given to broadening and mobilizing support for change • Crystallize the vision and focus for the organization, initially starting broad and then focusing down as an emerging strategy vision is identified • Recognize that the focus is on people and the process of change • Change is inevitable and continuous • The pace of change varies, sometimes smooth and linear, sometimes rapid and non-linear • Renewal may require destruction of some valued but outdated practices • Emerging structures and processes are a product of a multiplicity of factors, including constraints imposed by the environment • Change is a natural process of development
Ingredients for successful change management	• Provide clear evidence that change is both desirable and feasible • Develop clear strategic aims and identifiable objectives • Gain support at the highest level • Understand the role of corporate culture and corporate politics in key stakeholder organizations • Make senior managers accountable for change, but allow contribution from others • Make changes in the power structure of the organization(s) • Plan and manage implementation carefully • Communicate effectively and avoid rumour mongering • Gain ownership by key stakeholders of the outcomes and process • Manage stress levels and help stakeholders to cope with the changes • Where possible seek compatibility with existing systems, procedures, cultures and traditions • Plan the pace of change

Continued

Table 8.2. Continued

Issue	Explanation and actions to be taken
	• Build in systems to reward relevant behaviour • Create success early on, and provide positive feedback to build confidence • Provide examples for role modelling • Initiate appropriate, flexible and timely training and support • Monitor and evaluate the process, and make timely adjustments as required • Appreciate that managing change is a learning process for all • Empower key stakeholders, make them: *aware*, *capable* and *included*
Cycles of change	**Thinking:** • Diagnosis • Feasibility studies • Brainstorming • Communication of concerns • Problem recognition • Establish steering group **Addressing:** • Task forces • Training • Buying in new skills • Building support • Building coalitions • 'Pilot' trials **Doing:** • Creating change • Champions • Proposals for change • New structure and skills • Team building • Rewards and recognition • Sell change • Publicize success

possible methods of communication to broadcast the vision, and continually reinforce and repeat the message. Kotter makes the point that communication comes in both words and deeds; nothing undermines trust and thus change more than key players behaving in a manner that is inconsistent with their words.

Step 5 involves empowering others to act on the vision and identifying and removing obstacles to change. These may be people, organizational structures, procedures or policies. People can be key obstacles and not only those where it is known that they resist the change, but also those who are in key positions but do not champion the change. As the momentum for change grows so strength is gained to remove the blockages, though significant ones have to be confronted early on. Real transformation takes time, and as in any struggle short-term wins (Step 6) are needed to show progress towards the goal. In successful transformation managers look to provide clear milestones whose attainment can be measured, and for which people can be rewarded.

Kotter points out (Step 7) that there is a danger of declaring victory too early, letting go of the momentum and initiative and allowing any resisting forces to regroup and return to traditional ways. Leaders of successful transformation take pains to consolidate the gains made, to promote and support systems, structures and people that support the change process. They don't let up. Finally Kotter emphasizes (Step 8) the need to anchor the changes made into the corporate culture,

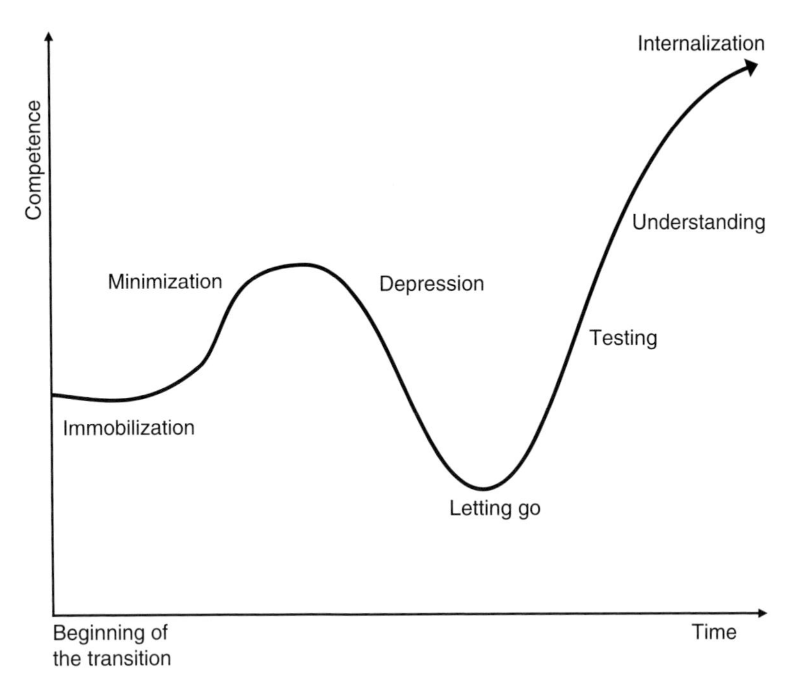

Fig. 8.1. The transition curve. (Modified from Adams *et al.*, 1976; de Vries and Miller, 1984; Sadler, 1995.)

such that the new behaviours are rooted in social norms and shared values. There is first a need to show how the new approaches, behaviours and attitudes have benefited the organization and the people in it, and second to ensure that the next generation of top management personify and are fully committed to the new approach. Kotter concludes with a simple statement:

> just as a relatively simple vision is needed to guide people through a major change, so a vision of the change process can reduce the error rate.

Leadership in change management

> Change, by definition, requires creating a new system, which in turn always demands leadership.
>
> (Kotter, 1995)

There is a considerable amount of literature on leadership. It is not the intention to review this literature in detail, rather to briefly summarize some of the developments in our understanding of leadership to date and then to look at the impact of leadership on the change management process.

Early studies of leadership in the 1930s focused on the belief that leadership is essentially a personal attribute (trait theory). This understanding developed in the later 1950s to an understanding of leadership as an interactive process. In the 1960s the concept of contingency/situational leadership emerged, where leadership is seen as a dynamic relationship between leaders, followers and tasks. In these situations the leaders have to adapt their style to the nature of the task and the people with whom they have to achieve the defined objectives. Contingency theory argues that a leader is temperamentally suited to a particular style, while situational leadership argues that the leader's style is adapted from a range of styles to suit given situations. In this context more recent work on emotional intelligence has emerged, which studies the role that a leader's moods and behaviours play in driving the behaviour of others. Goleman (1998) found that effective leaders are alike in one crucial way: they all have a high degree of 'emotional intelligence', comprising the five elements of self-awareness, self-regulation, motivation, empathy and social skills.

More recent theories focus on leadership in the context of a rapidly changing world. Organizations are seen as being in a constant state of flux; to survive organizations need leaders with vision and inspiration to gain trust and commitment to the strategy to achieve that vision. The concept of 'transformational' and 'transactional' leadership emerges, with the former providing the leadership for change and the latter for implementing measures to achieve the vision. The distinction is drawn between leaders and managers, where the two are seen as different but complementary in successful organizations. It was suggested that managers seek order, control and stability to address problems and achieve their objectives, without necessarily understanding their potential significance. In contrast, leaders tolerate chaos and are able to keep problems and potential answers in suspense, avoiding premature closure on key issues. In this context it was suggested that business leaders have more in common with artists, scientists and creative thinkers than with managers.

Kotter (1990; Table 8.3) summarized the differences between managers and leaders, and their interrelationship:

> Managers promote stability while leaders press for change, and only organisations that embrace both sides of that contradiction can thrive in turbulent times... leadership and management are two distinctive and complementary systems of action. Each has its own function and characteristic activities. Both are necessary for success in an increasingly complex and volatile business environment.

In analysing the role of the leader in setting the direction, Kotter makes it clear that the crucial issue with the vision is not that it is brilliantly innovative, rather how well it serves the interests of the key stakeholders – customers, shareholders and employees – and how effectively it can be implemented through a realistic strategy.

The key management role of organizing human systems within organizations is recognized, as is the need for standard and efficient procedures and systems. The need to set the direction and then to align people to obtain their agreement and acceptance of the vision is seen as a *leadership role*, integrated with the capability to identify both those who can support as well as block implementation of the vision. This is a major *communication* challenge, with *credibility* being a key component of the leadership effort. A major outcome of such leadership communication is seen as the *empowerment* of many stakeholders, enabling them to act towards the attainment of the vision in the knowledge of what it is, what it aims to achieve and the process by which it seeks to achieve its aims.

Motivating and inspiring people are seen as leadership roles in contrast with management systems for control and problem solving. Control systems are central to effective management, they must be fail-safe and risk free. They seek to help staff to efficiently and

Table 8.3. Differences between managers and leaders. (From Kotter, 1990.)

Managers	Leaders
About *coping with complexity* based on defined practices and procedures	About *coping with change* in a competitive and volatile world
Planning and budgeting, setting targets and goals for the future. Setting detailed steps for their achievement and allocating resources	*Setting a direction* and developing a vision of the future and formulating a strategy for its achievement
Organizing and staffing the organization to accomplish the necessary tasks of the organization	*Aligning people* through communication to accept and become committed to the new direction
Ensuring accomplishment of the plan through *controlling and problem solving*	*Motivating and inspiring* to achieve the attainment of the vision

successfully complete routine jobs day after day. Leadership requires a burst of energy to galvanize change and implement new ways of doing things, involving stakeholders in the process to develop ownership and participation. Through this process leadership is created across the organization at many levels as people work together to achieve the new reality. In this context the role of informal networks that exist within organizations needs to be recognized and understood, they comprise a formidable source of support or obstruction for change initiatives.

In well-performing organizations management and leadership are developed at multiple levels. In leading from the top, four key functions are:

- setting direction through setting a realistic vision and implementation strategy;
- aligning people through communication of the vision and strategy;
- inspiring action by motivating and influencing people;
- getting results by focusing energy on strategically appropriate goals.

The important role of leadership in successful irrigation management transfer can be seen in the case of Mexico and Turkey, where the government and the I&D agency were fully behind the transfer process. In Kyrgyzstan the leadership came initially from politicians and was then taken up by the implementation project and the WUA Support Unit, and after 7 years of establishment is now being taken up by individual WUA Chairmen and Executive Directors who are making their own changes and innovations within their WUAs.[2]

Management processes checklist for irrigation management transfer

The framework provided by Kotter with his eight steps to transforming an organization provides a sound base on which to build a checklist of the key management processes associated with irrigation management transfer. The fundamental components are identified as:

- building change management structures;
- effective leadership;
- vision;
- clear strategies for change;
- effective and widespread communication;
- commitment to change;
- building teams, networks and coalitions;
- careful planning and target setting;
- commitment of adequate resources;
- active monitoring and evaluation;
- reward systems for achievement;
- a flexible and supportive management control system.

The checklist presented in Fig. 8.2 was used by the author (Burton, 2003) to gather feedback from 23 professionals with experience in irrigation management transfer in a number of countries. The results of the survey are presented in Fig. 8.3, showing the areas where the IMT processes were considered by the respondents to have succeeded or failed. With each element of the checklist being marked out of 10, 11 of the 23 responses scored less than a total of 5.0, with a spread of stages receiving less than 5.0. In the better than average group, setting direction and making progress visible were found to be the weakest areas. The difficulty in making these assessments is shown by the different scores given for the same country – Nepal scores badly on three counts and well on one. This relates either to the time at which the assessor was involved with the IMT programme, or to particular schemes on which the IMT process was carried out.

The process provides a structured approach to assessing irrigation management transfer in the context of change management processes, and does highlight the weaknesses in the processes used in some countries to date. Better understanding of, and compliance with, fundamental change management processes is required if IMT programmes are to be successful.

Establishing and Supporting Water Users Associations

Overview

As discussed above, a number of countries have established water users associations,

A. Context and need for change

- ☐ The environment was conducive to IMT
- ☐ There were drivers for change
- ☐ The key stakeholders were ready for change
- ☐ There was limited resistance to change

B. Establishing a coalition for change

- ☐ Stakeholders were fully involved in the IMT process from the start
- ☐ There was an effective coalition for change
- ☐ There was clear leadership of the IMT process in the initial stages

C. Setting direction

- ☐ A vision was created for the IMT programme
- ☐ A vision statement was formulated
- ☐ Clear objectives were set for IMT
- ☐ A suitable, feasible and acceptable strategy and action plan was formulated
- ☐ A realistic timeframe was set for implementation
- ☐ The implementation strategy was well balanced in relation to technical, financial and institutional factors
- ☐ Clear leadership was displayed in setting the direction for change

D. Communicating direction and anticipated outcomes

- ☐ The vision was effectively communicated to stakeholders
- ☐ The strategy and action plan were effectively communicated to stakeholders
- ☐ Stakeholders had a clear understanding of how the changes would affect them
- ☐ Leadership continued to show commitment to change

E. Empowering action

- ☐ Change agents were appointed and were effective
- ☐ A sense of ownership was created of the change process among key stakeholders
- ☐ Stakeholders were encouraged and supported to adapt and improve the change process
- ☐ New leaders of the change process emerged, and were encouraged and supported
- ☐ Effective measures were taken to change institutional and organizational structures, systems and processes
- ☐ Sufficient resources were committed to the change programme
- ☐ An effective training programme was established

F. Making progress visible

- ☐ Short-term wins were planned for and incorporated in the IMT programme
- ☐ Short-term wins were identified and individuals/groups rewarded
- ☐ Progress was effectively communicated to stakeholders
- ☐ An effective monitoring and evaluation programme was established
- ☐ Stakeholder attitudes were assessed and acted upon

G. Sustaining and consolidating progress

- ☐ Progressive stakeholders were identified and supported
- ☐ The change process was flexible and adaptable, and incorporated emergent strategies
- ☐ Continuing resistance to change was identified and acted upon

H. Institutionalizing new approaches

- ☐ Links between new practices and beneficial outcomes of the change programme were identified and communicated to stakeholders
- ☐ New structures, systems and processes were institutionalized
- ☐ Future leaders were identified and continued to lead the change process
- ☐ Following IMT water users felt empowered and able to solve problems

Fig. 8.2. Checklist for assessing irrigation management transfer (IMT) change management. (From Burton, 2003.)

No.	Country	Stage A	Stage B	Stage C	Stage D	Stage E	Stage F	Stage G	Stage H	
		Establishing the context and need for change	Establishing a coalition for change	Setting direction	Communicating direction and anticipated outcomes	Empowering action	Making progress visible	Sustaining and consolidating progress	Institutionalizing new approaches	Overall
1	South Africa	5.3	0.7	1.9	1.0	0.6	0	0	0.5	1.2
2	Kazakhstan	5.0	1.7	2.1	4.3	3.4	1.0	2.0	3.3	2.8
3	Nepal	4.8	4.0	2.6	2.0	3.3	2.8	1.7	2.3	2.9
4	Nepal	5.5	2.0	2.3	2.0	4.1	2.0	3.7	2.8	3.0
5	Niger	5.3	2.0	3.0	3.5	3.3	1.6	3.3	4.0	3.2
6	Sri Lanka	6.3	6.7	6.1	2.0	4.9	1.4	2.0	2.5	4.0
7	Egypt	3.0	5.7	4.0	5.5	3.6	5.0	5.7	4.3	4.6
8	Albania	6.3	4.7	3.1	3.5	5.4	3.4	5.0	5.0	4.5
9	Yemen	4.3	3.0	3.7	6.0	5.0	3.2	6.3	5.5	4.6
10	Pakistan	4.8	2.7	4.1	6.3	4.7	3.2	6.0	5.3	4.6
11	Nepal	4.5	6.0	4.4	4.0	6.0	4.4	4.3	4.3	4.7
Average – Less well performing programmes		**5.0**	**3.5**	**3.4**	**3.6**	**4.0**	**2.5**	**3.6**	**3.6**	**3.7**
12	Maharashtra	5.5	4.3	3.9	6.0	4.9	3.2	6.7	5.5	5.0
13	Philippines	7.0	6.0	4.1	4.8	6.6	2.0	5.7	5.5	5.2
14	Kazakhstan	7.0	7.7	3.9	7.3	5.1	3.8	6.0	4.5	5.7
15	Viet Nam	7.0	5.3	6.0	6.3	6.3	1.8	5.3	6.3	5.5
16	Turkey	5.5	5.3	6.4	7.3	6.0	5.8	4.3	5.3	5.7
17	Andhra Pradesh	4.8	6.7	6.3	6.5	6.0	6.4	6.0	6.3	6.1
18	Mexico	6.5	8.3	5.1	6.8	6.6	5.2	8.3	6.0	6.6
19	Romania	6.8	7.0	5.4	7.0	7.0	7.2	7.3	6.3	6.7
20	Nepal	7.3	7.7	6.4	6.0	6.6	5.8	7.0	7.3	6.7
21	Kyrgyzstan	8.0	7.7	5.7	5.8	6.6	7.0	7.3	7.5	6.9
22	Kyrgyzstan	7.0	8.0	6.9	8.3	7.9	6.8	7.0	6.8	7.3
23	Mexico	6.3	8.3	8.0	8.3	6.9	6.8	8.0	7.3	7.5
Average – Better performing programmes		**6.1**	**6.4**	**5.1**	**6.1**	**5.9**	**4.7**	**6.2**	**5.8**	**5.8**
Average – All programmes		**5.6**	**5.0**	**4.3**	**4.9**	**5.0**	**3.7**	**5.0**	**4.7**	**4.8**

Legend
Relative failure (dark shading)
Relative success (light shading)

Fig. 8.3. Summary scorecard of irrigation management transfer programmes. (From Burton, 2003.)

with mixed success. Mexico and Turkey are generally recognized as examples of successful establishment of water users associations, though Mexico has had some difficulties with sustainability issues during a recent drought.

A fundamental problem in some countries has been the belief that I&D systems can be transferred to water users first in a short space of time, and second with limited resources put into the transfer process. The reason that Mexico has been successful is that significant political support was provided and significant resources were committed to making the transfer programme work. Creating awareness of the process and benefits, and training of WUA personnel and water users, requires significant effort and resources; where these have not been provided the transfer process has generally not gone well.

It is also important to realize that formation of water users associations is a *process*, not a one-off *activity*. It requires resources, planning and long-term commitment through the transformation process from a government-run I&D system to a farmer-run I&D system. In Turkey WUAs are still receiving advice and guidance from the government agency, some 10 years after the initial transformation process started.

The establishment of water users associations can be likened to a child growing up. In the early years a child needs guidance, tuition and support; as they grow up so they gain their own experiences and grow in confidence, until they reach adulthood when they can manage their own affairs (albeit with occasional reference/recourse to their parents!). As with a parent, the implementing agency can take considerable pride in seeing an independent, confident and competent WUA, providing good levels of service delivery to water users.

As noted previously in the section on change management, there is a danger of declaring victory too early. The WUAs will continue to need support from governments and politicians to protect their interests, particularly in relation to access to water supplies, taxation and access to good markets for their products.

Forming successful water users associations

Experience from several countries in forming successful water users associations shows the following.

- A comprehensive enabling environment is required, comprising:
  - strong, high-level political support and commitment;
  - clear policy direction;
  - a sound legal base, incorporating a legal basis for the new management entities and well-defined property and water rights.
- Functioning markets for agricultural produce.
- Roles, responsibilities and management functions need to be clearly specified and delineated for all parties (government agencies, WUAs and water users).
- It takes time and resources to establish fully functioning, effective and sustainable water users associations; 10–15 years is a typical time frame.
- Considerable effort is required in the early stages to establish the water users associations. Water users may initially be distrustful of these organizations, for a variety of reasons, and need to be convinced of the benefits of forming them. Time spent convincing, rather than coercing, farmers to form water users associations is well spent.
- Water users are unlikely to form associations of their own accord, external support is almost always required in the form of Community Organizers, Community Mobilizers, Support Units and the like. These personnel are provided through either a project or a government-supported programme.
- The Support Unit personnel need to carry out a significant amount of work in the early stages to explain and promote the association. The support units may also need to assist in grouping water users together in hydraulic units, in mapping these units, identifying landholding plots, etc. Unless this work is done, and done well, at this stage it is unlikely that sustainable associations will be formed.
- Training of WUA personnel and water users is a key activity once the association is formed and registered, and requires significant resources.
- Measures need to be in place for oversight of the association, both by the water users and by a government regulatory authority. All processes need to be transparent and accountable.
- Processes and procedures need to be in place for effective and timely conflict resolution.
- External support will continue to be needed for the associations until they are fully able to stand on their own feet.

A useful list of key success factors for management transfer was developed by Frederiksen and Vissia (1998). This list comprised eight key factors with associated tasks (Table 8.4).

Table 8.4. Key factors for management transfer. (From Frederiksen and Vissia, 1998 with permission.)

Factors	Tasks
1. Scope of transfer	• Define objectives • Define facilities and services to be transferred • Define responsibilities of all parties
2. Condition of facilities	• Ensure functionality of facilities to be transferred • Develop a rehabilitation plan if necessary
3. Ownership of facilities	• Place title of the facilities with the entity that will: ◦ Achieve proper O&M ◦ Reduce political interference ◦ Sustain the transfer programme
4. Water rights	• Stipulate legal rights to supplies for WUAs • Protect WUAs' water supplies from more powerful users • Exert control over appropriation • Allow conjunctive use of surface and groundwater • Reduce political pressure on water resources agencies
5. Service charges, funding and finance	• Obtain payment from beneficiaries for all services • Ensure WUAs have authority to: ◦ Set and collect fees ◦ Borrow money or issue bonds ◦ Levy taxes if necessary • Ensure the availability of loan programmes for WUAs
6. Form of water service entity	• Develop appropriate type of WUA • Provide the WUA with appropriate authority • Ensure that the WUA has all the critical characteristics
7. Preparation for and execution of transfer	• Address the present situation and forecast the future • Address the need for: ◦ Legislation ◦ Changing government roles and organizations ◦ Budgets ◦ Financial plans ◦ Asset management plans • Develop an execution programme
8. Follow-up support and oversight	• Provide government support to sustain the programme: ◦ Regulatory role ◦ Audit and oversight ◦ Water resources management ◦ O&M of major facilities • Construction and financing of major new projects

O&M, operation and maintenance; WUA, water users association.

Legal framework[3]

Experience has shown that a sound legal framework is essential in establishing successful and sustainable water users associations. Legislation is required to define the rules governing the operation of the WUA, and its relationship with its members and other individuals and organizations. The clear and unambiguous specification of these rules enables all parties concerned, from the water users, the association management through to government and government agencies, to function effectively and with the minimum of contention or dispute.

The main components of the legal framework are:

- an 'enabling law', which allows the WUA to be established and which describes its legal and organizational form;
- a constitution for each individual WUA, which can be amended by the water users subject to a specified majority in

favour of the proposed amendment and, in some cases, agreement by the state supervisory body;
- the operating rules prepared by the WUA in accordance with its constitution and the WUA law.

It is necessary in drafting the law to achieve a balance between the rules contained in the primary legislation, the WUA law, and those contained in the constitution (statutes) and operating rules (by-laws). There is some argument for most of the detail being provided in the constitution document, and less in the law. This may, however, have its problems in communities that are not familiar with legal forms, and has the potential for the constitution being drawn to favour certain parties. There is a case to be made in the public interest to specify minimum standards and conditions in the law to protect the rights of all potential members.

Some of the main components of the WUA legislation are outlined in the sections below (FAO, 2003, 2009).

Status, name and tasks of water users associations

The enabling law specifies that the WUAs are legal persons or enjoy legal personality. The law details the following.

- The generic name to be used by WUAs in order to identify it as a particular legal entity.
- The purpose or purposes of the WUA, its boundaries of operation and its permitted tasks, functions and responsibilities. These may include:
    - abstraction and delivery of irrigation water;
    - collection and disposal of drainage water;
    - operation and maintenance of the I&D system;
    - maintenance and repair of flood defence works;
    - recovery of costs from water users.
- The form of the organization operating in the public interest, or as a body of public law, on a non-profit or non-commercial basis.[4]

There is often discussion on whether the WUA should be able to carry out tasks other than those related to water delivery and removal. The consensus is that it should stick to its core function and leave other tasks, such as input supply and marketing, to other organizations, groups or individuals. Engagement in other activities dilutes the focus on water delivery and removal, and raises tax issues on the more commercially oriented functions of input supply and marketing.

Constitution of water users associations

The minimum content of the WUA constitution is generally specified in the WUA law and may include:

- the name of the WUA;
- the location of the WUA;
- description of the WUA service area by reference to plans and maps;
- objects and purposes of the WUA's activity;
- structure and competences of management organs of the WUA;
- the rights and duties of members of the WUA;
- procedures for joining the WUA and for termination of membership of the WUA;
- procedures for the calling of meetings of the General Assembly;
- provisions on the setting fees in the WUA;
- the responsibility of WUA members;
- procedures for compensation for damage to agricultural crops and agricultural plots of land to members of the WUA;
- conditions of termination activity (reorganization and liquidation) of the WUA.

Participation

There are a number of issues surrounding participation in WUAs, as outlined below.

- *Voluntary or compulsory?* A key issue for the legislation to clarify is whether participation in the WUA is voluntary or compulsory. In the case of land drainage or flood defence it is generally a requirement that all of those within the relevant command area are required to be members.

The case may be less clear for irrigation in some areas, where farmers may choose to rely solely on rainfall for their agricultural activities. In some countries coercion to join an association is strongly resented, and cannot be applied. In all cases a WUA based on voluntary, engaged and active participation is likely to be more effective than one based on coercion.

- *Membership*. Though it is common, it is not essential that participation in the WUA has to be by membership. Under the law the WUA could be given the right to distribute irrigation water or manage the drainage of the land, and to levy a charge for the service provided to the landowners. The landowners might have the right to participate in governance of the WUA, for example by election of WUA officials, but may not need to be members of the WUA per se. Membership is, however, the general form of participation in WUAs, and has benefits of defined rights and responsibilities for the parties concerned. It is important to specify who has a legal right to membership, and if membership is voluntary to allow non-members to have access to the services.[5]
- *Membership rights*. In some countries membership is open to landowners only; tenants can participate through their landlords. This is not appropriate in countries where there are large numbers of farmers who use land on the basis of leases or other rights. Allowance can be made in the legislation for those with long-term use rights to be eligible for membership, while short-term tenants may be members with the agreement of the landowner. It is important to make clear, however, that the landowner and tenant cannot both be members of the WUA at the same time.

Rights, duties and responsibilities

The rights, duties and responsibilities of the relevant parties need to be specified in the WUA legislation. This will cover all relevant stakeholders – the water users, the WUA, the government agency, the regulatory authority. There is good reason for specifying the minimum rights, duties and responsibilities in the WUA law rather than the constitution; these include the right to:

- access a fair share of water, or benefit from services provided by the WUA (such as drainage or flood protection);
- vote in WUA elections;
- stand for office;
- propose matters for discussion at meetings and the General Assembly;
- inspect the WUA books and records.

Associated with these rights are duties and responsibilities, which may include:

- compliance with provisions of the constitution and operating rules;
- payment of fees (on time);
- permitting water to pass to other users unhindered;
- access to land for operation and maintenance purposes;
- compliance with decisions of WUA officials, staff and/or the General Meeting.

Procedures for establishing water users associations

The procedures for establishing WUAs need to be established in the legislation. This means setting out the steps for establishment and the people and organizations involved. The process requires specification of the following.

- *The initiative group*: A small group of self-elected people wishing to form a WUA.
- *The Founding Committee*: A representative group covering the service area of the proposed WUA. The main task of the Founding Committee is to consult with all water users and to prepare the documentation required to establish the WUA.
- *Formulation of establishment documents*: This will include formulating the WUA constitution, identifying potential members of the WUA and their landholdings, obtaining maps of the service area and details of the infrastructure.
- *Approval of draft establishment documents*: There are genuine public interest reasons

for requiring that the draft establishment documents are first approved by the state supervisory body before being submitted to the WUA members for approval. This is to ensure that the public interest is adequately protected and to assess the viability and sustainability of the WUA proposed in the documentation (e.g. are paid staffing levels too onerous for water users?).

- *Establishment meeting*: This meeting is held with as many members as possible attending. The proposed WUA constitution and establishment documentation are outlined and discussed. It is typically a requirement that over 50% of the members approve the documentation, and that they own or use over 50% of the land in the WUA service area.[6] The meeting will also elect WUA office holders and approve the initial budget and work plan.
- *Formal legal establishment*: Following the establishment meeting the elected WUA officials lodge the establishment documents with the relevant authority. This is usually the civil courts or the supervisory authority.
- *Costs of establishment*: The legislation typically allows for reasonable costs incurred by the founders to be reimbursed by the WUA.

Governance structure of water users associations

The WUA law needs to outline the institutional arrangements for the WUA. This will cover the following.

- *The structure of the organization*: This includes the general or representative assembly, management board, WUA chairperson, WUA executive and WUA sub-committees, WUA representatives, etc. Care has to be taken to ensure sufficient guidance in the legislation while allowing adequate flexibility for WUAs to adapt the institutional framework to their specific needs.
- *Voting rights*: These need to be specified, are these to be one member–one vote, or based on landholding size?
- *Role of the general assembly, management board and chairperson*: In some cases the (elected) Chair of the Management Board is also the Executive Director of the Association, and therefore responsible for and involved in all matters related to the WUA. This has its benefits and drawbacks; the benefits being that a strong chairperson can drive the WUA forward and take action as required, the drawback being that it concentrates too much power in one person. A preferable situation is to have an elected Chairperson of the Board and a separate appointed Executive Director and staff.
- *The role and procedures of the General Assembly*: The legislation will stipulate the role and procedures of the WUA General Assembly, which include:
  - electing the WUA Management Board, committee members and other officers;
  - setting/approving the budget and service fees;
  - approving the annual work plan and irrigation plan;
  - receiving, reviewing and approving the annual report and accounts;
  - adopting the operational rules of the WUA;
  - amending the WUA constitution.

  The legislation will also specify the number of times the General Assembly should meet, the procedures for calling emergency meetings, the minimum numbers required to make a meeting quorate, voting procedures, etc. The legislation may also allow for representation of water users and procedures for a Representative Assembly.
- *The role and procedures for the Management Board and sub-committees*: The legislation should specify the procedures for the election of the Management Board and its duties and responsibilities. It is also helpful if the law specifies minimum and maximum number of members. The law may specify sub-committees for auditing WUA accounts and dispute resolution, the composition, procedures for appointment/election of members; duties and responsibilities need to be defined.

- *Provisions for elected officials and staff*: The legislation will detail the provisions for elected officials and staff of the association, with provisions for election and re-election, terms of office, etc.
- *WUA income*: The legislation will specify the possible sources of income for WUAs, such as:
  - membership fees;
  - service fees;
  - grants and subsidies;
  - gifts;
  - loans;
  - interest on savings.

  The list of possible sources on income will be specified in the law, the method of levying the membership and service fee will be detailed in the constitution or operating rules. There are a number of mechanisms for charging for the service fee, including on the bases of area, crop type and area, number of irrigations, number and duration of irrigations, etc.
- *Record keeping and accounts*: Keeping of transparent and accountable records and accounts is of paramount importance to the sustainability of a WUA. The legislation may specify the minimum level of record keeping that is required, which may include:
  - a register of all members with their names and landholding details;
  - a register showing water received and delivered;
  - accounts books detailing sums due and sums paid;
  - minutes of meetings (General or Representative Meetings, Management Board meetings, meetings with other organizations, etc.);
  - copies of any contracts entered into.
- *Sanctions*: The legislation should specify the nature of offences against which action will be taken by the WUA, the process by which they will be addressed and the nature of the sanctions.
- *Rights of WUAs*: The legislation will confer a number of additional legal rights on WUAs, which may include the right to:
  - use infrastructure;
  - impose sanctions and fines;
  - expel members;
  - acquire access rights over land;
  - recover outstanding fees and charges.
- *Liquidation*: The possibility of dissolution and liquidation needs to be covered by the legislation. The grounds for liquidation will be specified, as will the procedures following liquidation (e.g. who will take over the functions previously carried out by the WUA, who will own the infrastructure if it has been transferred to the WUA, etc.). To avoid the risk of political interference or coercion formal notice should be issued to the WUA members to enable them to take legal action if they are not in agreement with the proposed liquidation.

Supervision of water users associations

In the public interest it is important that the government provide some oversight and monitoring of WUAs. This should not be too imposing, but at the same time must be effective in being able to assess whether a WUA is functioning properly, and not failing. The WUA law will identify the supervisory body, typically a ministry or unit within a ministry, and will describe the extent of the supervisory body's powers and the conditions under which it may intervene in a WUA's affairs. The supervisory body will need to collect data and information on each WUA, and to report on a regular (annual) basis on WUA performance (see later section on monitoring of WUAs).

Merging of WUAs, Federations of WUAs and National Associations of WUAs

It is helpful if the legislation allows for the merging of WUAs to form larger WUAs, the formation of Federations of WUAs to manage higher-order infrastructure and the ability to form a National Association of WUAs. The Federation of WUAs will need to have an independent legal personality, with defined jurisdiction, roles and responsibilities; likewise for the National Association of WUAs.

Consideration of other issues

In formulating WUA legislation it will be necessary to take into account other related legislation, taxation and property rights. Water resources and irrigation acts will need to be amended to take account of the changes brought about by the formation of WUAs. Tax codes will need to be amended to allow for issues such as transfer of government property to WUAs with charge, payment of VAT on service fees and maintenance work. Land tenure status can be an issue in some countries.

Stages in formation of water users associations

Main stages in formation, establishment and support

The main stages in forming, establishing and supporting water users associations are shown in Box 8.1. *Formation* is the process of getting water users to agree to form an association, *establishment* is the process of getting the WUA going once the water users have agreed to its formation, and *support* is the process of hand-holding until the WUA is established and fully capable of running its own affairs.

The initial step is to review the existing legislation and to revise it to allow for the formation of water users associations. In most cases a WUA Law will be required, and adjustments will be required to the existing Water Law and Tax Code.

As mentioned above, the formation and establishment of WUAs takes time, effort and resources. One of the main resources is a dedicated team of trained personnel working with water users to form and establish the associations. In some countries a directive has been issued to form WUAs without any specialist team being formed, or any guidance for training of government agency personnel. Not surprisingly the transfer process has not been successful.

Once formed the Support Units work with water users to form and register the association. For this data need to be collected to define the boundaries of the association, the membership and the areas of land held by each member. At the same time procedures need to be put in place to form the WUA Regulatory Office. This body will be responsible for regulating and monitoring the WUA, and reporting to government on its progress and performance.

For larger WUAs it is advisable to break the WUA command area into zones with a representative for the water users in that zone. The zone area varies from 10 from 40 ha, with ten to 30 water users being typical. The representatives are elected by the water users and attend Representative Assembly meetings on behalf of these members, and report back to

Box 8.1. Main stages in Forming, Establishing and Supporting Water Users Associations

- Review existing legislation, update and enact legislation to support WUA formation.
- Establish Support Units.
- Train Support Unit staff.
- Promote formation of WUAs.
- Gather data, produce maps, identify canal ownership, etc.
- Form WUA.
- Register WUA.
- Establish WUA Regulatory Office.
- Establish if water users want a Representative Assembly or a General Assembly.
- Train WUA Representative Council members.
- Train WUA Executive staff.
- Identify and form representative zones.
- Hold Representative Assembly meetings.
- Carry out asset surveys.
- Identify performance measures.
- Provide advice and guidance to established and functioning WUA.

them. The alternative to the Representative Assembly is a General Assembly, in which all water users participate.

Once the WUA has been formed and the staff appointed, work can commence on establishing rules for water allocation and in carrying out asset surveys to establish the extent, type and condition of the WUA's assets (canals, drains, structures, etc.). The asset survey forms the basis for maintenance and possible upgrading or rehabilitation of the I&D system.

Finally procedures are put in place for monitoring and evaluation of WUA performance, both by WUA management, but also for the Regulatory Office or the project.[7]

Review water sector legislation

The first step in the formation of water users associations is to review the existing legal frameworks and establish if new legislation is required. In some countries, to speed up the formation process, WUAs have been established under existing legal frameworks, such as for cooperatives. This has generally not been successful, and the formation, establishment and sustainability of the WUAs have been adversely affected. As shown in the discussion above there are some very specific elements of WUA legislation, which require a dedicated legal framework.

Where new legislation is required specialist guidance is generally required, drawing on relevant international experience. The new WUA legislation needs to be drafted, reviewed, revised and then submitted to the legislature for ratification. This can be a lengthy process, but time and effort taken at this stage sets a solid base for the subsequent processes, and for the establishment of viable and sustainable water users associations.

Steps in establishing support units for water users associations

An important and sometimes overlooked part of the WUA formation and establishment process is the setting up of a WUA support team or unit. This team or unit can be established within the government I&D agency which is transferring the management of the I&D systems, or it can be set up as part of a project team engaged in the formation and establishment of water users associations.[8] Following training, this team/unit becomes the driving force behind the transfer process and can be the key element in the success or failure of the process.

The first step is to prepare a programme for the establishment of the Support Unit (Box 8.2). This will involve decisions on the level at which to provide Support Unit staff. In the example given here based on experience in several countries, three levels have been assumed, at Central, Regional and District level.[9] The Central Support Unit staff are appointed or recruited and trained. This training is crucial, and needs to be carried out by personnel with experience in forming, establishing and supporting water users associations. Often a key part of the training

Box 8.2. Steps in Establishing Support Units for Water Users Associations

- Prepare programme for establishment of Support Units.
- Appoint or recruit Central Support Unit (CSU) personnel.
- Establish Central Support Unit office.
- Train Central Support Unit staff.
- International study tours for Central Support Unit staff and senior government agency staff.
- Appoint or recruit Regional Support Unit (RSU) personnel.
- Appoint or recruit District Support Unit (DSU) personnel.
- Establish Regional and District Support Unit offices.
- Prepare training material and programme for Regional Support Unit and District Support Units.
- Train Regional and District Support Unit staff.
- Organize international study tours for Regional Support Unit and District Support Unit staff.
- Regional and District Support Unit staff work with water users to form and train WUAs.

programme will be study tours to countries such as the USA, Turkey and Mexico where successful WUAs have been established. The study tours will comprise the Support Unit staff and may include politicians and senior personnel from relevant government agencies. These visits are valuable in enabling these key players in the transfer process to understand the processes followed and see the outcomes of management transfer.

The next step is to appoint or recruit the Regional and District Support Unit staff, and to establish their offices. In many cases the Support Unit offices are established in the offices of the I&D agency, which has some benefits and some limitations. The benefits are that the Support Units can work closely with the I&D agency, and liaise with them on behalf of the WUAs; the constraint is that this relationship may be too close and not sufficiently independent.

Following the establishment of the Regional and District Support Units the Central Support Unit staff prepare training material for training of Regional and District Support Unit personnel, WUA personnel and water users. This training material will cover WUA governance, water management and system maintenance. The Central Support Unit will then train the Regional and District Support Unit in the principles and practices of WUA formation, establishment and support, and in the giving of training to WUAs and water users. If funds permit, international study tours to countries with well-established and functioning WUAs may be organized for some Regional and District Support Unit personnel, together with Regional and District irrigation and drainage agency staff, and WUA Chairmen and/or Board members.

Following their establishment and training, the Support Units are ready to start the process of WUA formation, establishment and support.

Steps for forming, establishing and supporting water users associations

The various steps for forming, establishing and supporting WUAs are shown in Box 8.3. The first step is to prepare information material for raising public awareness related to water users associations. This information can be in the form of leaflets, posters, booklets,

Box 8.3. Steps in Establishing and Supporting Water Users Associations

- Prepare public information/WUA formation material.
- Meet with local authorities and government agencies (water resources, irrigation and drainage, environment).
- Meet with local leaders.
- Conduct WUA awareness workshops with water users.
- Obtain and prepare WUA documentation (maps, landholding areas, etc.).
- Hold initial WUA formation meeting and appoint Founding Members.
- Prepare WUA Constitution and statutes.
- Hold General Meeting to approve WUA Constitution and statutes.
- Submit WUA documentation for legal registration.
- Hold elections for WUA Board.
- Interview and appoint WUA executive staff.
- Prepare WUA training material.
- Train WUA Board.
- Train WUA Accountant.
- Train WUA water management staff.
- Establish Representative Zones and then elect and train Zonal Representatives.
- Ensure participation of WUA in design and construction supervision of rehabilitation works.
- Prepare asset management plan for the I&D system.
- Prepare MOM manual for the WUA and I&D system.
- Organize maintenance awareness workshops with water users.
- Organize fee setting and fee recovery workshops with water users.

etc. or radio and television broadcasts. The next step is to meet with local authorities and government agencies to inform them on the approach and benefits of forming WUAs, and to gain their support in the process.

The engagement with the I&D scheme begins with meeting local leaders and explaining the benefits and process of WUA formation. With the support of the local leaders, awareness raising meetings are held with the water users with the aim to get them to sign up to the process. At this stage locally based community motivators may be engaged to promote the WUA message and to obtain feedback and address concerns of the water users.

Information needs to be collected on the extent of the I&D system, the landholdings and the ownership of the landholdings. A key principle is that the WUA has clearly established hydraulic boundaries, with the area supplied solely from one or more water sources. A preliminary register of water users and their landholdings is drawn up and forms the basis for gaining signatures from landowners to show their agreement to forming the association. In some countries it is a stipulation that more than a certain percentage of landowners agree to form the association (this can be as low as 51%), thus signatures are required as evidence of this agreement.

A group of Founding Members is appointed to organize and oversee the initial stages of the WUA formation process. This group is generally drawn from local leaders and respected water users within the proposed WUA command area. The Founding Members then work with the Support Unit to prepare the WUA Charter and statutes and to prepare the documentation required for registration of the WUA (such as a map of the command area, a list of WUA members with location and areas of their landholdings).

A General Meeting is held by the Founding Members to discuss the WUA Constitution and statutes with the members. Following agreement on the constitution and statutes the registration documents are formally submitted to the relevant authority (this is often the local magistrates court or Ministry of Justice office).

Following registration, the Founding Members with the assistance of the Support Unit organize and hold elections for the WUA Management Board/Council. The WUA Management Board generally comprises some five to 12 persons drawn from landowners within the WUA command area. In some cases the WUA charter allows or requires specific persons to be members of the Management Board, such as representatives or headmen from each village within the command area. The Management Board elects a Chairman/woman to chair the meetings and to represent the WUA. The Management Board will also appoint an executive to carry out the day-to-day tasks associated with the running of the association. This executive generally comprises an Executive Director, a Treasurer/Accountant, an O&M Engineer/Technician and Water Masters.

Following the establishment of the WUA the Support Unit commences the training programme. The training focuses on the different elements – governance, accounting, water management and maintenance. Training is carried out for: the WUA Management Board in duties and responsibilities; the WUA Accountant in financial and accounting procedures; the WUA Executive Director in his/her duties and responsibilities; the WUA O&M Engineer/Technician and Water Masters in water management and maintenance, and liaison/communication with water users. Additional training will be carried out for other elements of the management structure, such as the Audit Sub-Committee and the Conflict Resolution Committee.

In some cases a Representative Assembly rather than a General Assembly is formed by the WUA. If a Representative Assembly is formed, Representative Zones need to be identified and Zonal Representatives elected and trained. The zones should comprise discrete hydraulic units (e.g. comprising water users on one quaternary canal) with 30–40 water users in each zone.

The formation and establishment of WUAs may be associated with a programme or project for rehabilitation of the I&D system. If this is the case it is preferable that the WUA is formed before the rehabilitation works commence (particularly the planning and design phase) so that the WUA can participate in and guide the rehabilitation work.

This involvement of the WUAs from planning through to construction and implementation creates a strong sense of ownership for the physical infrastructure, and is a key part of the process of transferring the responsibility and sense of ownership for this infrastructure from government to water users.

Following the initial training there may be follow-up activities and training, such as that related to preparing asset management plans for the I&D system such that proper budgeting can be carried out for the maintenance of the system. Additionally, if the WUA formation and establishment is part of a project, WUA management, operation and maintenance manuals may be prepared. These manuals can form the basis for more detailed training of WUA personnel.

Periodically, additional training or awareness-raising meetings may be organized, such as meetings to inform water users how the WUA budget is prepared and the irrigation service fee set, and meetings on the importance and cost of maintenance work. These meetings may be run by the Support Unit staff, or by the WUA Executive Director or Chairman.

Establish a regulatory authority for the water users association

In the public interest government needs to maintain some oversight over the established water users associations. As discussed above in the section on legislation, the formation and outline structure of the supervisory body (hereafter referred to as the WUA Regulatory Authority) should be specified in the WUA law, and should be referred to in the constitution and operating rules of the association.

A typical process for establishing the WUA Regulatory Office is:

- prepare legal framework (for WUA Regulatory Authority);
- enact legislation/regulation to establish the authority;
- recruit staff;
- establish office;
- monitor and report on WUAs.

Typically the WUA Regulatory Authority comprises an office with two or three staff established in a relevant ministry, either agriculture, water resources or irrigation. The purpose of the Regulatory Authority is to:

- oversee the formation and establishment of water users associations;
- maintain a complete register of the establishment documentation for each WUA (this will be in parallel to the documentation submitted to the relevant local authority or court to register the WUA);
- collect data, usually annually, from WUAs (these data will be specified in the WUA law, and should be sufficient to monitor the progress and performance of each WUA);
- report on the performance of WUAs;
- support water users and WUAs where their rights are being threatened by other parties.

Other roles may include mediation in the event of disputes between water users and WUAs, and oversight of auditing of WUA accounts.

Due to the quantity of data that require processing each year the WUA Regulatory Authority will need to establish a computerized database. At the end of each cropping year the Regulatory Authority can send out standard data collection forms to be completed and returned by each WUA. Where a WUA Support Unit has been established they can assist the WUAs in the collection of data and the completion of these forms. Possible data that might be requested are presented in Table 8.5 and includes information on WUA meetings, WUA personnel, WUA budget, water use, and expenditure and income related to the water delivery service. Where the WUA provides drainage services as well there may be additional data requested.

Management structure of water users associations

There are a number of variations for the management structure of WUAs. A typical structure is provided in Fig. 8.4.

Table 8.5. Possible data to be collected by WUA Regulatory Authority.

A. General WUA information

- Year
- Region name
- District name
- WUA name
- WUA Registration Number
- Establishment date
- Registration date
- Number of members
- Service area (ha)
- Number of WUA staff
- Number of Management Board Members
- General or Representative Assembly?
- Number of water user representatives
- WUA Chairman name
- WUA Executive Director name
- WUA Accountant name
- Number of meetings this year of General Assembly or Representative Assembly
- Number of meetings this year of WUA Management Board

B. Irrigation area and cropping pattern

- Total cropped area (ha)
- Planned irrigated area (ha)
- Actual irrigated area (ha)
- Rainfed area this year (ha)
- Area not cultivated this year (ha)
- Major crops (crop name, area)
  - List of the type and area of the main irrigated crops grown
- Yield of crops (crop name, yield)
- Average yields of the main crops
- Crop market prices (crop, market price)
- Average market price of the main crops
- Crop area damaged and cause

C. Water supply

- Total from external sources (main system service provider, MCM)
- For each named main canal (MCM)
- Total from own sources (MCM)
  - For each named source (MCM)

D. Water distribution by WUA

- Contracted water delivery to water users (MCM)
- Actual water delivered to water users (MCM)
- Actual water use per ha (m^3/ha)

E. Income and expenditure for water delivery

- Total value of water delivery contracted with main system service provider (MU)
- Total value of water delivery contracted with water users (MU)
- Total value of payments received from water users for water delivery (MU)

E. Income and expenditure for water delivery (continued)

- Amount owed by water users to WUA at beginning of year for water (MU)
- Amount owed at end of year by water users to WUA for water delivery (MU)
- Amount owed to main system service provider at end of year (MU)
- Amount paid to main system service provider (MU)
- Method used for determining ISF and value of ISF charged
  - Per hectare basis (MU/ha)
  - Per irrigation basis (MU/irrigation)
  - Volumetric basis (MU/m^3)

F. Budget information (planned and actual)

- Recurrent costs (MU)
  - Salaries
  - Social fund
  - Temporary staff costs
  - Transport costs
  - Admin costs
  - O&M costs
  - Other operating costs
  - ISF payment to main system service provider
  - Maintenance expenditure
- Investment costs (MU)
  - Major repairs
  - Rehabilitation
  - Equipment and vehicles
  - Other acquisitions
- Financial and other costs (MU)
  - Reserve Fund
  - Repayments of loans and credit
  - Interest payments
  - Taxes
  - Contingencies
- WUA income
  - ISF
  - Fines and penalties
  - Interest income
  - Other income
- Income less expenditure
- Repayment for rehabilitation (if any)
- Total amount of accumulated reserve fund
- Amount in bank account
- Debts to main system service provider

G. Changes in WUA staff, charter or area

- Elections of new senior members
- Increase or decrease of WUA service area in last year (ha)
- Transfer of assets
- List of assets
- Key Management Board decisions and/or changes in Constitution or Internal Regulations

WUA, water users association; MCM, millions of cubic metres; MU, monetary unit; ISF, irrigation service fee; O&M, operation and maintenance.

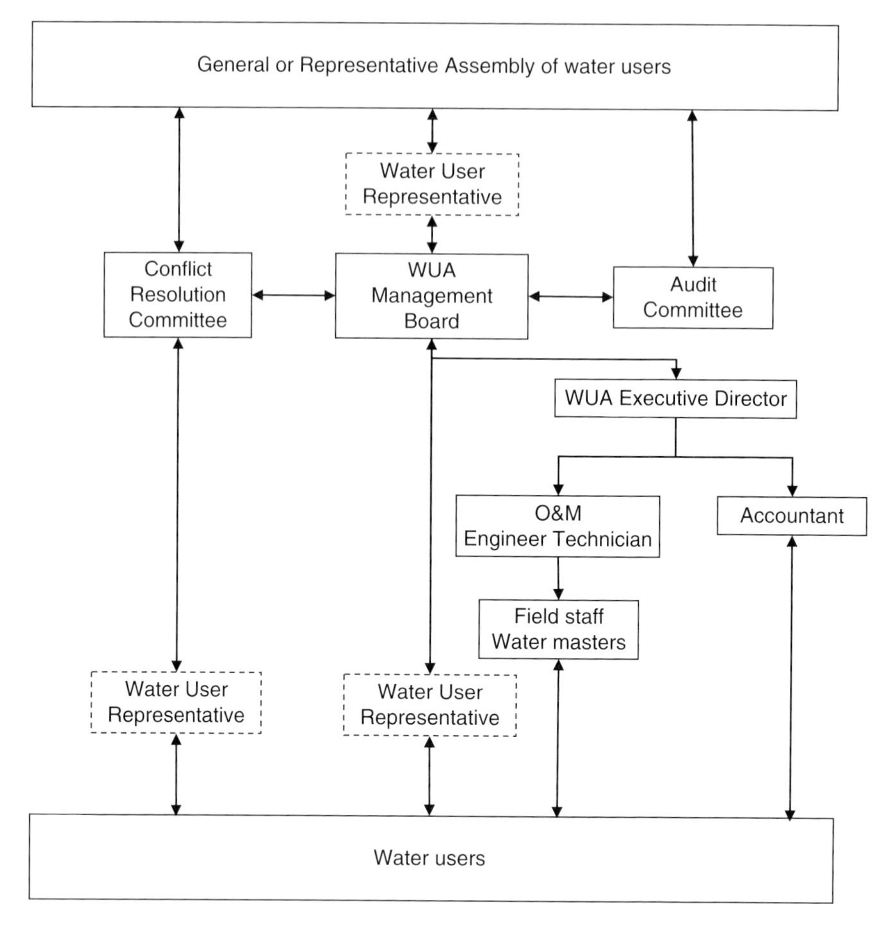

Fig. 8.4. Typical management structure of a water users association (WUA) (O&M, operation and maintenance).

When the WUA is formed the water users elect a Management Board comprising five to 12 members (the number depends on the water users, it is up to them to decide what a suitable size is for the Management Board). The Management Board elects a Chairman/woman, and then appoints the executive staff, generally comprising an Executive Director, an Accountant, an O&M Engineer/Technician and field staff (Water Masters). Out of the elected members the Management Board may also establish an Audit Committee of three or four people and a Dispute Resolution Committee with three or four people. To be effective it is important that the members appointed to these two subcommittees are respected and trusted by the water users. In larger systems a representative system may be used whereby groups of water users elect a representative to attend management meetings, represent their views and report back.

The Management Board members and the Water Users Representatives are generally not paid, but the executive staff are paid. Payment may be in cash, in kind (agricultural produce), or in the allocation of preferential access to water or land. The number of staff depends on the size of the irrigated area, its complexity, and the resources available to the water users. In systems with a good market, growing profitable crops, then staff can be paid in cash. In systems with subsistence agriculture staff may be allocated some land or preferential access to water, or maybe paid in kind with agricultural produce.

The roles and responsibilities of the various personnel associated with water users associations are summarized below. Note that these duties and responsibilities are linked to the organizational structure outlined in Fig. 8.4. There are variations on this structure; for example, some WUAs may not have an O&M Engineer/Technician, the duties and responsibilities for this position might then be divided between the WUA Executive Director and the Water Masters.

WUA Chairman/woman and Management Board members

The WUA Chairman and Management Board members are responsible for the proper functioning of the WUA and for the performance of the WUA staff. Their main tasks include:

- attending and chairing General or Representative Meetings;
- attending and chairing WUA Management Board meetings;
- receiving, checking and approving reports on WUA activities, membership and performance from the WUA Executive Director;
- submitting the Annual Report to the members, including reporting on WUA performance and finances, together with proposals for service fee levels for the coming year;
- liaising with external agencies on behalf of the WUA;
- reviewing and adapting WUA by-laws where requested by the General or Representative Assembly;
- agreeing on terms and conditions and employing WUA staff.

WUA Conflict Resolution Committee members

The WUA Conflict Resolution Committee members are responsible for resolving disputes that may arise between the WUA Executive and water users, and disputes between water users. Their main tasks include:

- bringing together the parties, taking evidence, allowing presentations of information and facilitating discussion between affected parties;
- making and enforcing a judgement;
- recording the procedures followed and the judgement made;
- reporting to the WUA Management Board and the WUA Annual General or Representative Assembly.

WUA Audit Committee members

The WUA Audit Committee members are responsible for auditing the WUA financial records on behalf of the membership. Their main tasks include:

- receiving and checking the accounts each year;
- verifying the accounts;
- reporting to the WUA Management Board and the WUA Annual General or Representative Assembly.

WUA Executive Director

The WUA Executive Director is responsible for the proper day-to-day functioning of the water user association. He/she reports to the WUA Chairman and Management Board and the members. His/her main tasks include:

- representing the WUA in all matters related to its day-to-day activities;
- preparing the annual work plan and budget and proposed service fee;
- organizing regular meetings with the Management Board, preparing the agenda and ensuring that minutes are taken;
- organizing and chairing seasonal meetings with farmers to discuss operational issues including seasonal cropping plans, water supply availability, system operation and maintenance, service fee setting and payment, and dispute resolution;
- liaising with water users on timely payment of service fees, and enforcement of penalties and sanctions for non-payment;

- liaising with the main system service provider on water delivery and service fee collection and payment;
- overseeing water distribution and maintenance as managed by the WUA O&M Engineer/Technician and/or Water Masters;
- conducting seasonal walk-through inspection of the I&D system with the WUA O&M Engineer/Technician, Water Masters and WUA Representatives;
- investigating farmers' complaints and taking appropriate action;
- as and when required, applying sanctions in line with by-laws and regulations;
- liaising with governmental and non-governmental bodies on issues related to WUA activities and water users' concerns;
- engaging local contractors for maintenance work (as sanctioned by the Management Board);
- assessing scheme performance and reporting back to the Management Board and water users;
- monitoring and oversight of WUA accounts and financial administration;
- providing information as requested by the WUA Regulatory Authority and other legitimate government agencies.

WUA O&M Engineer/Technician

The WUA O&M Engineer/Technician is responsible for the management, operation and maintenance of the I&D system. He/she reports to the WUA Executive Director. His/her main tasks include:

- preparing a seasonal water allocation plan for the service area based on the proposed cropping;
- advising water users on water availability and allocation;
- receiving and recording requests for water from water users;
- keeping operational records (e.g. discharges, pump operating hours, water use and allocation, etc.);
- in association with the Water Masters, preparing daily irrigation schedules;
- identifying maintenance requirements on a regular basis and through seasonal walk-through inspections;
- presenting matters relating to system operation and maintenance to the Management Board and General/Representative Meetings/Assembly;
- assessing maintenance requirements, estimating costs and preparing the maintenance budget;
- coordinating and supervising maintenance work;
- organizing communal labour for periodic maintenance work;
- monitoring and evaluating performance related to the operation and maintenance of the system.

WUA Accountant

The WUA Accountant is responsible for the financial management of the WUA. He/she reports to the WUA Executive Director and to the Management Board. His/her main tasks include:

- collecting and recording membership registrations and fees;
- collecting and recording irrigation service fees,[10] donations and fines;
- keeping and issuing receipts;
- keeping cash and bank accounts;
- keeping the WUA cheque book;
- making payment on authorized procurement of goods and services;
- keeping and recording invoices;
- keeping the stock book;
- preparing monthly accounts;
- preparing seasonal and annual statements of accounts;
- preparing annual and seasonal budgets;
- arranging for seasonal or annual auditing;
- reporting on financial matters to the Management Board and General/Representative Meetings/Assembly.

WUA field staff (Water Masters)

The WUA field staff (Water Masters) are responsible for operation and maintenance of the system under the direction of the WUA O&M Engineer/Technician, liaising with water users on a daily basis, and where possible resolving conflicts arising at field level. The Water Masters report to the WUA Executive Director. Their main tasks include:

- receiving and recording requests for water from water users;
- implementing the daily irrigation schedule;
- advising farmers of their turns and durations;
- monitoring and controlling irrigation supplies according to the schedule;
- resolving disputes at field level, where possible;
- liaising with water users in the field and reporting back to the WUA O&M Engineer/Technician as appropriate;
- keeping field operational records (e.g. discharges, pump operating hours, water use and allocation, etc.);
- inspecting the I&D system on a daily basis and reporting any problems;
- assisting in organizing communal activities for periodic maintenance.

Water Users Representatives

The Water Users Representatives are responsible for representing the needs of farmers within their designated command area. They report to the water users in their zone or block. Their main tasks include:

- liaising on behalf of zone/block water users with the WUA Executive Director, O&M Engineer/Technician and Water Master on issues related to management, operation and maintenance of the irrigation system;
- representing the zone/block user group at WUA Management Board and other meetings, and reporting back to water users;
- collecting data from landowners and tenants within their zones/blocks on behalf of the WUA;
- communicating and liaising with zone/block water users on behalf of the WUA;
- attending training sessions;
- organizing and running information and training sessions for farmers;
- in association with WUA O&M Engineer/Technician and Water Master, assisting farmers in the organization of water distribution within the zone/block service area;
- in association with WUA O&M Engineer/Technician and Water Master, identifying maintenance needs affecting the zone/block area;
- keeping records of water shortages, water supply failures, power cuts, etc. affecting water delivery to farmers;
- assessing and maintaining records of agricultural performance within the zone/block command area (this may be an optional requirement).

Training programmes for water users associations

Training is a central element in any programme to establish water users associations. Key areas for training are summarized in Table 8.6. To establish and carry out these training programmes requires significant financial and human resources. Without these resources the training cannot be implemented, and without the training it is unlikely that the water users and the WUA personnel will fully understand the processes involved and be capable of carrying out the functions required of them.

Where the transfer programme is funded the training is typically implemented by the Support Unit, with training of the Central Support Unit being carried out by personnel experienced in WUA formation and support. As an adjunct to this training these experienced personnel will often also assist of the Central Support Unit in the preparation of the training material for the Regional and District Support Units, and that for the WUAs and water users. Once trained the Central Support Unit staff will train the Regional and District Support Unit staff, who in turn will train the WUA personnel and the water users.

Procedures for preparing and carrying out training are detailed in Chapter 7. It suffices to mention here that the training for

Table 8.6. Key target groups and topics for awareness raising and training in a water users association establishment programme.

Target group	Training topics
WUA Support Unit/ Team staff	• Purpose and objectives of WUA formation • Organization, functions and tasks of Support Units • Processes and procedures for formation and establishment of WUAs • Legal aspects related to WUA formation • WUA duties and responsibilities • WUA processes and procedures, functions and tasks • Principles and practices of system management, operation and maintenance at on-farm level • Supporting WUAs
WUA members	• Purpose and benefits of WUA formation • Procedures for formation and establishment of WUAs • WUA legal framework – WUA law, statutes and by-laws • Purpose and role of Water Users Representatives and election procedures • Purpose of ISFs • Setting and paying the ISF • WUA management processes and procedures • Operation processes and procedures • Maintenance processes, procedures, costs and benefits
WUA Managing Board members and sub-committees	• Purpose and functions of WUA Management Board and sub-committees • Management Board processes and procedures • Management Board duties and responsibilities • WUA legal framework – WUA law, statutes and by-laws • Dispute resolution procedures • Financial management and auditing for WUAs
WUA Executive Director and staff	• Purpose and functions of WUA Executive Director and staff • WUA legal framework – WUA law, statutes and by-laws • WUA management, operation and maintenance processes and procedures • Liaison with water users • Fee collection processes and procedures – fee setting and collection • Dispute resolution
Water Users Representatives	• Purpose and role of Water Users Representatives • Water Users Representatives' duties and responsibilities • WUA legal framework – WUA law, statutes and by-laws • Water users' rights and responsibilities • Liaising with and representing water users • Dispute resolution procedures • Fee collection processes and procedures – fee setting and collection
I&D agency staff	• Purpose and role of WUAs • WUA organizational structure, roles and functions • WUA legal framework – WUA law, statutes and by-laws • Liaising and working with WUAs and water users • Dispute resolution procedures • Fee collection processes and procedures – fee setting and collection • Operation and maintenance procedures at the on-farm level
Local administrators and politicians	• Purpose and role of WUAs • WUA legal framework – WUA law, statutes and by-laws • Benefits of WUAs • Liaising and working with WUAs • Supporting WUAs • WUAs and local government

WUA, water users association; I&D irrigation and drainage; ISF, irrigation service fee.

the WUA personnel and water users *must* be practical, with the use of exercises and field work wherever possible. Water users will have limited tolerance for training material and methods which are too theoretical.

Factors to consider in establishing a water users association

The following are some key factors to consider in establishing effective and sustainable water users associations.

- If there is an existing (traditional) water users associations assess its performance. If it is functioning effectively, do not try to impose (external) structures, rules and regulations.
- Develop a phased development approach if establishing new WUAs; do not try to do everything at once. The priority is to improve the water management, followed by fee collection, followed by improving maintenance.
- Keep the water users association as a single-function organization focused on improving water management. If the association is successful, consideration may be given to branching out into provision of inputs, marketing, etc., but be careful not to lose sight of the core water management function. Consider carefully if these other functions might be better provided by other mechanisms.
- Ensure that the WUA management remains open and transparent, and responsible to the membership at all times.
- In association with water users and WUA management, set measurable and achievable targets, and provide feedback on progress made. This creates a feeling of achievement, generates confidence and increases expectations and ambition.
- Allow sufficient time for WUAs to become fully self-sustaining; 10–12 years is a reasonable timeframe.
- Key factors in the success of a WUA are the membership, the leadership and developing a sense of ownership. Good (or bad) leadership can make (or break) any organization; WUAs are no exception. Ownership of the I&D system by the members is central to success. Particular effort is required if the system has been designed, built and then managed by a government agency before being transferred to the water users.
- Continuing agency support is essential, particularly in the early stages of WUA formation and establishment. Successful international examples of WUA establishment following transfer (Mexico, Turkey) have all involved significant involvement by the government agency responsible for irrigation and drainage.

Reasons for failure of water users associations

At an international conference on irrigation management transfer held in Wuhan, China in September 1994 it was generally concluded that IMT, if properly executed, could benefit both farmers and the government, though a number of issue were identified that could adversely affect WUA formation and establishment (Kloezen and Samad, 1995). Given below are some of the reasons why water users associations might fail.

- *Lack of time*: IMT needs to be carried out in carefully managed stages, and requires considerable time and supporting effort.
- *Lack of support/commitment*: Although most governments find IMT attractive there is often only partial support for the process, especially among politicians and senior government personnel. Mixed messages and lack of support at crucial times can cripple IMT programmes.
- *Failure to provide institutional support for the IMT process*: In some countries the

IMT process has taken the form of a government instruction, with the expectation that water users will spontaneously form water users associations. This is unlikely to happen, specialist WUA support teams are required to initiate and support the IMT process.

- *Inadequate legal framework*: Governments have not always formulated the requisite policies and legal frameworks for IMT.
- *Failure to upscale*: In many countries IMT has not progressed beyond the pilot stage.
- *Focus on cost reduction, rather than improved performance and service delivery*: IMT is often initiated by government with a focus on reducing costs in the irrigation and drainage sector, rather than on improving performance and service delivery. This results in a failure to invest adequately during the transfer process and to take advantage of the opportunities offered by IMT.
- *Lack of funds for system rehabilitation*: In many cases prior to IMT the I&D systems are in a poor condition as government has not provided sufficient funds for system management, operation and maintenance. Newly formed water users associations need external support to repair or rehabilitate their systems if they are to provide adequate levels of service to water users.
- *Failure to take account of farmers' needs*: Often farmers' needs, aspirations and capability to take over management are not adequately considered.
- *Profitability of irrigated agriculture*: For IMT to be sustainable, irrigated agriculture needs to be profitable for farmers.
- *Need to focus on sustainability*: Initially the focus is on the IMT process, attention then needs to be focused on ensuring the sustainability of the management of the transferred systems, especially in terms of service fee recovery and thereby system maintenance.
- *Failure to adequately consider government agency staff*: A major issue during the transfer process is the retrenchment of irrigation agency personnel, and the need for strategic reorientation of the irrigation agency from the role of service provider to a regulatory organization. Failure to adequately address this issue results in resentment, resistance to the IMT process, and possible sabotage of the process.
- *Context specific nature of management*: It is recognized that post-turnover management systems are context-specific and dependent on a mixture of social, political, economic and technical factors.
- *Transparency and accountability*: Management accountability, financial autonomy, water rights and property rights are vital ingredients to successful IMT.
- *Evolutionary process*: IMT should be seen as a long-term evolutionary process, rather than a structural adjustment programme.

Monitoring progress with the formation and establishment of water users associations

A list of typical activities involved in establishing water users associations, together with associated performance indicators, is outlined in Table 8.7. While progress with these activities will be monitored at the lower level a useful approach for senior management is to monitor progress through the use of milestones. Table 8.8 shows the milestones used for monitoring progress in Kyrgyzstan, while Fig. 8.5 provides a graphical plot of the data recorded. In this example, WUAs achieving Milestone 4 were eligible for rehabilitation work under the project.

Appendix 2 provides a checklist for making a rapid performance assessment of WUAs. The checklist covers the three main areas of institutional, financial and technical components of WUAs, and is intended for use where there is a large number of WUAs. It is important to note that the performance standards given are indicative and should be reviewed and adjusted to the standards expected in a given country.

Table 8.7. List of possible activities and indicators for implementation monitoring.

No.	Activity	Measures and indicators
1.	Enact new, or upgrade existing, legislation for establishing WUAs and Federations	Status of legislation – drafted, enacted, in use
2.	Formation of WUA Support Units	Number of Support Units formed (each quarter, year) Number and types of staff Training events carried out (for Support Unit staff)
3.	Formation and establishment of WUAs	Number of WUAs formed (each quarter, year) Milestones achieved – formed, staff hired, O&M plan prepared, etc. Area covered by WUAs (area and as a percentage of the total irrigable area in the country) Number of WUAs formed in each region Assets transferred from government to WUA account
4.	Publicity, communication and awareness campaigns	Status of campaigns – needs identified, material produced, campaign started, activities done, etc. Number and types of people, communities, agencies, etc. contacted through the campaigns Impact evaluation – pre- and post-campaign awareness assessment
5.	Training and capacity building programmes	Status of programmes – needs identified, training plan produced, training material produced, trainees identified, training course run,etc. Number and types of training courses carried out Number and types of people trained Training evaluation – pre- and post-training knowledge tests, pre- and post-training assessment of understanding, knowledge and skills
6.	Development of management capability, including record keeping and performance monitoring	Status – identification of needs, development plan, management systems functioning, etc. Implementation of plan – training, preparation of maps, records, filing system, etc. Performance monitoring – meetings held, level of attendance, complaints, issues arising, etc.
7.	Development of financial management capability	Status – identification of needs, development plan, financial systems functioning, etc. Implementation of plan – training, preparation of recording systems, bills and receipts, etc. Performance monitoring – fee level set, budget, expenditure, fee collection, results of annual audit, etc.
8.	Development of technical management capability (system operation and maintenance)	Status – identification of needs, development plan, O&M systems functioning, etc. Implementation of plan – training, preparation of scheduling systems, water delivery records, etc. Performance monitoring – water abstracted and used, amount of water invoiced and paid for, crops grown, yields, maintenance work carried out, complaints, issues arising, etc.
9.	Support for the purchase of maintenance machinery and equipment	Status – identification of needs, plan, purchased, etc. Implementation of plan – training, purchase, cost, etc. Performance monitoring – work completed each year, expenditure on fuel, maintenance, etc.
10.	Formation and establishment of Federations of WUAs	Number of Federations formed (each quarter, year) Milestones achieved – formed, staff hired, O&M plan prepared, etc. Area covered by Federations (area and as percentage of total area) Number of Federations formed in each region
11.	Development of processes and procedures for WUA Regulatory Authority	Status – formed, staff trained, etc. Activities being carried out – reporting forms prepared, reports received, database designed and operational, etc.

Continued

Table 8.7. Continued.

No.	Activity	Measures and indicators
		Number of complete sets of records for WUAs Performance monitoring – reported on in Annual Report: WUA status, budget, expenditure, fees set and recovered, maintenance expenditure, etc.
12.	Formation and establishment of National Association of WUAs	Status – discussions, National Association formed, meetings held, etc. Number of members and total area covered

WUA, water users association; O&M, operation and maintenance.

Table 8.8. Framework for development monitoring of water users associations using milestones. (From OIP, 2008.)

	Number of WUAs in category by region				
Milestone	Region A	Region B	...	Region Z	Total
1. Formerly established					
Last reporting period (no.)	84	31			450
Current reporting period (no.)	84	33			454
Changes (no.)	0	2			4
2. Staff hired and training started					
Last reporting period, current reporting period, changes (no.)					
3. O&M plan prepared					
Last reporting period, current reporting period, changes (no.)					
4. Irrigation service fee paid					
Last reporting period, current reporting period, changes (no.)					
5. Rehabilitation alternatives developed					
Last reporting period, current reporting period, changes (no.)					
6. Rehabilitation alternative selected					
Last reporting period, current reporting period, changes (no.)					
7. WUA is ready for cooperation					
Last reporting period, current reporting period, changes (no.)					

WUA, water users association; O&M, operation and maintenance.

Restructuring and Modernizing Irrigation and Drainage Agencies

The need for restructuring and modernization

There are a number of drivers bringing about the need for change in I&D agencies, including the following.

- Increasing pressure on water resources, particularly from other sectors, including the environment.
- Increasing demands for transparency and accountability in relation to water resource abstraction.
- The introduction of charging systems for water users and the subsequent demand, and right, by water users for greater

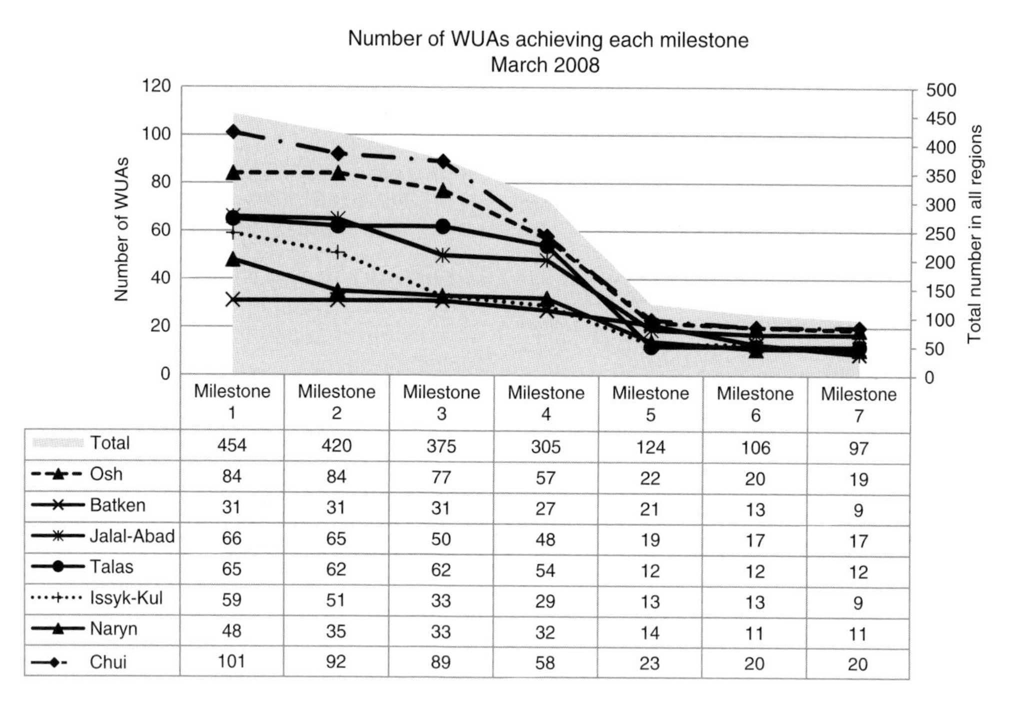

	Milestone 1	Milestone 2	Milestone 3	Milestone 4	Milestone 5	Milestone 6	Milestone 7
Total	454	420	375	305	124	106	97
Osh	84	84	77	57	22	20	19
Batken	31	31	31	27	21	13	9
Jalal-Abad	66	65	50	48	19	17	17
Talas	65	62	62	54	12	12	12
Issyk-Kul	59	51	33	29	13	13	9
Naryn	48	35	33	32	14	11	11
Chui	101	92	89	58	23	20	20

Fig. 8.5. Example of graphical representation of milestones achieved (WUA, water users association). (From OIP, 2008.)

transparency and accountability with regard to:
- the use of funds;
- irrigation water allocation and distribution.

- Increasing pressure to define, agree with users and then provide a responsible level of service.
- Changes in society, with a greater expectation and demand for accountability by government agencies to the wider society.
- Turnover/transfer of all or sections of I&D systems to management by water users.
- Changes brought about by reforms in government policy, with moves towards decentralization and reduction in government ownership and funding.
- Reduced disparities between the educational and economic status of the irrigated farmer and the irrigation agency professionals.
- Reducing importance of the role of the design and construction engineer and increasing importance of the O&M engineer and water management specialist.

Significant changes are happening in many societies, with a perceived need for government to disengage from activities that might be equally well or better performed by the private sector or by communities. Funding is also an issue, with government wishing to increase the financial contribution made by beneficiaries, particularly where beneficiaries are deriving financial benefit from the services provided. In many countries the government monopoly on a range of services including electricity, telecommunications and potable water supply is being relinquished, with private sector organizations taking on the responsibility for their management.

While an I&D agency may not be privatized, it will need to change with the times and become more transparent and accountable to its clients, as well as needing to justify and raise its standard of service delivery.

Areas where change is required

There are a number of areas where change may be required. These may include changes to:

- the legal status of the entity, from a line agency to a parastatal or private company;
- the charter of the organization;
- the staffing structure, staffing numbers, employment procedures, training and capacity building;
- attitudes and behaviours of agency personnel;
- financial and accounting systems;
- management information systems;
- accountability and reporting requirements.

Approach to change

The first step in the process is an acknowledgement that change is required, and an appreciation that doing 'business as usual' is no longer a feasible option. The next step is to commission a study to objectively review the organization's performance, identify the current and possible future operating environment and pressures, and make recommendations for change. Assistance may also be required to implement the change management process.

The approach and content of an institutional and organizational analysis will vary depending on the circumstances. The following general steps are proposed based on experience with such analysis of several organizations.

Step 1: Specify clear terms of reference

It is important at the outset to specify clear terms of reference for the study. These should state the purpose and objectives of the study, and the boundaries.

Step 2: Put together a suitable team

Due to the diverse nature of functions within an I&D agency it is likely that a team of specialists will be required to carry out the study. Specialist functions will include: institutional development; management, operation and maintenance; finance and economics; and human resource development, training and capacity building.

Step 3: Inception, data collection, processing and analysis

The first step in the implementation of the study is to formulate an approach to the work. A broad framework is to ask the following questions.

- Where are we now?
- Where do we want to be?
- How are we going to get there?

The study can then be broken down into three parts, as summarized below and discussed in the following sections.

1. Strategic analysis:
   - analysis of the external environment within which the organization operates;
   - analysis of the organization's current internal functions;
   - analysis of combined external and internal environment.
2. Strategic planning:
   - formulation of desired future function and structure;
   - formulation of options for future direction for the organization.
3. Organizational transformation:
   - detailed analysis of changes required, pathways and costs;
   - preparation of an implementation plan.

STRATEGIC ANALYSIS *Analysis of the external environment within which the organization operates.* An initial starting point for the study is to understand the current environment within which the organization operates, and the possible future environment. A PESTLE analysis is useful for this purpose.

- *Political*: An analysis of the political environment within which the organization has to operate. It will include an analysis of political trends and aspirations.
- *Economic*: Assessment of the current and future economic environment, both

within and outside the irrigated agricultural sector. It will include investigation of economic trends within the country.

- *Social*: An assessment of the social norms and expectations, particularly within the rural community and future possible changes.
- *Technical*: An assessment of the current and possible future technical capability and context within which the organization must operate (level of technology, etc.).
- *Legal*: An assessment of the current and future possible legal environment. This will in particular cover legislation related to the organization, and associated organizations such as line ministries and water users associations.
- *Environmental*: An assessment of current and future environmental factors, particularly in relation to water resources and agriculture.

The analysis will look at the key stakeholders with whom the organization interacts, their role and relative importance to the functioning of the organization.

Analysis of the organization's current internal functions. A detailed analysis will be carried out of the organization's current:

- roles and responsibilities;
- organizational structure;
- staffing (roles, responsibilities, numbers, salaries, etc.);
- operations procedures (water delivery, maintenance, etc.);
- assets (infrastructure, equipment, etc.);
- income, expenditure and financial resources.

The analysis of essential water resources management functions (as outlined in Chapter 2, Fig. 2.5) has a role to play in identifying the functions of the I&D agency and other key stakeholders.

Analysis of the combined external and internal environment. This analysis will draw together the external and internal analyses, and highlight the current strengths and weaknesses of the organization, and the opportunities for, and threats to, its future. This SWOT analysis will form the basis for formulating future plans.

STRATEGIC PLANNING *Formulation of desired future function and structure.* Having gained an understanding of the current situation (*Where are we now?*), options for future directions (*Where do we want to be?*) will need to be investigated. This process will be based around discussions with senior personnel within the organization and other key stakeholders to formulate broad criteria and objectives. Key factors to consider will include the following.

- What will be the future functions of the organization?
- What will be the duties and responsibilities (and what will they not include)?
- How will the organization be financed?
- What staffing and resources will it require?

Formulation of options for future direction for the organization. Following the discussions above a number of options may evolve and require further investigation. Scenario formulation and analysis can be used to analyse each of these options. The scenario formulation will look at options related to the key variables, including:

- legal status – whether a public, parastatal or private entity;
- relationship with external bodies, including line agencies, WUAs and water users;
- duties and responsibilities;
- income and expenditure profiles;
- physical system and organizational modernization;
- staffing and salary levels;
- working practices.

Following discussion of the possible scenarios with the organization's senior management and key stakeholders, the preferred future option will be selected.

ORGANIZATIONAL TRANSFORMATION *Detailed analysis of changes required, pathways and costs.* Following selection of the future option, strategic planning will be carried out to detail how the organization will achieve that state (*How are we going to get there?*). The analysis will cover:

- required legal framework;
- structure and functions at different levels;
- staff duties and responsibilities and working practices;
- staffing levels and salaries;
- human resource development (capacity building and training);
- linkages with related organizations and individuals (e.g. WUAs and water users);
- income, expenditure and risk analysis;
- differentiation between state-funded functions (flood protection, drainage) and privately funded functions (irrigation water supply, on-farm drainage and water removal).

Preparation of implementation plan. The identified changes will be incorporated into an implementation plan detailing:

- the actions required;
- the costs involved;
- the resources required;
- measures to gain acceptance of the changes among the organization's staff and key stakeholders;
- training and capacity building requirements and approaches.

Associated with this formulation a list of the work required (Table 8.9) and a work programme (Fig. 8.6) will be formulated. As shown in Table 8.9, information will be required on a wide range of topics, ranging from the organizational structure to the income-generating potential of irrigated agriculture and water users' organizations. In situations where the organization is expected to be self-financing it is essential that a thorough analysis is carried out of the potential for water users to finance the services to be provided.

A useful tool in analysing the current situation in relation to the organization is a SWOT analysis (Table 8.10). Summarizing the organization's key strengths and weaknesses in the SWOT analysis provides information on the internal environment, while summarizing the key opportunities and threats provides information on the external environment. From the example shown the major issues are a lack of finance for the organization; recently formed, weak water users associations; and poorly developed markets and low returns to irrigated agriculture.

Restructuring an organization under these conditions is difficult, for this example the challenges include:

- severely reduced budget allocation from central government;
- badly deteriorating hydraulic systems due to lack of adequate maintenance over the last 15 years;
- severely depleted mechanical maintenance machinery and equipment, much of it past its useful life;
- low staff salaries;
- ageing staff profile, with many staff nearing retirement and few new, younger staff joining the organization;
- low morale and motivation at some levels within the organization.

Balanced against these challenges are some positive aspects:

- the Agency is a professional organization with a wealth of experience;
- despite the recent difficulties the Agency retains much of its professionalism;
- though some staff have left or retired, the organization still has very experienced personnel in key positions;
- the Agency has retained all its records during periods of turmoil;
- though it could benefit from modernization through computerization, the Agency has a well-developed management information system.

Step 4: Formulation of options

The previous section has outlined how the analysis identifies the challenges facing an organization. In the case of the Agency represented in the SWOT analysis above, the areas where changes may be beneficial include:

Table 8.9. Data requirements for a reorganization study.

No. Description	Why needed
Organizational structure, facilities and finances	
1. Current organization structure diagram	To understand the organizational structure
2. Current staffing details to include: • Location of staff by office • Staff designation/position • Duties and responsibilities, job descriptions • Qualifications (key staff) • Age (key staff) • Salary (for position) • Recruitment plan • Interviews with staff	To know the current level of staffing, their distribution and cost To be able to match staff with current and future job functions To understand the current levels of knowledge, understanding, skills together with attitudes, aspirations and levels of motivation
3. Inventory of offices, including: • Number of rooms • Approximate floor area (m^2)	To know the extent of the organization's office assets To match the office assets with current and future needs
4. Inventory of office equipment	To know the extent of the organization's office equipment assets To match the office equipment assets with current and future needs
5. Inventory of vehicles, maintenance machinery and workshop equipment, including: • Type and make • Average age • Average condition	To know the extent of the organization's plant, machinery and workshop equipment assets
6. Inventory of physical infrastructure (per scheme), including: • Type of irrigation and drainage scheme (gravity, pumped, etc. • Type of key structures (canal, drain, head regulator, cross regulator, etc.) • Extent/number/size of infrastructure (canals, drains, head regulators, aqueducts, etc.) • Date scheme built • General condition (good, moderate, poor) • Operation and maintenance staff per scheme • Key data related to reservoirs (type, live storage, dead storage, elevation, etc.) • Key data for other infrastructure	To know the extent of infrastructure for which the organization is responsible To estimate the annual O&M costs for these physical assets, and hence budget requirements To update the estimate of the value of the assets (using MEA value valuation approach)
7. Details of management, operation and maintenance procedures, including: • General management processes and procedures • Processes and procedures for system operation, including scheduling, performance monitoring and evaluation, communication and liaison with water users, etc. • Processes and procedures for maintenance of the I&D systems, including maintenance identification, planning and budgeting, prioritization, implementation and monitoring and evaluation	To understand the management processes related to general management and administration, operation and maintenance In general, to be able to make an assessment of current practices, and propose recommendations for improvement To identify key processes that are fundamental to good service provision, identify how they might be improved, and establish key performance indicators for management monitoring and evaluation

Continued

Table 8.9. Continued.

No.	Description	Why needed
8.	Budget allocations for the last 10 years to include: • Total budget requested • Total funds allocated • Total annual irrigation command area • Total annual area actually irrigated • Divisions of funds (salaries, office costs, operation, maintenance, machinery costs, machinery maintenance, etc.) • Source of funds (Government, donors, fee payment, etc.)	To understand the historical budget and actual funding allocation To ascertain the gap between funds actually required, those budgeted for and actually received
Legal aspects		
9.	Legislation (laws and ordinances) related to current functions, organizational strucutre, staffing, financing, etc.	To understand current legal framework, opportunities, restrictions, roles and responsibilities
10.	Planned or drafted legislation related to future possible functions, organizational structure, staffing, financing, etc.	To understand future possible legal framework, opportunities, restrictions, roles and responsibilities
11.	Planned, drafted or enacted legislation related to WUA's roles and responsibilities	To understand current legal framework, roles and responsibilities
12.	Planned, drafted or enacted legislation related to associated agencies' roles and responsibilities (Environment Agency, Ministry of Agriculture, etc.)	To understand current legal framework, roles and responsibilities, impact on the organization and WUA activities
Tax aspects		
13.	Tax implications for the organization now and in the future for different possible organizational structures	To understand current tax implications for the organization
Farm income and market trends		
14.	Farm budgets for different agro-climatic zones and range of farm sizes	To estimate the likely levels of income from water users now and in the future To estimate the ability of water users to pay water fees
15.	Market prices for key products over last 5 years	Trend analysis to understand how the market is changing over time
16.	Other relevant market indicators relevant to agricultural sector over last 5 years (imports, exports, etc.)	Trend analysis to understand how the market is changing over time
17.	Details for typical agro-climatic zones (climate, soils, land slopes, etc.)	To understand the irrigation and drainage needs in each zone
Water user associations and farmers		
18.	Data related to WUAs, including: • Total command area • Total number of WUAs planned • Current number of WUAs formed, date formed and status in categories (Functioning well, Functioning, Not functioning) • Command area of each WUA • Total number of farmers in each WUA command area • Total number of WUA members in each WUA command area • Service fee levels and amounts collected • Expenditure on system maintenance	To know how many WUAs there are and their status To determine current and future performance
19.	Farm family data, including: • Landholding sizes • Family size • Educational levels • Family income and sources, etc.	To understand the socio-economic status of the farming community

WUA, water users association; MEA, modern equivalent asset.

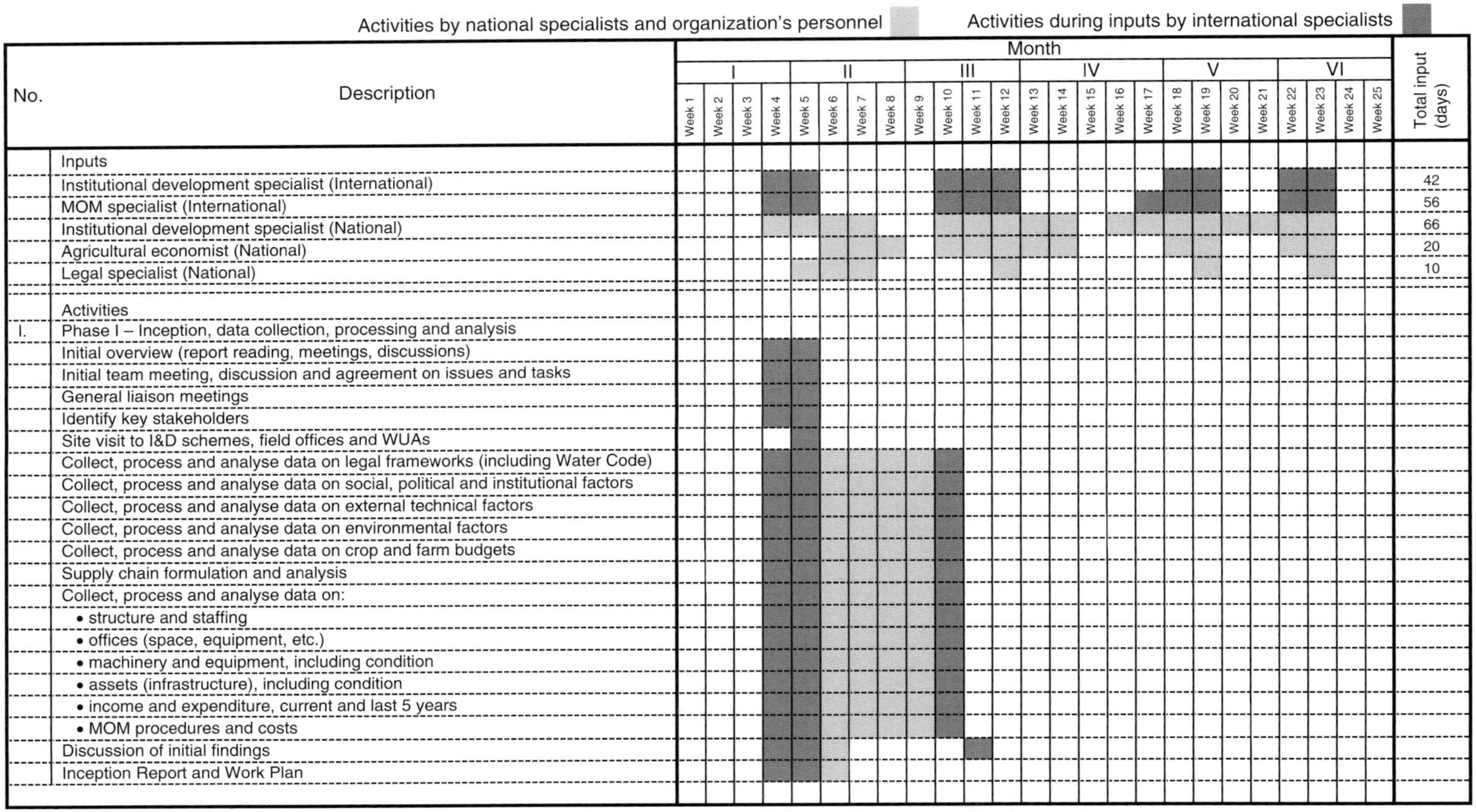

Activities by national specialists and organization's personnel (light shading) | Activities during inputs by international specialists (dark shading)

No.	Description	Month																									Total input (days)
		I				II				III				IV					V				VI				
		Week 1	Week 2	Week 3	Week 4	Week 5	Week 6	Week 7	Week 8	Week 9	Week 10	Week 11	Week 12	Week 13	Week 14	Week 15	Week 16	Week 17	Week 18	Week 19	Week 20	Week 21	Week 22	Week 23	Week 24	Week 25	
	Inputs																										
	Institutional development specialist (International)																										42
	MOM specialist (International)																										56
	Institutional development specialist (National)																										66
	Agricultural economist (National)																										20
	Legal specialist (National)																										10
	Activities																										
I.	Phase I – Inception, data collection, processing and analysis																										
	Initial overview (report reading, meetings, discussions)																										
	Initial team meeting, discussion and agreement on issues and tasks																										
	General liaison meetings																										
	Identify key stakeholders																										
	Site visit to I&D schemes, field offices and WUAs																										
	Collect, process and analyse data on legal frameworks (including Water Code)																										
	Collect, process and analyse data on social, political and institutional factors																										
	Collect, process and analyse data on external technical factors																										
	Collect, process and analyse data on environmental factors																										
	Collect, process and analyse data on crop and farm budgets																										
	Supply chain formulation and analysis																										
	Collect, process and analyse data on:																										
	• structure and staffing																										
	• offices (space, equipment, etc.)																										
	• machinery and equipment, including condition																										
	• assets (infrastructure), including condition																										
	• income and expenditure, current and last 5 years																										
	• MOM procedures and costs																										
	Discussion of initial findings																										
	Inception Report and Work Plan																										

Fig. 8.6. Example of a restructuring study programme (I&D, irrigation and drainage; WUA, water users association; MOM, management, operation and maintenance; O&M, operation and maintenance; ISF, irrigation service fee).

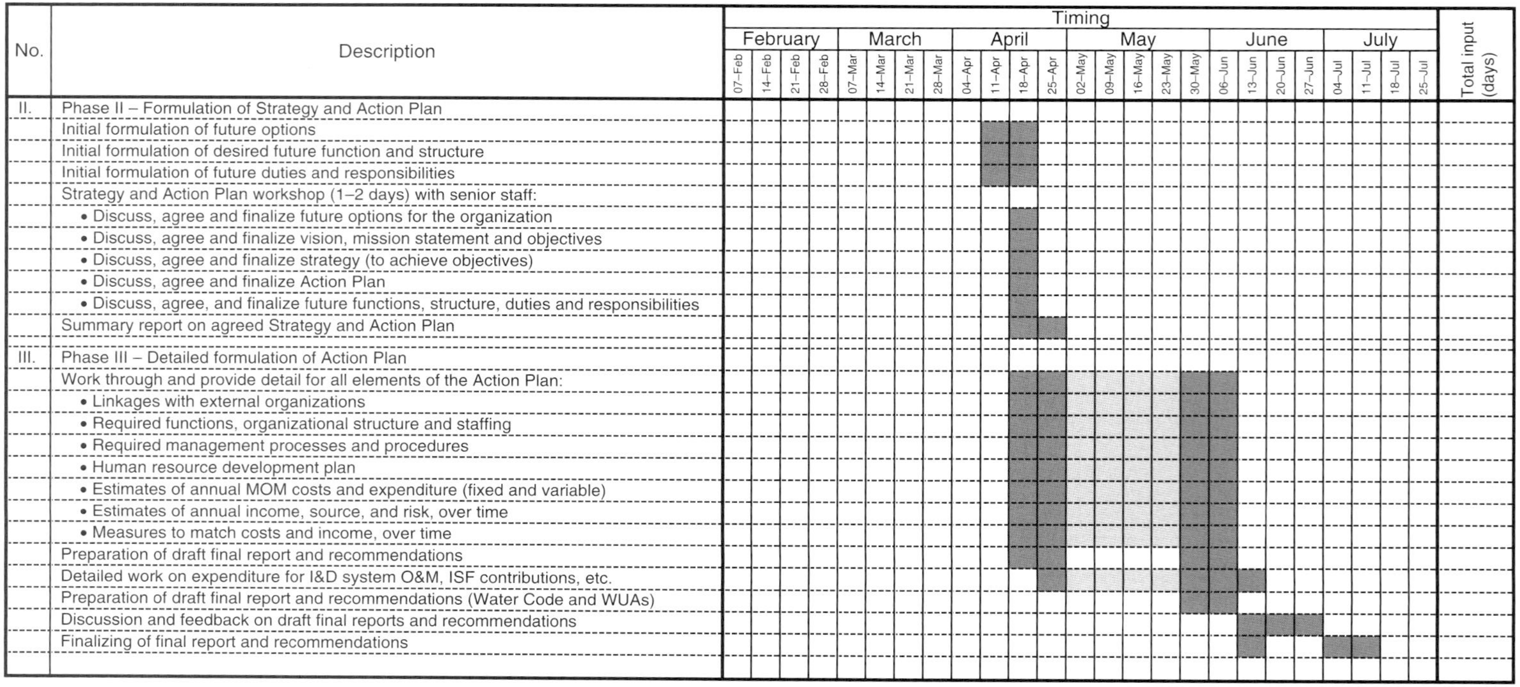

No.	Description	Timing																									Total input (days)
		February				March				April				May					June				July				
		07–Feb	14–Feb	21–Feb	28–Feb	07–Mar	14–Mar	21–Mar	28–Mar	04–Apr	11–Apr	18–Apr	25–Apr	02–May	09–May	16–May	23–May	30–May	06–Jun	13–Jun	20–Jun	27–Jun	04–Jul	11–Jul	18–Jul	25–Jul	
II.	Phase II – Formulation of Strategy and Action Plan																										
	Initial formulation of future options																										
	Initial formulation of desired future function and structure																										
	Initial formulation of future duties and responsibilities																										
	Strategy and Action Plan workshop (1–2 days) with senior staff:																										
	• Discuss, agree and finalize future options for the organization																										
	• Discuss, agree and finalize vision, mission statement and objectives																										
	• Discuss, agree and finalize strategy (to achieve objectives)																										
	• Discuss, agree and finalize Action Plan																										
	• Discuss, agree, and finalize future functions, structure, duties and responsibilities																										
	Summary report on agreed Strategy and Action Plan																										
III.	Phase III – Detailed formulation of Action Plan																										
	Work through and provide detail for all elements of the Action Plan:																										
	• Linkages with external organizations																										
	• Required functions, organizational structure and staffing																										
	• Required management processes and procedures																										
	• Human resource development plan																										
	• Estimates of annual MOM costs and expenditure (fixed and variable)																										
	• Estimates of annual income, source, and risk, over time																										
	• Measures to match costs and income, over time																										
	Preparation of draft final report and recommendations																										
	Detailed work on expenditure for I&D system O&M, ISF contributions, etc.																										
	Preparation of draft final report and recommendations (Water Code and WUAs)																										
	Discussion and feedback on draft final reports and recommendations																										
	Finalizing of final report and recommendations																										

Fig. 8.6. Continued

Table 8.10. Example of a SWOT analysis for an irrigation and drainage agency.

Internal factors	
Strengths	**Weaknesses**
• Agency is a professional organization, with considerable experience and expertise • Dynamic senior management committed to change and reorganization, with a clear vision for the future role of the Agency • Robust procedures, which have withstood the rapid decline in Agency fortunes over the last 15 years • Despite the difficulties, the Agency has retained its records, maps and knowledge of schemes	• Significantly reduced budget allocation over last 10 years • Poor condition of I&D infrastructure (due to lack of maintenance funding) • Poor office and operational facilities (due to lack of funding) • Loss of some key staff to other organizations/ projects • Some among the older generation looking backwards, rather than forwards to the future • Lack of experience in working with WUAs and with customer-focused service delivery • Low salaries, acting as a barrier to young professionals joining the organization • Loss of corporate experience as experienced staff retire before being able to pass on their knowledge to the next generation • Using outdated manual administrative procedures rather than computer-based approaches

External factors	
Opportunities	**Threats**
• Reformation of the Agency as a legal entity under public law • Water resources and I&D legislation has been revised to permit WUAs and Federations to manage I&D systems • Formation of some active and functioning WUAs, providing good levels of service to water users • Donor-funded rehabilitation of some I&D schemes, allowing the Agency the opportunity to provide good levels of service on rehabilitated schemes • Changing attitudes in society, greater perception of the relationship between service delivery and fee payment • Prioritization of certain schemes to allow full MOM funding and thus high levels of service delivery and fee recovery • In some areas, good returns to irrigation	• Lack of full support from Government, especially in relation to adequate levels of funding until water users are able to meet the real costs of service provision • Supplementary nature of irrigation in many regions • Poorly developed markets, and inadequate returns to irrigated/drained agriculture • Small, fragmented landholdings, many landowners, subsistence farming • Lack of knowledge and farming skills among farming community • Lack of interest in irrigation by some farmers and low levels of reported irrigated area • Low levels of fee recovery, freeloading by some water users • Time taken to form effective WUAs, capable of delivering services to water users for which they are prepared to pay (a portion of which is to be transferred to the Agency)

I&D, irrigation and drainage; WUA, water users association; MOM, management, operation and maintenance.

Table 8.11. Example of elements of a restructuring study Strategy and Action Plan.

New name: National Water Management Agency (NWMA)

Vision: *Our vision is to become a leading service provider in the water sector. We will lead in terms of the quality of our service provision, and the level of fee recovery we obtain from our customers. We will be a professional, modern organization, responsive to the needs of our customers, employing well-trained and highly motivated staff.*

Mission: *Our mission is to provide high-quality and cost-effective services to support efficient, productive and sustainable use of irrigated and drained land. We will achieve this through the adequate and timely delivery of irrigation water, and the adequate and timely removal of excess water, together with measures for protection of irrigated land from flooding and maintenance of the quality of such land.*

Objectives:
Short-term (2005–2008)
- Change the legal status to become a parastatal organization
- Review the feasible extent of Agency-designated responsibilities, given the current and future anticipated funding
- Review office facilities and staffing requirements and restructure the organization to match feasible extent of designated responsibilities, including staff training
- Review maintenance equipment needs and purchase limited quantity of new equipment
- Review irrigation and drainage service delivery tariff rates with a view to increasing them to levels sufficient for sustainable MOM
- Accelerate the transfer of irrigation schemes smaller than 3000 ha to WUAs

Medium-term (2010–2015)
- Capital costs are estimated and financed for selected schemes
- Agency to assist in the formation of Federations of WUAs to enable Federations to take over the management of medium-sized irrigation/drainage systems
- Complete transfer of irrigation schemes smaller than 5000 ha to WUAs or Federations of WUAs

Long-term (2015–2020)
- Fully functioning hydraulic systems that are under Agency control
- Full MOM costs covered on I&D systems through water user fees
- Full MOM costs covered for reservoirs, flood embankments and drainage crossings/pathways, from water users' fees (20%) and government subvention (80%)
- Modern organization, with a staffing of between 500 and 600 staff, paid at rates comparable with the private sector

Strategy: *Our strategy is to reform the Department to become a modern, customer-focused organization (the National Water Management Agency), deriving the most part of our income from services provided to water users. We will achieve this by changing our legal status to that of a government-owned parastatal accountable to a Management Board, reforming the organizational structure, procedures and staffing so that we are able to provide a responsive, reliable and efficient service to our customers. We will seek to reduce the extent of the hydraulic infrastructure for which we are responsible to the maximum sustainable under anticipated levels of funding.*

Action Plan:
1. Change legal status
2. Carry out an inventory and prioritization of I&D systems and their infrastructure condition and serviceability
3. Identification and agreement with government on extent of Agency's designated responsibilities and associated financial requirements
4. Study of maintenance needs, and associated machinery and equipment requirements
5. Restructure the Department to become the National Water Management Agency
6. Train staff
7. Improve management information systems
8. Carry out irrigation and drainage tariff study (ability and willingness to pay, and level of service provision)

Continued

Table 8.11. Continued

9. Monitor and support WUAs and water users for enhanced performance
10. Revise maintenance budgeting and implementation procedures
11. Implement performance-based management

MOM, management, operation and maintenance; WUA, water users association; I&D, irrigation and drainage.

- type of legal entity (government, parastatal, private);
- making a commitment to service delivery;
- recovering full MOM costs from water users;
- balanced budgeting;
- full maintenance on rehabilitated schemes.

The SWOT analysis assists in the formulation of the options available for the organization and the possible direction it might take in the future. These options need to be identified, preferably though discussion with, and presentation of, ideas to senior management. Such discussion is important at this stage if the senior management in the organization is to take ownership of proposed changes and be supportive in their implementation. In addition, the time required to implement the changes and the resources required (in particular, political and financial support) need to be identified.

Step 5: Formulation of strategy and action plan

From the discussion of the different options the agreed option is selected and a Strategy and Action Plan formulated to implement it. The Strategy will identify the vision, mission, objectives and the strategy for implementing the change. The details of the strategy will be formulated in an Action Plan, which will provide the actions required to attain the objectives, and the time and resources needed.

An example of these elements of the Strategy and Action Plan is provided in Table 8.11.

Step 6: Implementation

The preparation of the Strategy and Action Plan is a relatively easy task compared with its implementation. At this level there is significant interest by government, and changes often have to be debated and discussed at the highest level. The ability to make the proposed changes will depend on the level of support in the line agency to which the organization reports. It will also be susceptible to changes and policies proposed by government for reforming the public sector in general.[11]

Finance will also be a key factor. Though the organization might be self-financing in the long run, in the short run it will most likely require continued government support to allow it to restructure, retrain, re-equip and reorganize itself.

Endnotes

[1] *Effectiveness*: the ability to meet goals, objectives or needs. *Efficiency*: the manner in which the goals are met, at as low a cost as possible without negative impact. *Accountability*: institutionalized responsiveness to those who are affected by one's actions.

[2] This innovation and adaptation by WUA management is an excellent indicator for success of an IMT programme.

[3] This information is drawn from two FAO publications written by Stephen Hodgson, a lawyer specializing in water users association legislation: (i) FAO Legislative Study No. 79, *Legislation on Water Users' Organizations: A Comparative Study* (FAO, 2003); and (ii) FAO Legislative Study No. 100, *Creating Legal Space for Water User Organizations: Transparency, Governance and The Law* (FAO, 2009).

[4] This statement can have important implications for tax purposes.

[5] In this case the service charges to non-members may be greater than those for members, up to a maximum specified level.
[6] Though this may be done by a show of hands, it is preferable and sometimes obligatory to obtain signatures from those who agree to the proposals.
[7] If the WUAs have been established as part of a project.
[8] Often WUA formation and establishment is part of a rehabilitation project.
[9] The terminology for these different administrative levels will vary from country to country.
[10] In some systems the Water Masters collect the service fee from water users, either at the time of irrigation or later. In general, in the interests of transparency and accountability it is preferable that the service fee payment is made in the WUA office to the accountant.
[11] Any such changes should have been picked up during the study and factored into the analysis and choice of option proposed.

9

Performance Assessment, Monitoring and Evaluation

This chapter provides a framework for performance assessment, emphasizing the importance of identifying the purpose and objectives of the performance assessment programme and going on to provide details of how such assessment can be carried out. Key performance indicators are detailed for management, operation and maintenance at the main system, on-farm and field level. This is followed with an outline of processes and procedures for project monitoring and evaluation (M&E), with details of two commonly used approaches, the logical framework and the results framework. The final section of the chapter introduces benchmarking, an increasingly widely used tool for comparative assessment of performance of irrigation and drainage schemes (I&D schemes), systems and processes. A detailed example is provided to show how a benchmarking study can be implemented.

Introduction

Performance assessment is an integral component of management. In the context of irrigation and drainage systems (I&D systems) performance assessment is carried out at different levels by different entities. At the main system level the main system service provider is interested in knowing how well water is being delivered, and whether the required fees are being recovered. At the on-farm level the water users association is interested in knowing how much water it is receiving, how well it is distributing the water and the level of fee recovery from water users. At the field level water users measure the performance of the association in delivering water, the output from their fields in terms of agricultural produce, and the income that it generates when sold.

By measuring performance, at whatever level, those responsible for management at that level are able to assess whether performance is satisfactory, or whether it can be improved. Through an assessment of the performance (and the associated process of diagnostic analysis) the manager is able to identify areas where performance can be enhanced.

Monitoring and evaluation is a part of performance assessment, and is generally used in the context of project,[1] rather than scheme, management. Monitoring is an integral part of the management of an irrigation and drainage project, seeking to ensure that the project is on track to complete the assigned activities on time, within budget and to the quality required. Evaluation is carried out once the project has been completed and is used to assess whether the project has been successful in achieving its objectives.

Information gained from evaluation of the project can then be fed back into the planning, design and implementation of future projects.

Framework for Performance Assessment[2]

Overview

Prior to establishing procedures for assessing performance it is important to think through the various components of the process and to establish a framework for the assessment. The framework serves to define why the performance assessment is needed, what data are required, what methods of analysis will be used, who will use the information provided, etc. Without a suitable framework the performance assessment programme may fail to collect all the necessary data, and may not provide the required information and understanding.

The framework is based on a series of questions (Fig. 9.1). The first stage, purpose and scope, looks at the broad scope of the performance assessment – who it is for, from whose viewpoint it is undertaken, who will carry it out, its type and extent. Once these are decided, the performance assessment programme can be designed, selecting suitable criteria for the performance assessment, the performance indicators and the data that will be collected. The implementation of the planned programme follows, with data being collected, processed and analysed. The final part of the programme is to act on the information provided, with a variety of actions possible, ranging from changes to long-term goals and strategy, to improvements in day-to-day procedures for system management, operation and maintenance.

Purpose and scope

The initial part of formulating a performance assessment programme is to decide on the purpose and scope of the performance assessment. Key issues relate to who the assessment is for, from whose viewpoint, the type of assessment and the extent/boundaries. It is important that adequate time is spent on this part of the work as it structures the remaining stages.

Purpose

As with any project or task it is essential that the purpose and objectives of the performance assessment be defined at the outset. Three levels of objective setting can be identified:

- rationale;
- overall objective;
- specific objectives.

The *rationale* outlines the reason why a performance assessment programme is required. The *overall objective* details the overall aim of the performance assessment programme, while *specific objectives* may be required to provide further detail on how the overall objective will be achieved (Table 9.1).

Establishing the rationale and identifying the overall and specific objectives of the performance assessment programme are not always straightforward; care needs to be taken at this stage of planning to ensure that these objectives are clearly defined before proceeding further.

For whom?

The performance assessment can be carried out on behalf of a variety of stakeholders. These include:

- government;
- funding agencies;
- irrigation and drainage service providers;
- I&D system managers;
- farmers;
- research organizations.

Who the assessment is for is closely linked to the purpose of the assessment.

From whose viewpoint?

The assessment may be carried out on behalf of one stakeholder or group of stakeholders

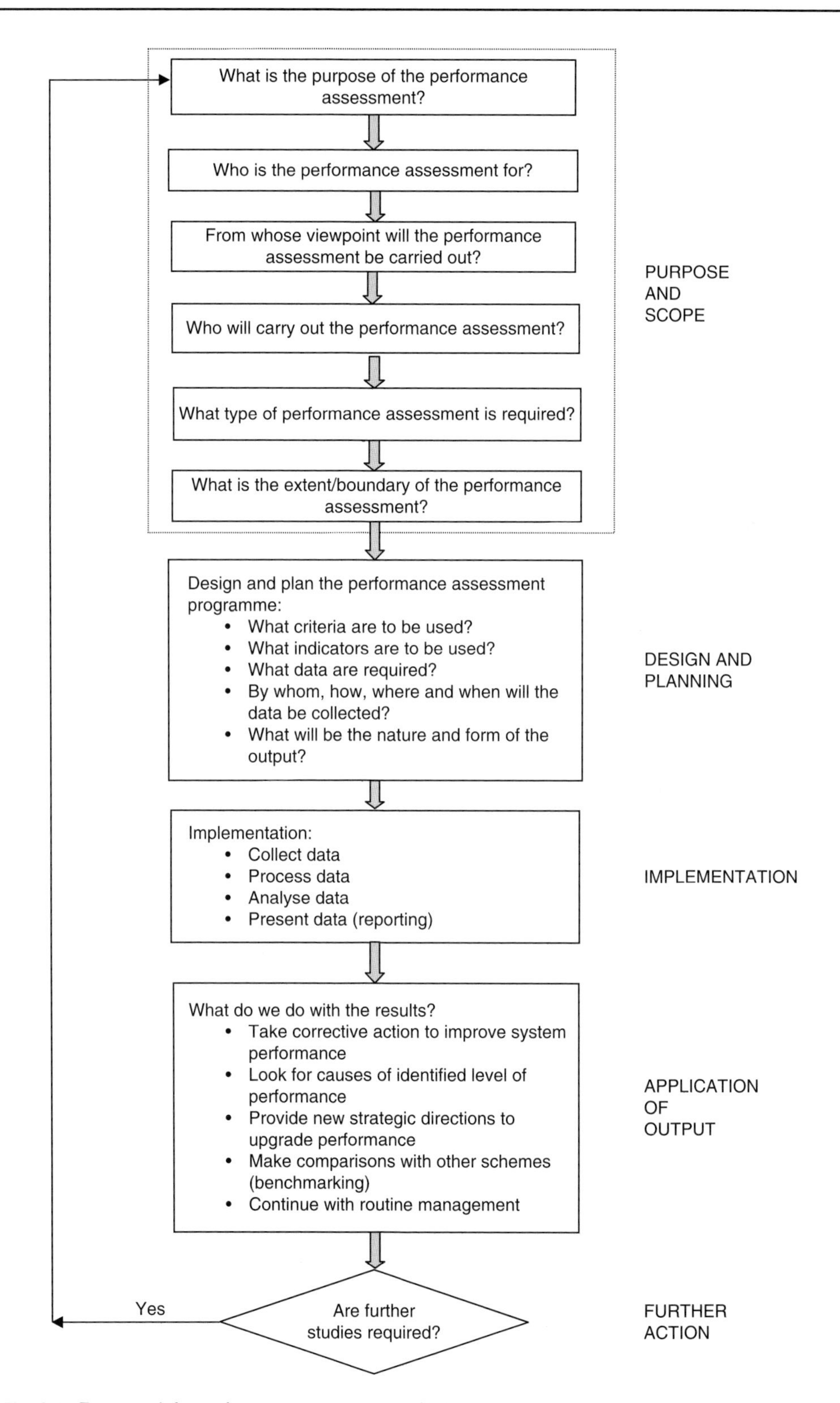

Fig. 9.1. Framework for performance assessment of irrigation and drainage schemes.

Table 9.1. Example of the rationale and a set of objectives for a performance assessment programme.

Rationale	Water management needs to be improved if all farmers within the scheme are to obtain adequate livelihoods
Overall objective	To identify feasible and sustainable water management practices that lead to improved crop production and thereby income for the farming community
Specific objectives	1. Identify how irrigation water is currently used *Activities:* • Monitor water demands and water supply • Analyse match between demand and supply 2. Identify areas where improvement can lead to increased productivity *Activities*: • Understand and monitor processes and procedures used for data collection, processing and analysis of demands, and allocation of supplies • Identify shortfalls and areas where improvements can be made

but may be looking at performance assessment from the perspective of another stakeholder or group of stakeholders (Fig. 9.2). Government may commission a performance assessment, for example, to be carried out by a research institute to study the impact of system performance on farmer livelihoods. Farmers might commission a study of the irrigation service provider in order to ascertain if they are receiving an adequate return for service fees paid.

By whom?

Different organizations or individuals have different capabilities in respect of performance assessment, and different types of performance assessment will require different types of organization or individuals to carry out the assessment (Table 9.2). A scheme manager might establish a performance assessment programme using existing operation and maintenance (O&M) personnel to be able to monitor and evaluate scheme performance.

Fig. 9.2. Good performance from a farmer's viewpoint.

Table 9.2. Examples of for whom, from whose viewpoint and by whom performance assessment might be carried out.

For whom?	From whose viewpoint?	By whom?
Scheme manager	The scheme management	Scheme manager and staff
Government	Government (return on proposed investment)	Consultant
Government	Society in general, but specifically water users	Government regulatory authority
Funding agency	Farmers (livelihood)	Consultant
Scientific community	The management of the system	Research institute/university
Farmers	Farmers	Consultant

A government agency might employ a consultant to carry out performance assessment of a scheme with a view to further investment, while a university research team might carry out a research programme to identify and understand generic factors that affect system performance.

Type

Small and Svendsen (1992) identify four different types of performance assessment, to which a fifth – diagnostic analysis – can be added:

- operational;
- accountability;
- intervention;
- sustainability;
- diagnostic analysis.

The type of performance assessment is linked with the purpose; in fact Small and Svendsen refer to these categories as the rationale for performance assessment.

Operational performance assessment relates to the day-to-day, season-to-season monitoring and evaluation of scheme performance. *Accountability* performance assessment is carried out to assess the performance of those responsible for managing a scheme. *Intervention* assessment is carried out to study the performance of the scheme and, generally, to look for ways to enhance that performance. Performance assessment associated with *sustainability* looks at the longer-term resource use and impacts. *Diagnostic analysis* seeks to use performance assessment to track down the cause, or causes, of performance in order that improvements can be made or performance levels sustained.

Internal or external assessment

It is important to define at the outset whether the performance assessment relates to one scheme (internal analysis) or comparison between schemes (external analysis).

A significant problem with performance assessment of I&D schemes is the complexity and thus variety of types of scheme. This makes comparison between schemes problematic. Some schemes are farmer-managed, some are private estates with shareholders, some are gravity-fed, some fed via pressurized pipe systems, etc. There is as yet no definitive methodology for categorizing I&D schemes, therefore there will always be discussion as to whether one is comparing like with like. A short list of key descriptors for I&D schemes is presented in Table 9.3. This list of descriptors can be used as a starting point to select schemes with similar key characteristics for comparison; other important characteristics can be added as necessary.

It is important to understand, however, that comparison between different types of scheme can be equally valuable; as for instance might be the case for governments in comparing the performance of privately owned estates with smallholder irrigation schemes. The two have different management objectives and processes, but their performance relative to criteria based on the efficiency and productivity of resource use (land, water, finance, labour) would be of value in policy formulation and financial resource allocation.

Benchmarking of I&D systems is a form of comparative (external) performance assessment that is increasingly being used (see later sections for more detail). Benchmarking seeks

Table 9.3. Key descriptors for irrigation and drainage schemes.

Descriptor	Possible options	Explanatory notes	Example
Irrigable area	–	Defines whether the scheme is large, medium or small in scale	8567 ha (net)
Annual irrigated area	Area supplied from surface water Area supplied from groundwater	Shows the intensity of land use and balance between surface or groundwater irrigation	7267 ha 4253 ha surface 3014 ha groundwater
Climate	Arid; semi-arid; humid tropics; Mediterranean	Sets the climatic context, influences the types of crops that can be grown	Mediterranean
Average annual rainfall (P)	–	Associated with climate, sets the climatic context and need for irrigation and/or drainage	440 mm
Average annual reference crop evapotranspiration (ET_0)	–	Associated with climate, sets the climatic context and need for irrigation	780 mm
Water source	Storage on river; groundwater; run-of-the river; conjunctive use of surface and groundwater	Describes the availability and reliability of irrigation water supply	Over-year storage reservoir in upper reaches, groundwater aquifers
Method of water abstraction	Pumped; gravity; artesian	Influences the pattern of supply and cost of irrigation water	Gravity-fed from rivers, pumped from groundwater
Water delivery infrastructure	Open channel; pipelines; lined; unlined	Influences the potential level of performance	Open channel, lined primary and secondary canals
Type of water distribution	Demand; arranged on-demand; arranged; supply-oriented	Influences the potential level of performance	Arranged on-demand
Availability and type of water storage	River storage; in-system (online/offline); on-farm; over-year, seasonal, night storage	Influences the availability (reliability, quantity and timing) of water supply	Over-year reservoir on river upstream, no in-system storage
Predominant on-farm irrigation practice	Surface: furrow; level basin; border; flood; ridge-in-basin Overhead: raingun; lateral move; centre pivot; drip/trickle Subsurface: drip	Influences the potential level of performance	Predominantly furrow, with some sprinkler and (increasingly) drip
Major crops (with percentages of total irrigated area)	–	Sets the agricultural context; separates out rice and non-rice schemes, monoculture from mixed cropping schemes	Cotton (53%) Grapes (27%) Maize (17%) Other crops (3%)

Continued

Table 9.3. Continued

Descriptor	Possible options	Explanatory notes	Example
Average farm size	–	Important for comparison between schemes, whether they are large estates or smallholder schemes	0.5–5 ha (20%) >5–20 ha (40%) >20–50 ha (20%) > 50 ha (20%)
Type of management	Government agency; private company; joint government agency/ farmer; farmer-managed	Influences the potential level of performance	River system: Government Primary and secondary systems: water users associations

to compare the performance of 'best practice' schemes with less well performing schemes, and to understand where the differences in performance lie. Initially performance assessment might be focused on a comparison of output performance indicators (water delivery, crop production, etc.), followed by diagnostic analysis to understand: (i) what causes the relative difference in performance; and (ii) what measures can feasibly be taken to raise performance in the less well performing scheme(s).

The selection of performance assessment criteria will be influenced by whether the exercise looks internally at the specific objectives of an irrigation scheme, or whether it looks to externally defined performance criteria. Different schemes will have different objectives, and different degrees to which these objectives are implicitly or explicitly stated. It may well be that when measured against its own explicitly stated objectives (e.g. to provide a specified number of families with secure livelihoods) a scheme is deemed a success. However, when measured against an external criterion of crop productivity per unit of water used, or impact on the environment, it may not perform as well. This reinforces the point made earlier that assessment of performance is often dependent on people's perspective – irrigation is seen as beneficial by farmers, possibly less so by fishermen and downstream water users.

Extent/boundaries

The extent of the performance assessment needs to be identified and the boundaries defined. Two primary boundaries relate to spatial and temporal dimensions. *Spatial* relates to the area or number of schemes covered (whether the performance assessment is limited to one secondary canal within a system, to one system, or to several systems); *temporal* relates to the duration of the assessment exercise and temporal extent (one week, one season, or several years).

Other boundaries are sometimes less clear-cut, and can relate to whether the performance assessment aims to cover technical aspects alone, or whether it should include institutional and financial aspects. How much influence, for example, does the existence of a water law on the establishment of water users associations have on the performance of transferred I&D systems?

The use of the *systems approach* advocated by Small and Svendsen (1992) can add to the definition and understanding of the boundaries and extent of the performance assessment programme. The systems approach focuses on *inputs*, *processes*, *outputs* and *impacts* (Fig. 9.3). Measurement of outputs (e.g. water delivery to tertiary unit intakes) provides information on the effectiveness of the use of inputs (water abstracted at river intake), while comparison of outputs against inputs provides information on the efficiency of the process of converting inputs into outputs. The process of transforming inputs into outputs has impacts down the line – the pattern of water delivery to the tertiary intake has, for example, an impact on the level of crop production attained by the farmer.

Measurements of canal discharges will provide information on how the irrigation system (network) is performing, but tell us

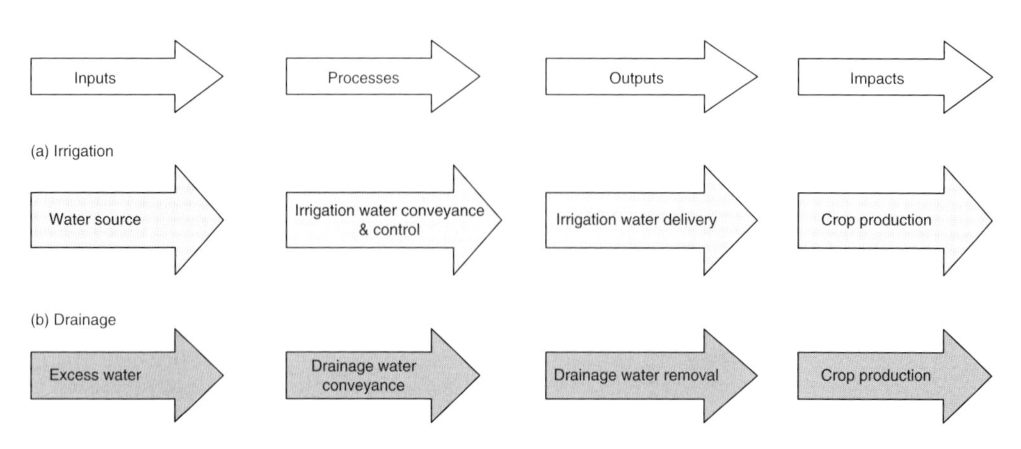

Fig. 9.3. Inputs, processes, outputs and impacts in irrigation and drainage.

little about the performance of the I&D scheme as a whole. To obtain this information we need to collect data within the irrigated agriculture system (see Fig. 2.2 and discussion in Chapter 2) and the agricultural economic system to set the performance of the irrigation system in context. Care is needed here in relating the performance of the irrigation system (e.g. adequate and timely water supply) to that of the agricultural economic system (e.g. farmer income), as many variables intervene between the supply of the irrigation water and the net return to the farmer for the crops produced.

Design of the performance assessment programme

Having specified the approach to the performance assessment programme in terms of the purpose and scope, the performance assessment programme can be designed. The following are key issues to consider.

- What criteria are to be used?
- What performance indicators are to be used?
- What data are required?
- By whom, how, where and when will the data be collected?
- What is the required form of output?

Performance criteria and scheme objectives

In the literature the terms *performance criteria, performance indicators* and *performance measures* are used by different authors to mean different things. The following definitions are proposed in order to clarify the terms *performance criteria, objectives, performance indicators* and *targets*.

1. *Objectives* are made up of *criteria*:
- 'to maximize agricultural production';
- 'to ensure equity of water supply to all farmers';
- 'to optimize the efficiency of water distribution'.

2. Criteria can be measured using *performance indicators*.
3. Defined *performance indicators* identify data requirements.
4. Data can then be collected, processed and analysed.
5. If *target, standards, reference* or *benchmark* values of performance indicators are set or known then performance can be assessed.

In selection of criteria for performance assessment it is necessary to define whether the assessment will be made against the scheme's stated objectives and criteria, or against an alternative set of performance objectives or criteria. An example of where a scheme's objectives and target values are stated is shown in Table 9.4. In this case the targets for cropped area and crop production (in terms of crop production and value) can also be monitored over time to assess the sustainability of the scheme.

While an irrigation scheme may have stated objectives, its performance may need to be assessed against different criteria

Table 9.4. Example of linkage of objectives, criteria, performance indicators and targets. (From Calculations for Mogambo Irrigation Scheme, Somalia, in Burton, 1993.)

Objective	Criterion	Performance indicator	Target value
Maximize area harvested	Productivity	Cropping intensity	2052 ha (100%)
Maximize total crop production	Productivity	Total production	7600 t
Maximize total value of agricultural production	Productivity	Total value of production	$1,067,238
Maximize productivity of water	Productivity	Water productivity	0.16 kg/m^3
		Value of production per unit water	$0.023/m^3
Maximize equity of water supply	Equity	Area planted/area harvested	1.0
		Delivery performance ratio	SD<10%

SD, standard deviation.

(Table 9.5). For example, a government might assess a scheme's performance in relation to the country's economic needs, or environmental sustainability and impact. Simply because these criteria are not stated in the objectives for the scheme does not mean that the scheme cannot be assessed against such externally stipulated criteria. For example, a scheme may not have stated objectives about pollution loading, but an environmental regulatory agency may have its own standards against which the scheme's performance is assessed.

In some of the literature on performance assessment authors have stated that performance should be assessed against objectives set for a given scheme. This is an obvious starting point, but is more difficult to apply when there are no explicitly stated objectives for the scheme.

As outlined in Murray-Rust and Snellen (1993) the setting of objectives is a crucial part of the management process, and much has been written on the subject in the context of business management. Some key points in relation to objective setting for irrigation management and performance assessment are outlined below.

1. *Explicit or implicit*: Objectives can be *explicit*, where they are clearly stated, or *implicit*, where they are assumed rather than stated. For example, for the Ganges Kobadak irrigation scheme in Bangladesh the explicit objective is food production but an (essential) implicit objective is flood protection to prevent the irrigation scheme being inundated by the waters of the Ganges River. In performance assessment it is important to identify both types of objective.

2. *Hierarchy of objectives*: Objectives occur at different levels within a system or systems. As discussed in Chapter 3 a hierarchy of objectives for irrigation development, identified by FAO (1982), was, in ascending order:

- appropriate use of water;
- appropriate use of agricultural inputs;
- remunerative selling of agricultural products;
- improvement in social facilities;
- betterment of farmers' welfare.

Table 9.5. Criteria for good system performance according to type of person. (Modified from Chambers, 1988.)

Type of person	Possible first criterion of good system performance
Landless labourer	Increased labour demand, days of working and wages
Farmer	Delivery of an adequate, convenient, predictable and timely water supply
Irrigation engineer	Efficient delivery of water from headworks to the tertiary outlet
Agricultural economist	High and stable farm production and incomes
Economist	High internal rate of return
Political economist	Equitable distribution of benefits, especially to disadvantaged groups
Environmental scientist	Low levels of fertilizer and pesticide contamination in drainage water

Table 9.6. Comparison of objectives, weightings and rankings for a state farm and a settlement scheme. (From Burton, 1993.)

	State farm		Settlement scheme			
Objective	Weighting[a]	Ranking[a]	Weighting[a]	Ranking[a]	Performance indicator	Target value
Maximize area harvested	6	(v)	10	(ii)	Area harvested	2052 ha (100%)
Maximize total production	10	(iv)	6	(iii)	Total production	7600 t
Maximize total value of agricultural production	10	(i)	6	(iv)	Total value of production	$1,067,238
Maximize productivity of land (kg/m^3)	10	(ii)	10	(v)	Water productivity	0.16 kg/m^3
Maximize productivity of water ($/m^3)	10	(iii)	10	(vi)	Value of production per unit water	$0.023/m^3
Maximize equity of water supply	0	(vi)	10	(i)	Area planted/area harvested	1.0
					Delivery performance ratio	SD<10%

SD, standard deviation.
[a]For weightings 1 is low, 10 is high; for ranking (i) is highest, (vi) lowest.

Each of these objectives is important at its own system level, satisfying the objectives at one level means that those at another (higher) level might also be satisfied. This hierarchy of objectives is an integral part of the *logical framework* project planning tool (see 'Framework for Monitoring and Evaluation' below), moving from outputs to purpose to satisfy the overall goal.

3. *Ranking or weighting of objectives*: Within a system there may be several, sometimes competing, objectives. For performance assessment these may need to be ranked or weighted and assessments made to evaluate how well individual and collective objectives are satisfied. This process is commonly termed multi-criteria analysis (see Appendix F in Snell, 1997 for more details). An example of the weightings and rankings attached to individual objectives depending on whether the irrigation scheme is run as a state farm or settlement scheme is presented in Table 9.6. Objectives to maximize equitable distribution of water might be favoured for a settlement scheme, while objectives to maximize value of production might be favoured for a state farm.

Performance indicators

Performance is measured through the use of indicators, for which data are collected and recorded. The analysis of the indicators then informs us on the level of performance.

The linkage between the criteria against which performance is to be measured, and the indicators that are to be used to measure attainment of those criteria, is important. Using the nested systems outlined in Fig. 2.2, for example, performance criteria and indicators for the irrigation system, the agricultural system and the agricultural economic systems can be defined (Table 9.7). Note that a

Table 9.7. Examples of linkages between performance criteria and performance indicators.

	Performance indicator[a]		
Criterion	Irrigation and drainage system[b]	Irrigated agriculture system[b]	Agricultural economic system[b]
Command adequacy	Water level ratio Overall consumed ratio Delivery performance ratio	– Crop production relative to family food needs	– Cash value of crop production relative to defined poverty level
Equity	Overall consumed ratio Delivery performance ratio	Spatial distribution within scheme of: • Crop type • Crop yield • Cropping intensity	Spatial distribution within scheme of farm income
Reliability	Overall consumed ratio Delivery performance ratio	Number of years crop production is adequate	Number of years income from crop production is adequate
Efficiency	Overall consumed ratio Field application ratio Outflow over inflow ratio	Crop yield	O&M fraction
Productivity	–	Crop yield	Crop gross margin Internal rate of return
Profitability	–	–	Farm profit Return on investment (EIRR)
Sustainability	Efficacy of infrastructure Groundwater depth Indicator value on salinity	Sustainability of irrigable area	Financial self-sufficiency O&M fraction Fee collection ratio

O&M, operation and management; EIRR, economic internal rate of return.
[a]See Table 9.9 for more detail on these indicators
[b]As detailed in Fig. 2.2.

performance criterion, such as equity, can be defined differently depending on the system to which it relates.

In some instances it is useful to consider indicators for the inputs and outputs across a number of systems, examples are presented in Table 9.8.

Target values may be set for these indicators, or the values obtained at a particular location or time can be compared with values of the indicator collected at other locations (spatial variation) or time (temporal variation). Thus values of performance indicators can be compared within or between schemes.

Data requirements

Following on from identification of the performance criteria and indicators to be used in the performance assessment programme, the data needs can be identified (Table 9.9).

Data collection (who, how, where and when)

During the design stage of the performance assessment programme it will be necessary to identify *who* will collect these data, and *how*, *where* and *when* they will be collected.

All or some of the required data, such as crop areas, may already be available or there may be a need for additional data collection procedures or special equipment to collect data (such as automatic water level recorders to gather detailed information on canal discharges day and night). Allowance will need to be made in the performance assessment budget for the costs associated with the data collection and handling programme.

To understand the performance of an I&D scheme it is not necessary, economic or time-efficient to collect data for every location in a scheme. The performance assessment programme should be designed to take representative samples to enable an adequate analysis to be carried out in keeping with the prescribed needs. It is, for example, common to take sample tertiary units from the head, middle and tail of irrigation systems when studying irrigation water management performance.

When the data needs have been decided a data collection schedule can then be drawn up. An example schedule for a performance assessment programme by a scheme manager is presented in Table 9.10.

In addition a matrix can be drawn up (Table 9.11) showing the performance indicators to be used and the data to be collected. As can be seen in the example provided, some data apply to a number of indicators.

Form of output

At the planning stage for the performance assessment programme it is helpful to think about the form of the report output. Preparing a draft annotated contents list of the report, and a list of tables and figures and their anticipated content, helps focus thinking and ensures that data are collected to match. An example is given below (Table 9.12) for a study to gain a broad understanding of performance related to irrigation water supply throughout a scheme.

Simple sketches of the form of the expected output are helpful, as is thinking about the form of data presentation that the users of the performance assessment report and data would find most useful. Non-technical personnel might be interested, for example, in a graph showing the trend in the decline in water quality over time, without requiring too much detail on the actual numbers. Technical personnel, however, would

Table 9.8. Examples of indicators using inputs and outputs across different systems.

Criterion	Indicator example	Systems covered
Productivity	Water productivity (kg/m^3)	Irrigation system Irrigated agriculture system
Productivity	Land productivity (kg/ha)	Irrigation system Agricultural economic system

Table 9.9. Linking performance indicators to data requirements. (From Bos *et al.*, 2005, Chapter 3.)

Indicator	Definition	Units	Data required
Cropping intensity	$\frac{\text{Actual cropped area}}{\text{Irrigable area}}$	%	Actual cropped area (ha) Irrigable area (ha)
Crop yield	$\frac{\text{Crop production}}{\text{Area cultivated}}$	kg/ha	Crop production (kg) Area cultivated (ha)
Sustainability of irrigable area	$\frac{\text{Average cropped area}}{\text{Initial total irrigable area}}$	–	Average cropped area (ha) Initial total irrigable area (ha)
Overall consumed ratio	$\frac{\text{Crop water demand} - \text{effective rainfall}}{\text{Volume of water supplied to command area}}$	–	Crop water demand (mm) Effective rainfall (mm) Irrigation water supply (mm)
Delivery performance ratio	$\frac{\text{Actual flow of water}}{\text{Intended flow of water}}$	–	Actual volume delivered (m^3) Intended/planned volume to be delivered (m^3)
Water productivity	$\frac{\text{Yield of harvested crop}}{\text{Volume of supplied irrigation water}}$	kg/m^3	Crop production (kg) Volume of irrigation water supplied (m^3)
Water level ratio	$\frac{\text{Actual water level}}{\text{Design water level}}$	–	Actual water level (m) Design water level (m)
Field application ratio	$\frac{\text{Crop water demand} - \text{effective rainfall}}{\text{Volume of water delivered to the fields}}$	–	Crop water demand (mm) Effective rainfall (mm) Irrigation water supply (mm)
Efficacy of infrastructure	$\frac{\text{Functioning part of infrastructure}}{\text{Total infrastructure}}$	–	Number of functioning structures Total number of structures
Groundwater depth	Depth to groundwater	m	Depth to groundwater (m)
Indicator value on salinity	$\frac{\text{Actual concentration of salinity}}{\text{Critical concentration of salinity}}$	–	Actual concentration of salinity (mmho/cm) Critical concentration of salinity (mmho/cm)
O&M fraction	$\frac{\text{Cost of MOM}}{\text{Total budget for sustainable MOM}}$	–	Cost of MOM ($) Total budget for sustainable MOM ($)
Fee collection ratio	$\frac{\text{Irrigation service fees collected}}{\text{Irrigation service fees due}}$	–	Irrigation service fees collected ($) Irrigation service fees due ($)

O&M, operation and maintenance; MOM, management, operation and maintenance.

Table 9.10. Example of a data collection schedule – who, how, where and when.[a]

Data required	Units	Who	How	Where	When
Irrigable area	ha	Scheme manager	From design drawings or scheme database	In office	–
Crop production	kg	Scheme agronomist	Interviews with farmers	In selected sample tertiary units	At end of season
Actual cropped area	ha	Scheme agronomist	Data returns from farmers, and/or spot checks in field	For whole scheme but field checks made on selected sample tertiary units	During the irrigation season
Crop yield	kg/ha	Scheme agronomist	Crop cuttings	In selected sample tertiary units	At harvest time
Crop water demand	mm/day	Scheme agronomist or irrigation engineer	By calculation using standard procedures (e.g. CROPWAT or CRIWAR)	In selected sample tertiary units	During the season
Rainfall	mm/day	Water masters	Using rain gauge	At locations within the scheme area	Daily
Actual discharge	m^3/s	Water masters	Reading of measuring structure gauges	At primary, secondary and tertiary unit intakes	Daily
Actual duration of flow	h	Water masters	Reading of measuring structure gauges	At primary, secondary and tertiary unit intakes	Daily
Intended discharge	m^3/s	Scheme manager	From indents submitted by farmers	In office	Each week
Intended duration	h	Scheme manager	From indents submitted by farmers	In office	Each week
Crop market price	$/kg	Scheme agronomist	Interviews with farmers and traders	Villages and markets	At end of season

[a]The example given is for a performance assessment programme carried out by a scheme manager for the whole scheme with a view to understanding overall scheme performance.

Table 9.11. Linking performance indicators to data collection.[a]

Data required	Units	Indicator						
		Cropping intensity (%)	Crop yield (kg/ha)	Overall consumed ratio	Water productivity (kg/m^3)	Delivery performance ratio	Output per unit cropped area ($/ha)	Output per unit irrigation supply ($/m^3)
Irrigable area	ha	✓						
Crop production	kg		✓		✓		✓	✓
Actual cropped area	ha	✓	✓				✓	
Crop yield	kg/ha		✓					
Crop water demand	mm			✓				
Rainfall	mm			✓				
Actual discharge	m^3/s			✓	✓	✓		✓
Actual duration of flow	h			✓	✓	✓		✓
Intended discharge	m^3/s					✓		
Intended duration of flow	h					✓		
Crop market price	$/kg						✓	✓

[a]The example given is for a performance assessment programme carried out by a scheme manager for the whole scheme with a view to understanding overall scheme performance.

Table 9.12. Example of planned figures and tables for a performance assessment programme.

	Content
Figure no.	
1	Layout of irrigation system
2–10	Histogram plots of discharge versus time (daily) at primary, secondary and selected tertiary head regulators
11–16	Histogram plots of irrigation depth applied to a sample number of individual (sample) fields
17–22	Histogram plots of delivery performance ratio for a sample number of individual fields
Table no.	
1	Summary table of performance at head regulator level, including: total command area, irrigated area, total flow (MCM), total days flowing during season, average unit discharge (l/s/ha)
2–6	Summary tables of cultivable command area, cropped areas, crop types, cropping intensities for primary and a sample number of secondary and tertiary command areas
7–12	Summary tables of data collected at field level, including, for each sample field: area, crop type, number of irrigations, irrigation depths, irrigation intervals, maximum soil moisture deficit, total water supply, total estimated water demand, crop production and crop market price
13–18	Summary tables of results of calculation showing: yield per unit area (kg/ha), yield per unit irrigation supply (kg/m^3), output per cropped area ($/ha) and output per unit irrigation supply ($/m^3)

MCM, millions of cubic metres.

require the numbers to be presented, perhaps in a table associated with the graph.

Implementation

The performance assessment programme design phase is followed by the implementation phase, covering the actual collection, processing, analysis and reporting of the data. Depending on the nature of the performance assessment programme, implementation may be over a short (1 week) or long period (several years). In all cases it is worthwhile to process and analyse some, if not all, of the data collected as the work progresses in order to detect errors in data and take corrective action where necessary.

Application of output

The use of the information collected from a performance assessment study will vary depending on the purpose of the assessment. The use to which the results of the performance assessment are put will depend on the reason the performance assessment was carried out.

Possible actions following the conclusion of the performance assessment study might include the following.

- Redefining strategic objectives and/or targets.
- Redefining operational objectives and/or targets.
- Implementing corrective measures, for example:
  - training of staff;
  - building new infrastructure;
  - carrying out intensive maintenance;
  - developing new scheduling procedures;
  - changing to alternative irrigation method(s);
  - rehabilitation of the system;
  - modernization of the system.

Further action

Further studies may be required as a result of the performance assessment programme. Performance assessment is closely linked with diagnostic analysis and it is often the case that an initial performance assessment programme identifies areas where further measurements and data collection are required in order to identify the root causes of problems and constraints.

Where performance assessment identifies the root cause of a problem or constraint, further studies may be required to implement measures to alleviate the problem, such as, for example, field surveys for the planning and design of a drainage system to relieve waterlogging.

Performance Assessment at Different Levels

As discussed previously, performance assessment can take place at different levels:

- at the sector level when assessing how irrigation and drainage is performing in comparison with the objectives set for the sector, and in comparison with other uses of water;
- at the scheme level when assessing how individual schemes are performing against their own explicitly or implicitly stated objectives, or when assessing the performance of different schemes against themselves;
- at main system level where the performance of the water delivery service is assessed;
- at the on-farm level where the performance of the on-farm water delivery, water use and water application is assessed.[3]

The purpose of assessment at these different levels, and possible indicators to be used to assess performance, are briefly outlined in the following sections. It is important to note that the approach adopted here is that water delivery and water removal are taken as the primary function; other functions such as maintenance, fee recovery and the like are subsidiary to the prime function. Fee recovery, for example, is important in order that management staff can be employed and maintenance work carried out, with the end product that water is delivered to the crops' root zone at the right time and in the right quantity to match the crops' needs.

Sector level

At the sector level performance assessment is focused on the productivity of financial investment in the irrigation and drainage sector and on the productivity and efficiency of water use. In many countries and river basins there is increasing pressure on the available water resources, and an increasing need to justify the use of water for agricultural use against other uses, such as for domestic, industrial, environmental or navigation use. Assessment at this level is generally carried out by government, either through the water resource agency or by consultants.

Scheme level

At the scheme level performance assessment is focused on the outputs, outcomes and impacts of the I&D scheme. Outputs will generally focus on crop production, while outcomes will generally focus on protecting livelihoods and financial benefits to the farming community. The interest in impacts may range from the environmental impact of the scheme to its wider impact on the rural and national economy.

Table 9.13 presents key indicators that can be used for performance assessment at this level, with indicators covering a range of domains, including agricultural production, irrigation water delivery, drainage water removal, finance, and environmental protection.

Main system level

At the main system level performance assessment is focused on water delivery, which will depend on the management, operation and maintenance processes and procedures of the main system service provider. Table 9.14 summarizes the key indicators that can be used for assessing main system water delivery performance.

Table 9.13. Key indicators for assessing the scheme level management, operation and maintenance performance.

Indicators	Definition	Notes[a]
Agricultural production		
Total seasonal[b] area cropped per unit command area (cropping intensity)	$\frac{\text{Total seasonal area cropped}}{\text{Total command area of system}}$	A
Total seasonal crop production (t)	Total seasonal crop production by crop type within command area	A
Total seasonal crop production per unit command area (crop yield, kg/ha)	$\frac{\text{Total seasonal crop production}}{\text{Total command area of system}}$	A
Total seasonal value of crop production ($)	Total seasonal value of agricultural crop production received by producers	A
Total seasonal value of crop production per unit command area ($/ha)	$\frac{\text{Total seasonal value of crop production}}{\text{Total command area of system}}$	A
Total seasonal crop production per unit water supply (kg/m^3)	$\frac{\text{Total seasonal crop production}}{\text{Total seasonal volume of irrigation water supply}}$	A
Total seasonal value of crop production per unit water consumed ($/m^3)	$\frac{\text{Total seasonal value of crop production}}{\text{Total seasonal volume of crop water demand } (ET_c)}$	A
Total seasonal value of crop production per unit water supplied ($/m^3)	$\frac{\text{Total seasonal value of crop production}}{\text{Total seasonal volume of irrigation water supply}}$	A
Irrigation water delivery		
Total seasonal volume of irrigation water supply (MCM)	Total seasonal volume of water diverted or pumped for irrigation (not including diversion of internal drainage)	A
Seasonal irrigation water supply per unit command area (m^3/ha)	$\frac{\text{Total seasonal volume of irrigation water supply}}{\text{Total command area of system}}$	A
Main system water delivery efficiency	$\frac{\text{Total seasonal volume of irrigation water delivery}}{\text{Total seasonal volume of irrigation water supply}}$	B
Seasonal relative irrigation water supply	$\frac{\text{Total seasonal volume of irrigation water supply}}{\text{Total seasonal volume of crop water demand}}$	A
Water delivery capacity	$\frac{\text{Canal capacity at head of system}}{\text{Peak irrigation water demand at head of system}}$	–

Financial		
Total seasonal MOM expenditure[c] per unit command area ($/ha)	$\frac{\text{Total seasonal MOM expenditure}}{\text{Total command area of system}}$	C
Total seasonal MOM expenditure per unit irrigation water supply ($/m^3)	$\frac{\text{Total seasonal MOM expenditure}}{\text{Total seasonal volume of irrigation water supply}}$	C
Total seasonal maintenance expenditure per unit command area ($/ha)	$\frac{\text{Total seasonal maintenance expenditure}}{\text{Total command area of system}}$	C
Total seasonal maintenance expenditure fraction	$\frac{\text{Total seasonal maintenance expenditure}}{\text{Total seasonal MOM expenditure}}$	C
MOM funding ratio	$\frac{\text{Actual annual income}}{\text{Budget required for sustainable MOM}}$	D
Fee collection ratio	$\frac{\text{Irrigation (and drainage) service fees collected}}{\text{Irrigation (and drainage) service fees due}}$	D
Farm profit	Total farm income – total farm expenditure	E
Drainage water removal		
Average depth to groundwater (m)	Average seasonal depth to groundwater calculated from water table observations over the irrigation area	F
Environmental protection		
Salinity of soil water (mmho/cm)	Electrical conductivity of soil water	F
Soil salinity (mmho/cm)	Electrical conductivity of soil	F
Salinity of water in open drain (mmho/cm)	Electrical conductivity of water in open drains	F
Drainage water quality: biological (mg/l)	Biological load of drainage water expressed as biological oxygen demand (BOD)	F
Drainage water quality: chemical (mg/l)	Chemical load of drainage water expressed as chemical oxygen demand (COD)	F

MCM, millions of cubic metres; MOM, management, operation and maintenance.
[a]Location and sampling interval: A=determine for total command area and individual tertiary units; B=discharges measured at the main canal intake and tertiary unit intakes; C=determine for total command area, main system only and individual water users associations; D=determine for individual service providers (government agency or water users associations); E=for individual water users; F=periodic sampling at selected locations.
[b]May be seasonal or annual, depending on the circumstances. If there is more than one season and there are marked differences between the seasons' cropping patterns and water availability it is preferable to consider each season separately.
[c]Costs for irrigation water delivery and drainage water removal may be kept separate or combined; it depends on whether there is a separate drainage authority.

Table 9.14. Indicators used for assessing different performance criteria related to water delivery. (Adapted from Bos *et al.*, 2005.)

Criterion	Performance indicators	Definition	Notes
Reliability	RWS	$\frac{\text{Volume of irrigation water supply}}{\text{Volume of irrigation water demand}}$	Variation of the RWS at the main canal intake and at tertiary intakes during the season indicates the level of reliability of water supply and delivery
	DPR	$\frac{\text{Volume of irrigation water supplied}}{\text{Target volume of irrigation water supply}}$	Variation of the DPR at tertiary unit intakes during the season indicates the level of reliability of water delivery
Adequacy	RWS	$\frac{\text{Volume of irrigation water supplied}}{\text{Volume of irrigation water demand}}$	Measured at main canal intake and each tertiary unit intake. Target value = 1.0, a value less than 1.0 indicates water shortage
	DPR	$\frac{\text{Volume of irrigation water supplied}}{\text{Target volume of irrigation water supply}}$	Measured at main canal intake and each tertiary unit. Target value = 1.0. If there is a water shortage the target supply may be less than the actual irrigation water demand.
Timeliness	Dependability of irrigation interval	$\frac{\text{Actual irrigation interval}}{\text{Planned/required irrigation interval}}$	The planned/required interval between irrigations is either that planned (such as in a planned irrigation rotation regime) or that dictated by the crop's soil moisture status
	Timeliness of irrigation water delivery	$\frac{\text{Actual date/time of irrigation water delivery}}{\text{Planned/required date/time of irrigation water delivery}}$	Compares the actual date and time of delivery (planned in the rotation or requested by the farmer) with the actual delivery date and time
Equity	RWS	$\frac{\text{Volume of irrigation water supply}}{\text{Volume of irrigation water demand}}$	Variation of the RWS at tertiary intakes indicates degree of equity or inequity
	DPR	$\frac{\text{Volume of irrigation water supplied}}{\text{Target volume of irrigation water supply}}$	Variation of the RWS at tertiary intakes indicates degree of equity or inequity
Efficiency	RWS	$\frac{\text{Volume of irrigation water supply}}{\text{Volume of irrigation water demand}}$	Comparison of the RWS at the main canal intake and the tertiary unit intakes indicates the level of losses
	Overall scheme efficiency	$\frac{\text{Volume of water needed by crop}}{\text{Volume of water diverted/pumped from source}}$	Useful indicator. Relatively easy to obtain a meaningful value. Estimate crop irrigation water demand at the field (using FAO CROPWAT program, or similar) and measure actual discharge at main canal intake

	Main system water delivery efficiency	$\frac{\text{Volume of water delivered (to tertiary unit)}}{\text{Volume of water diverted/pumped from source}}$	Measure discharges at main canal intake and offtakes to tertiary units. Value may change due to the seasons (wet/dry), with drainage inflow possible in wet season
	Crop production per unit water supply	$\frac{\text{Total crop production}}{\text{Volume of water diverted/pumped from source}}$	A measure of efficiency use to determine change in production per unit of water diverted at source. Useful for monoculture schemes
Productivity	Crop production per unit water delivered	$\frac{\text{Total crop production}}{\text{Volume of water delivered (to tertiary unit or field)}}$	Increasingly important indicator. Need to be careful where there is mixed cropping
	Value of crop production per unit water delivered	$\frac{\text{Total value of crop production}}{\text{Volume of water delivered (to tertiary unit or field)}}$	Increasingly important indicator. Use the value of crop production where there is mixed cropping
Cost-effectiveness	Ratio of ISF collected to GVP	$\frac{\text{Total ISF collected}}{\text{Total GVP}}$	Assesses the cost of the ISF compared with the total GVP. A broad indicator only, as other costs are involved
	ISF to total crop input costs ratio	$\frac{\text{ISF due for the crop}}{\text{Total input costs for the crop}}$	Assesses the costs of the ISF as a fraction (or percentage) of the total input costs for planting, harvesting and marketing the crop. Often found to be in the range of 4–10% of total input costs where the ISF is set at adequate levels to recover sustainable MOM costs

RWS, relative water supply; DPR, delivery performance ratio; ISF, irrigation service fee; GVP, gross value of production; FAO, Food and Agriculture Organization of the United Nations; MOM, management, operation and maintenance.

On-farm level

At the on-farm level performance assessment is focused on water delivery from the tertiary unit intake to the farmers' field(s), and water application by the farmer to the crops in the field. In some cases the performance assessment can be subdivided into the water delivery function, often carried out by the water users association (WUA), and the water application function, generally carried out by the farmer. In these cases the assessment will look separately at the performance of the WUA and the performance of the farmer. Output from the field may be constrained by the performance of the farmer, or the WUA, or both, and might also be constrained by the water delivery pattern in the main system.

An example of a scoring system used for assessing the performance of water users associations is presented in Table 9.15. The indicators are divided into categories covering institutional, financial and technical performance of the WUA. Scores are applied by the assessment team to each of the indicators based on the achievement against stated target values.[4]

One of the most detailed guides for assessing irrigation performance at field level is by Merriam and Keller (1978). A subsequent publication under the FAO Irrigation and Drainage Paper series (FAO, 1989) built on this work and provided computer models to assist in the design and evaluation of surface irrigation methods. The performance indicators are relatively straightforward (Table 9.16), assessing the water actually applied against the water required in the root zone. Measurement and determination of the value of the indicators is, however, less straightforward.

Framework for Monitoring and Evaluation[5]

Purpose and definition

Monitoring and evaluation are distinct but related activities, as can be seen from the following definitions (Casley and Kumar, 1987; OECD, 2002).

- *Monitoring* is the continuous collection of data on specified indicators to assess for a development intervention (project, programme or policy), its *implementation* in relation to activity schedules and expenditure of allocated funds, and its *progress and achievements* in relation to its objectives.
- *Evaluation* is the periodic assessment of the *design*, *implementation*, *outcomes* and *impact* of a development intervention. It should assess the relevance and achievement of objectives, the implementation performance in terms of *effectiveness* and *efficiency*, and the *nature*, *distribution* and *sustainability* of impacts.

The linkage between monitoring and evaluation takes various forms:

- monitoring can raise issues for evaluation, while evaluation results can indicate where new processes or activities need to be monitored;
- monitoring and evaluation are used together by managers to identify and then diagnose problem areas;
- monitoring and evaluation often use the same data, but use these data in different ways.

Monitoring compares actual progress with that planned for a project, and provides managers and others with regular updates on the progress made towards the final outputs and outcomes of the project. Good monitoring is dependent on an effective management information system, the design and implementation of which is one of the first tasks when implementing a project.

Evaluation can take place either during project implementation or at the end. Project managers will need to evaluate the progress of a project and establish why targets are, or are not, being met. Formal evaluations may be required, such as are carried out by funding agencies, for mid-term or final reviews to establish project progress and achievements against the stated targets. Mid-term evaluations can be important in identifying problem areas and measures to address such problems in good time. Ex-post evaluations can be carried out some while after completion of the project in order to measure the full impacts of

Table 9.15. Example of key indicators used to monitor the performance of water users associations (WUAs).

Indicator	Definition	Value	Scoring	Score
Formation				
Area transferred to WUA	Area transferred to WUA / Total gross area serviced by the system		2 = 100% 1 = 50–99% 0 = <50%	
Membership, representation and accountability				
WUA membership ratio	Total number of WUA members / Total number of irrigators in service area		2 = >50% 1 = 25–50% 0 = <25%	
AGMs	AGM held		2 = yes 0 = no	
AGM attendance	Number of WUA members attending AGM / Total number of WUA members		2 = >50% 1 = 30–50% 0 = <30%	
Administrative Council meetings held	Number of meetings held during the year (January–December)		2 = >5 1 = 1–5 0 = 0	
Administrative Council elections	Elections for members of Administrative Council held in last 2 years		2 = yes 0 = no	
Women members of Administrative Council	Number of women members of Administrative Council		2 = 1 or more 0 = none	
Area irrigated				
First irrigation crop area ratio (of total service area)	Total annual recorded (first) irrigation crop area / Total gross area serviced by the system		2 = >50% 1 = 30–50% 0 = <30%	
Crop audit correction factor	Reported area of first irrigation / Crop area measured from crop area audit survey		2 = >90% 1 = 75–90% 0 = <75%	
Financial				
Employment of Accountant	Accountant employed and duration of employment		2 = yes, >4 months 1 = yes, <4 months 0 = no	
ISF collection per hectare of service area	Total ISF collected / Total gross area serviced by the system		2 = >$13[a]/ha 1 = $7–13/ha 0 = <$7/ha	

Continued

Table 9.15. Continued.

Indicator	Definition	Value	Scoring	Score
ISF collection as a percentage of target	Total ISF collected / Target total annual ISF		2 = >90% 1 = 60–90% 0 = <60%	
ISF collection per hectare irrigated	Total ISF collected / Total annual irrigated crop area		2 = >$18[a]/ha 1 = $7–18/ha 0 = <$7/ha	
Financial audit of WUA	Level of approval of WUA financial affairs by independent auditors		2 = accounts approved 1 = no audit undertaken 0 = accounts qualified/rejected	
Operation				
Area managed by water masters	Total gross area serviced by the system / Number of water masters employed by WUA		2 = <250 ha 1 = >250 ha 0 = no water masters	
Degree of flow measurement	Level of flow measurement at the head of the system (either primary canal or secondary canals)		2 = full water measurement record 1 = some water measurement 0 = no measurement	
Maintenance				
Annual maintenance planning	Extent of annual maintenance planning, costing and implementation		2 = inspection undertaken and detailed plan produced 1 = maintenance plan produced without proper inspection 0 = no plan produced	
Maintenance expenditure per unit of total service area	Maintenance cost / Total gross area serviced by the system		2 = >$7[a]/ha 1 = $4–7/ha 0 = <$4/ha	
Maintenance expenditure to revenue ratio	Maintenance expenditure / Gross revenue collected		2 = >70% 1 = 40–70% 0 = <40%	
Total score	Sum of scores for performance indicators. Top scores indicate that WUAs need no further support		2 = >32 1 = 20–32 0 = <20	

AGM, Annual General Meeting; ISF, irrigation service fee.
[a]Adjusted to current values.

Table 9.16. Measures of in-field performance for surface irrigation. (From Merriam and Keller, 1978; FAO, 1989.)

Indicator	Definition
Application uniformity	
Christiansen coefficient (C_u)	$100(1.0 - \Sigma x/mn$, where x is the absolute deviation from the mean application, m, and n is the number of observations
Distribution uniformity (DU)	$\frac{\text{Average depth infiltrated in the lowest one quarter of the area}}{\text{Average depth of water infiltrated}}$
Application efficiency (E_a)	$\frac{\text{Volume of water added to the root zone}}{\text{Volume of water applied to the field}}$
Water requirement efficiency (E_r)	$\frac{\text{Volume of water added to root zone storage}}{\text{Potential soil moisture storage volume}}$
Deep percolation efficiency (DPR)	$\frac{\text{Volume of deep percolation}}{\text{Volume of water applied to the field}}$
Tailwater ratio (TWR)	$\frac{\text{Volume of runoff}}{\text{Volume of water applied to the field}}$

the project; in the case of an irrigation rehabilitation project for example, this may be 2–3 years after completion of the physical works.

Definition of terms used in monitoring and evaluation

There are a number of terms used in M&E which have specific meanings; these are explained in Table 9.17.

Framework

There are two widely used frameworks for M&E:

- the logical framework;
- the results framework.

These two frameworks are related but different in their focus. The logical framework is more focused on M&E of project activities and outputs; the results framework is more focused on outcomes and impacts.

Both the frameworks rely on the 'causal chain', the hierarchy of links between inputs–activities–outputs–outcomes as shown in Figs 9.4 and 9.5. The funding of inputs (means) enables activities (ends) to be carried out, which in turn enables outputs to be achieved leading to attainment of desired outcomes and objectives. At all levels there are 'necessary conditions' that need to be satisfied; for example that there is sufficient skilled labour to convert the activities into the required outputs, or that other factors in the national economy enable the desired outputs of the project (e.g. rehabilitated and functioning I&D system) to be converted into desired outcomes (e.g. increased crop production through improved water supply and other inputs) or objectives (e.g. improved livelihoods through selling of surplus crop production). While rehabilitation of the physical components of the I&D system might be within the control of the project, hoped-for increases in crop production and farmer livelihoods are not, as these depend on resources and actions outside the control of the project. If at the project design stage key necessary conditions are not in place then measures may be taken to include additional activities within the project which will satisfy these conditions (e.g. including a component to provide farm machinery where this is required but not available).

Project monitoring can be usefully divided into results monitoring and implementation monitoring, as shown in Fig. 9.6. Implementation monitoring is more suited to project managers who are focused on achieving the required outputs from the

Table 9.17. Terms used in monitoring and evaluation.

Term	Explanation
Higher-level development objectives	The longer-term objective, change of state or improved situation to which achievement of the *project development objective(s)* is intended to contribute. Sometimes referred to as the higher-level development goal
Project development objective (also termed project goal or purpose)	The combination of one or more project component *outcomes* which make up the physical, financial, institutional, social, environmental or other development changes which the project is designed and expected to achieve
Outcomes	The effects of project components in terms of observable change in performance, behaviour or status of resources
Outputs	The products, capital goods and services resulting from a development intervention and which are necessary for the achievement of *project component outcomes*
Activities	The actions taken by project implementers that deliver the *outputs* by using the *inputs* provided
Inputs	The human and material resources financed by the project

Means–ends chain	*equals*	Logical project design	*subject to*	Required conditions being in place
End		Higher level development objectives		
↑		↑		necessary conditions
end (means)		Project development objective(s)		
↑		↑		necessary conditions
end (means)		Project component outcomes		
↑		↑		necessary conditions
end (means)		Outputs		
↑		↑		necessary conditions
end (means)		Activities		
↑		↑		necessary conditions
Means		Inputs		

Thus:

- **IF** inputs are provided, **THEN** activities can take place;
- **IF** activities are successfully completed, **THEN** planned outputs should result;
- **IF** outputs are used as intended, **THEN** the project component outcomes should be realized;
- **IF** the outcomes are achieved, **THEN** the project development objective(s) (PDO) should be achieved; and
- **IF** the PDO is achieved then the expected contribution should be made to higher level developmental objectives.

Fig. 9.4. Logical hierarchy for project design.

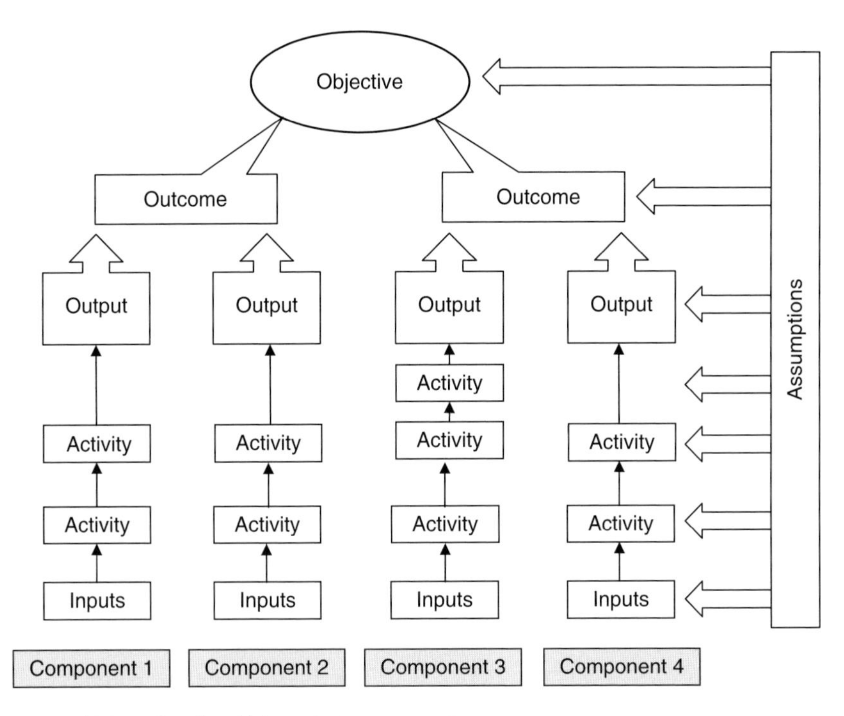

Fig. 9.5. Logical hierarchy of multiple project components.

Project logic	Types of indicator	*Focus of M&E*	Characteristics	Evaluation criteria	
Objectives	Impact	*Results monitoring*	Long-term widespread improvement in society	Relevance and impact	Sustainability
Outcomes	Outcome		Intermediate effects for beneficiaries		
Outputs	Output	*Implementation monitoring*	Capital goods, products and services produced	Effectiveness and efficiency	
Activities	Process		Tasks undertaken to transform inputs to outputs		
Inputs	Input		Human and material resources		

Fig. 9.6. A logical structure for project monitoring and evaluation (M&E).

various inputs and activities, while results monitoring is suited to senior project managers such as development agency task team leaders or supervising government personnel who are more interested in the long-term impacts of the project on society and the target beneficiaries.

For evaluation there are five commonly used criteria for assessing the performance of a project, as follows.

1. *Impact*: The effect of the project on its wider environment, and its contribution to the wider policy, sector or Country Assistance Strategy development objectives.
2. *Relevance*: The appropriateness of project objectives to the problems intended to be addressed, and to the physical and policy environment within which the project operates.
3. *Effectiveness*: How well the outputs contributed to the achievement of project component outcomes and the overall project development objective(s), and how well assumed external conditions contributed to project achievements.
4. *Efficiency*: Whether project outputs have been achieved at reasonable cost, i.e. how well inputs have been used in activities and converted into outputs.
5. *Sustainability*: The likelihood that benefits produced by the project continue to flow after external funding has ended.

The role of these five criteria in relation to the project logic and types of indicator are shown in the last two columns of Figure 9.6.

The logical framework

The logical framework was developed in the 1960s as a tool to improve project planning and implementation, and has been adopted by a number of development agencies, including the World Bank, as a project planning and management tool. At the core of the process is the logical framework matrix (Table 9.18), which is used to summarize the thinking that has occurred in the planning of the project based on problem and stakeholder analysis. The matrix comprises four columns and six rows which show:

- the hierarchy of project objectives (the causal chain or project logic);
- the indicators and sources of data to show how the project and its results will be monitored and evaluated;
- the assumptions and risks faced at each level showing the necessary external conditions that need to be satisfied if the next level up is to be achieved.

The stages followed in project planning contributing to the logical framework analysis are outlined in Box 9.1, while the sections below summarize the steps that are followed in formulating the logical framework matrix.

Identification of the target group

The first step is to identify the target group that the project intends to benefit, influence or change the behaviour of. The choice of the target group influences the approach of the project, the level of technology employed and the institutional and organizational arrangements that are required. Issues of status, access to resources, caste, ethnic status, gender, occupation/form of livelihood need to be considered and specified where appropriate.

Setting objectives

An objective states the desired state that is to be achieved through implementation of the project. There are three key objectives: (i) the higher-level development objective; (ii) the project development objective[6] (PDO); and (iii) the intermediate objectives identified for each project outcome. The PDO needs to specify the changes that can be expected in the target group, organization or location if the project is completed successfully, and must be a specific statement whose achievement can be verified. It is important that the PDO is realistic, and does not overstate the aims of the project. For example, a rehabilitation project will improve irrigation water delivery and thereby agricultural production; its impact on poverty eradication is less clear-cut, though this might be the higher-level objective in association with other interventions.

Identifying project outputs

Outputs are the result of the conversion of project inputs through the various project activities, and are a precondition for achievement of the project objectives. Importantly the achievement of the specified outputs is within the control of the project management, for which they should be held accountable. It is important that outputs are: identified; quantified (in terms of quantity, quality, time and place); realistic; and feasible within the resources available. Often there are several

Table 9.18. Structure of the logical framework matrix.[a] (From Burton *et al.*, 2008.)

Project logic	Indicators	Sources of verification	Assumptions and risks
Higher level development objective(s): the longer-term objective(s), change of state or improved situation to which achievement of the **project development objective(s)** is intended to contribute	How the objective(s) is to be measured; specified in terms of quality, quantity and timeframe	Data sources that exist or that can be provided cost-effectively through the completion of surveys or other forms of data collection	If the PDO(s) is achieved, what conditions beyond the project's direct control need to be in place to ensure the expected contribution to the higher-level development objectives?
Project development objective(s) (PDO): the combination of one or more **project component outcomes** which make up the physical, financial, institutional, social, environmental or other development changes which the project is designed and expected to achieve	How the PDO is to be measured in terms of its quality, quantity and timeframe	Details of data sources, how the data will be collected, by whom and when	If the project component outcomes are achieved, what conditions beyond the project's direct control need to be in place to achieve the PDO?
Project component outcomes: the effects of project components in terms of observable change in performance, behaviour or status of resources	Specification of how each project component outcome is to be measured in terms of its quality, quantity and timeframe	Details of data sources, how the data will be collected, by whom and when	If the outputs are produced, what conditions beyond the project's direct control need to be in place to achieve the project component outcomes?
Outputs: the products, capital goods and services resulting from a development intervention and which are necessary for the achievement of **project component outcomes**	How the outputs are to be measured in terms of their quality, quantity and timeframe	Details of data sources, how the data will be collected, by whom and when	If the activities are completed, what conditions beyond the project's direct control need to be in place to produce the outputs?
Activities: the actions taken by project implementers that deliver the **outputs** by using the **inputs** provided *(this level is not specified in some versions of logical framework analysis)*	*(a summary of the activities and resources may be included in this cell)*	*(a summary of the costs and budget may be provided in this cell)*	If the inputs are provided in full and on time, what conditions beyond the project's direct control need to be in place to ensure completion of the activities?
Inputs: the human and material resources financed by the project			What preconditions are necessary for input provision and project commencement?

[a]The structure and terminology used in the matrix may vary from that used by some organizations.

Box 9.1. Stages of Project Planning (European Commission, 2004)

Analysis stage

- *Stakeholder analysis*: identifying and characterizing key stakeholders and assessing their capacity.
- *Problem analysis*: identifying key problems, constraints and opportunities; determining cause-and-effect relationships.
- *Objective analysis*: developing solutions from the identified problems; identifying means-to-ends relationships.
- *Strategy analysis*: identifying different strategies to achieve solutions; selecting the most appropriate strategy.

Planning stage

- *Developing logical framework matrix*: defining the project structure, testing its internal logic and risks, formulating measurable indicators of achievement.
- *Activity scheduling*: determining the sequence and dependency of activities; estimating their duration and assigning responsibility.
- *Resource scheduling*: from the activity schedule, developing input schedules and a budget.

outputs contributing to the achievement of the PDO; it is important that the causal chain linking these outputs to achievement of the PDO is clearly identified.

Defining activities

An activity converts project inputs into output(s) over a specified timeframe. Activities need to be carefully specified such that their implementation progress can be measured and verified in terms of quantity, time and place. It should be clear who is responsible for implementation of each activity, and that all activities required to achieve a specified output are included in the project. Likewise no activity should be specified that does not contribute to a project output.

Identifying inputs

Inputs are the goods, personnel, services and other resources required for carrying out project activities. It is important at the planning stage to look at the inputs required for a project; these will include the purpose and type (personnel, equipment, materials, vehicles, etc.), the quantity required, duration, timing, cost and availability.

Assessing external conditions, assumptions and risks

The logical framework approach requires that proper attention is given to the environment within which the project is set and the assumptions and risks related to implement the project within this environment. False assumptions or failure to adequately take account of inherent risks has led to the failure of far too many projects. By considering assumptions and risks at the project design stage proper assessment can be made of their impact on the project outcome, and, where feasible, action taken to mitigate or remove them. Thoughtful filling in of this column at an early stage can help to define the project and its boundaries better. If an assumption about an aspect initially considered as outside the project turns out to be crucial to the project's success, it may be necessary to bring this aspect into the project, i.e. shifting it from the fourth to the first column.

The results framework

The results framework is a simplified version of the logical framework with a focus on the PDO and the intermediate outcomes (results) expected from the implementation of each project component.

The results framework comprises:

- a statement of the PDO, outcome indicators and the use of the indicators (Table 9.19);
- a table showing the intermediate results, results indicators and the use of these indicators in results monitoring (Table 9.20);

Table 9.19. Example of a results framework with project development objective and outcome indicators.

PDO	Outcome indicators	Use of outcome information
To improve irrigation and drainage service delivery and land and water management in order to sustainably increase agricultural productivity in irrigation and drainage schemes	Water distribution by main system service providers to WUAs in 75% of the irrigated area matches irrigation water demands	An enhanced irrigation and drainage service delivery will provide more reliable, timely and adequate water delivery and drainage water removal, providing improved crop production opportunities and livelihoods for farmers and their families
	Water distribution by WUAs to farmers in 75% of the rehabilitated systems closely matches irrigation water demands	
	Collection rates by WUAs at least 75% of total assessed fees following establishment of WUAs	Service fee collection rates provide indication of sustainability of management, operation and particularly maintenance
	Number of farmers in sub-project areas more knowledgeable and applying recommended irrigated agricultural practices	Improved agricultural practices combine with good water management to improve crop production
	Increase in average crop yields in sub-project areas after completion of rehabilitation works	Due to other contributory factors, care needs to be taken with attribution of increases in crop yields to project activities

PDO, project development objective; WUA, water users association.

- for both the outcome indicators and intermediate results indicators, a table showing the indicators, the target values for each year of the project, and details of the data collection and reporting to include the frequency of measurement and types of report, the data collection instruments and who is responsible for data collection (Table 9.21).

It is important that the PDO is clear and concise, and that it identifies the change in status to be brought about by the project. In the agricultural water management sector this change in status is usually expressed in terms of technology, agricultural productivity and value of agricultural production contributing to an increase in farmer income. The PDO should make clear:

- who are the beneficiaries and where they are located;
- what problem will have been addressed by the project;
- what will be the nature and scale of the change brought about by the project.

It is important that the PDO is expressed at the right level, that it is not set at too high a level (e.g. to reduce poverty in the rural sector) or too low a level (e.g. at activity level, such as to rehabilitate the physical infrastructure). The PDO should be realistic in terms of what it can achieve given its focus, resources and duration; it should be measurable and should summarize the achievements of the project as a whole, rather than reiterate the individual component outputs or outcomes.

It should be clear from the results table how the individual components of the project link together to achieve the project development objective. This can be shown through the intermediate results and results indicators stated in Table 9.20, with an explanation of how the results indicators will be used to monitor the progress of the project. This table is supported by Table 9.21, which specifies the annual targets, data collection, reporting and dissemination arrangements that will enable management to track and report on project progress.

Selection and specification of the indicators at the various levels is important; it may take several iterations until they are finalized. These indicators should measure and summarize the results of the work carried out, and it should be clear by whom the data will be collected and where they will be reported.

Table 9.20. Example of results table with intermediate results and indicators.

Intermediate results	Results indicators	Use of results monitoring
Component 1: Rehabilitation and modernization of I&D systems		
(1.1) Water users in 20 sub-project area command areas are provided with an improved, more reliable and more manageable water supply as a result of rehabilitation of main system and on-farm infrastructure	Number of systems with the ability to supply controlled and measured volumes of water to match water users' requests Number of systems with ability to manage and control groundwater levels within acceptable ranges	Results to be based on a system-by-system basis taking account of the phasing of the physical works
Component 2: Sustainable management, operation and maintenance of I&D systems		
(2.1) WUAs formed and functioning effectively in sub-project areas	Number of WUA Support Units functioning effectively and providing adequate levels of training and support to WUAs	Assess changing nature of Support Unit's role over the project period
	Number of WUAs setting ISF rates which match sustainable MOM needs based on asset management assessments	A leading indicator of understanding and acceptance of need for sustainable MOM, in particular system maintenance
(2.2) Main system service delivery functioning effectively	Number of main system service providers who are following updated procedures for preparation of seasonal water allocation plans and achieving target values for actual against planned delivery	Assesses the ability to plan water allocations to closely match demand and thereby conserve water supplies
	Number of WUAs receiving irrigation water supplies which closely match irrigation demands throughout the irrigation season	Assesses the capability to manage and operate the main system to provide reliable, timely, adequate and equitable water supply
	Annual maintenance expenditure on main system at least 75% of levels of expenditure assessed by maintenance studies	Assesses the level of adoption of the maintenance studies, and the commitment to sustainable MOM for the main system
(2.3) On-farm service delivery functioning effectively	80% of water users in each WUA receiving irrigation water supplies which match their requests	Assesses the WUA management's capability in operating the system
Component 3: Agricultural services and support		
(3.1) Capacity of farmers and farm managers strengthened	Number of farms who have successful and continued access to external technical and financial services Service Centres	Assumes that improved understanding, knowledge and skills combined with adequate and reliable financial services will lead to enhanced agricultural production
	Number of independent and well-functioning Service Centres measured in terms of delivery against Level of Service Agreements	Measures service delivery against stated criteria and farmer (customer) satisfaction with service delivered
(3.2) Farm mechanization improved	Number of farmers requesting farm machinery	Checks that a key pre-project constraint has been removed
	Number of WUAs requesting maintenance machinery	Checks that a key pre-project constraint has been removed

I&D, irrigation and drainage; WUA, water users association; ISF, irrigation service fee; MOM, management, operation and maintenance.

Table 9.21. Arrangements for results monitoring.

Indicator	Target values							Data collection and reporting		
	Baseline	YR1	YR2	YR3	YR4	YR5	YR6	Frequency and reports	Data collection instruments	Responsibility for data collection
Outcome indicators										
Water distribution by main system service providers to WUAs in 75% of the irrigation area in sub-project areas matches irrigation water demands (number of systems and area in ha)	0	0	0	0	3 12,000	5 21,000	12 51,000	Annually from YR4 as rehabilitation completed Quarterly and Annual Reports; Implementation Completion Report (YR6)	General project monitoring Water demand and supply data from main system service providers Baseline (YR1) and impact survey (YR6)	Project M&E team WUA Support Unit staff Main system service providers' staff Baseline and impact study contractor
Water distribution by WUAs to farmers in 75% of the rehabilitated systems closely matches the irrigation water demands (number of WUAs and area in ha)	0	0	0	0	5 21,000	10 42,000	30 124,000	Annually from YR4 as rehabilitation completed Quarterly and Annual Reports; Implementation Completion Report (YR6)	General project monitoring WUA records Baseline (YR1) and impact survey (YR6)	Project M&E team WUA Support Unit staff Baseline and impact study contractor

Continued

Table 9.21. Continued.

Indicator	Target values							Data collection and reporting		
	Baseline	YR1	YR2	YR3	YR4	YR5	YR6	Frequency and reports	Data collection instruments	Responsibility for data collection
Collection rates by WUAs at least 75% of total assessed fees (based on agreed annual budgets for MOM) after establishment of WUAs (number of WUAs)	0	0	30	50	70	90	110	Annually from YR2 following formation of WUAs	General project monitoring	Project M&E team
								Quarterly and Annual Reports; Implementation Completion Report (YR6)	Baseline (YR1) and impact survey (YR6)	WUA Support Unit staff Baseline and impact study contractor
Number of farmers in sub-project areas more knowledgeable and applying recommended irrigated agricultural practices (number of farmers)	0	0	0	2,500	4,000	5,500	7,000	Annually from YR3 onwards as farmer training completed	General project monitoring	Project M&E team
								Quarterly and Annual Reports; Implementation Completion Report (YR6)	Baseline (YR1) and impact survey (YR6)	WUA Support Unit staff Baseline and impact study contractor
Increase in crop yields in sub-project areas after completion of rehabilitation works (%)	0	0	0	0	15 (rice)	20 (rice)	30 (rice)	Annually from YR4 as rehabilitation completed	General project monitoring	Project M&E team
					15 (wheat) 10 (cotton)	20 (wheat) 15 (cotton)	30 (wheat) 25 (cotton)	Quarterly and Annual Reports; Implementation Completion Report (YR6)	Baseline (YR1) and impact survey (YR6)	WUA Support Unit staff Baseline and impact study contractor

WUA, water users association; MOM, management, operation and maintenance; M&E, monitoring and evaluation.

Benchmarking

Benchmarking originated in the corporate business sector as a means for companies to gauge, and subsequently improve, their performance relative to key competitors. By studying key competitors' outputs, and the processes used to achieve those outputs, many organizations have been able to adopt best management practices and enhance their own performance (Box 9.2). In some cases organizations have done so well that they have, in turn, become the organization that others use as a benchmark.

There are many reasons why organizations may be interested in the benchmarking activity. The private sector is primarily driven by a desire to improve return on investment or return to shareholders; in the public sector the aim is to improve the effectiveness and efficiency of the organization and the level of service provision. In the irrigation and drainage sector service providers are responding to a variety of drivers, including the following.

- Increasing competition for water, both within the irrigated agriculture sector, and from other sectors.
- Increasing demand on the irrigation sector to produce more food for growing populations. Coupled with the pressure on available water resources, this has resulted in the 'more crop per drop' initiative promoted by the International Water Management Institute (IWMI) and the Food and Agricultural Organization of the United Nations (FAO).
- Growing pressure to effect cost savings while increasing the productivity and efficiency of resource use (see Burton *et al.*, 2005; World Bank, 2005).
- Turnover and privatization of I&D schemes to water users, leading to more transparent and accountable (to users) management practices.
- Increasing interest by the wider community in productive and efficient water resource use and the protection of aquatic environments.
- Increasing need for accountability to both government and water users in respect of water resource use and the price paid for water.

Different drivers will apply in different situations; it is important at the outset of a benchmarking programme to identify the key drivers that are forcing change within the irrigation and drainage sector.

Benchmarking is about moving from one level of performance to another (Fig. 9.7). It is about changing the way in which systems are managed and about raising the expectations of all parties as to the level of achievable performance. It is a change management process that requires identification of shortcomings, and then acceptance by key stakeholders of the need and pathways for achieving the identified goals. Benchmarking is part of a strategic planning process, which asks and answers such questions as *'Where are we now?'*, *'Where do we want to be?'* and *'How do we get there?'*

Benchmarking uses performance assessment procedures to identify levels of performance and will use M&E procedures to see how actions taken to close identified performance gaps are progressing.

Benchmarking Stages

There are six key stages to benchmarking, as shown in Fig. 9.8.

Box 9.2. Definition of Benchmarking

Benchmarking can be defined as:

> A systematic process for securing continual improvement through comparison with relevant and achievable internal or external norms and standards.
>
> (Malano and Burton, 2001)

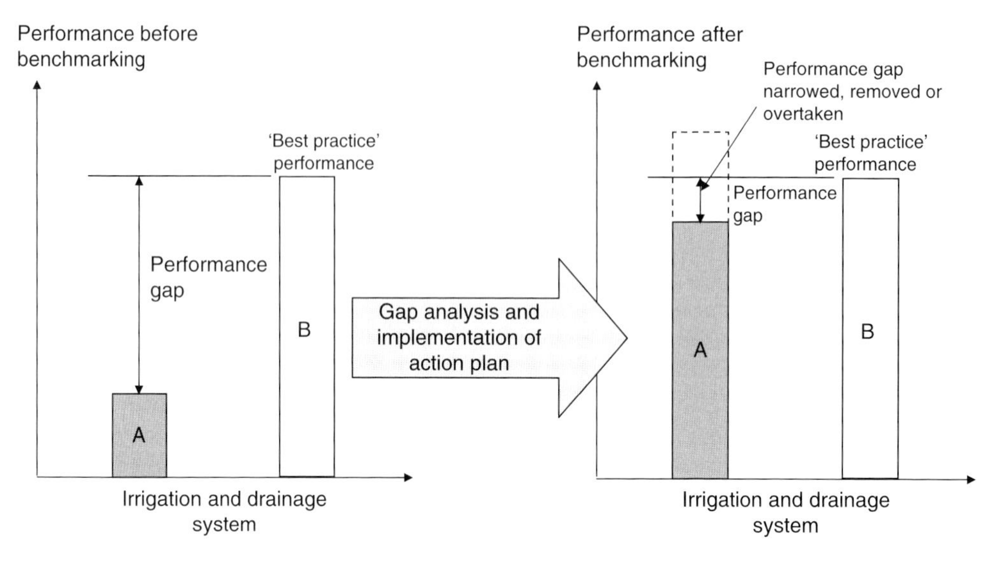

Fig. 9.7. Benchmarking – comparative performance against best practice.

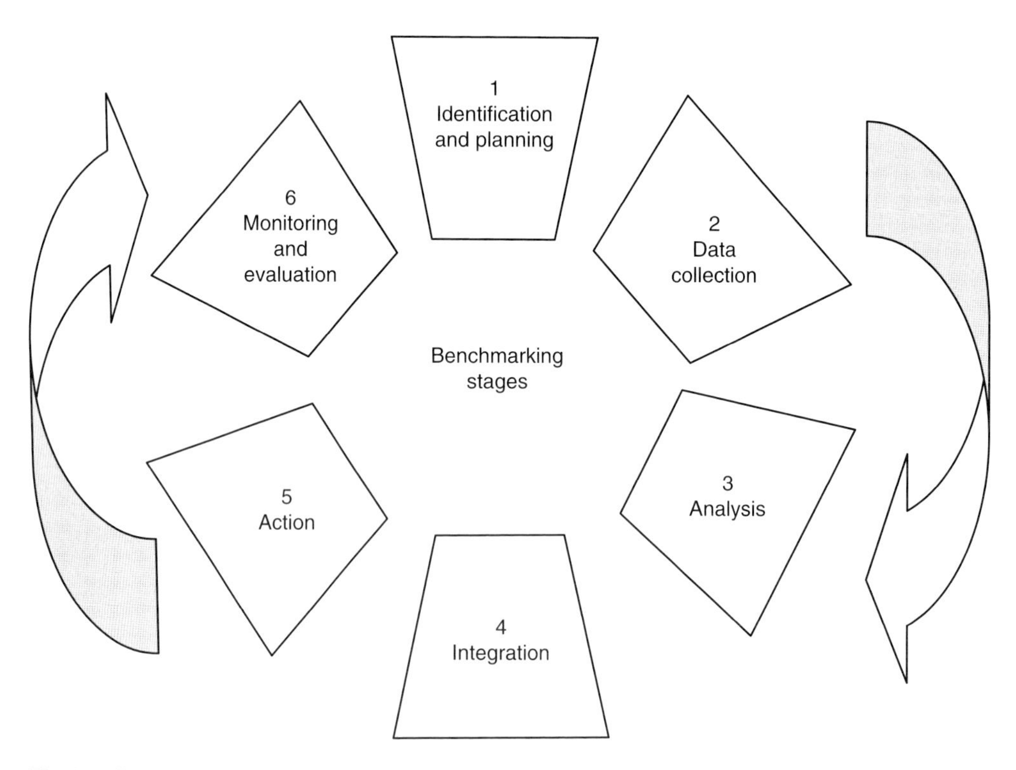

Fig. 9.8. Benchmarking stages. (From Malano *et al.*, 2004.)

Stage 1: Identification and planning

This stage identifies:

- the objectives and boundaries of the benchmarking programme;
- who the benchmarking is for;
- the key processes;
- the related performance indicators;
- the data requirements.

As discussed in earlier sections in this chapter, it is important at the outset to identify the objectives and boundaries of the benchmarking exercise. Is the objective to improve the efficiency and productivity of water alone, or irrigated farming as a whole? Is the benchmarking for the individual farmer, the service provider, the regulator or government? Having decided on these key issues, it is necessary to identify the processes involved within the identified boundaries and the related performance indicators and data needs.

A key part of the process is to identify successful organizations or I&D systems with similar processes. Use of key descriptors (Box 9.3) enables similar systems and processes to be identified and enables meaningful comparison to take place. For example, the water use on a rice scheme will be significantly different from that on a cotton scheme.

In identifying the key processes (Fig. 9.9) the following questions can be asked.

- What are the objectives of the enterprise?
- How is success measured? What are the outputs and desired outcomes?
- What are the processes that contribute to the attainment of these outputs and outcomes?
- How can these processes be measured?

It is also important to consider the impact of the key processes; the consequences of water abstraction from rivers and pollution from agricultural drainage water are key considerations in this respect.

Possible key processes and indicators include the following.

- Irrigation water abstraction, conveyance and application:
  - volume of water abstracted for irrigation;
  - irrigation water abstraction per unit area;
  - relative irrigation water supply (abstraction/demand).
- Crop production:
  - irrigated area;
  - cropping intensity;
  - crop yield;
  - value of crop production per unit area;
  - value of crop production per unit water abstracted.

Box 9.3. Descriptors for Irrigation and Drainage Schemes (Malano and Burton, 2001)

- Irrigable area.
- Drained area.
- Annual irrigated area.
- Climate.
- Water resources availability.
- Water source.
- Average annual rainfall.
- Average annual reference crop potential evapotranspiration (ET_0).
- Method of water abstraction.
- Water delivery infrastructure.
- Type of water distribution.
- Type of drainage.
- Predominant on-farm irrigation method.
- Major crops (with percentages of total irrigated area).
- Average farm size.
- Type of irrigation system management.
- Type of drainage system management.

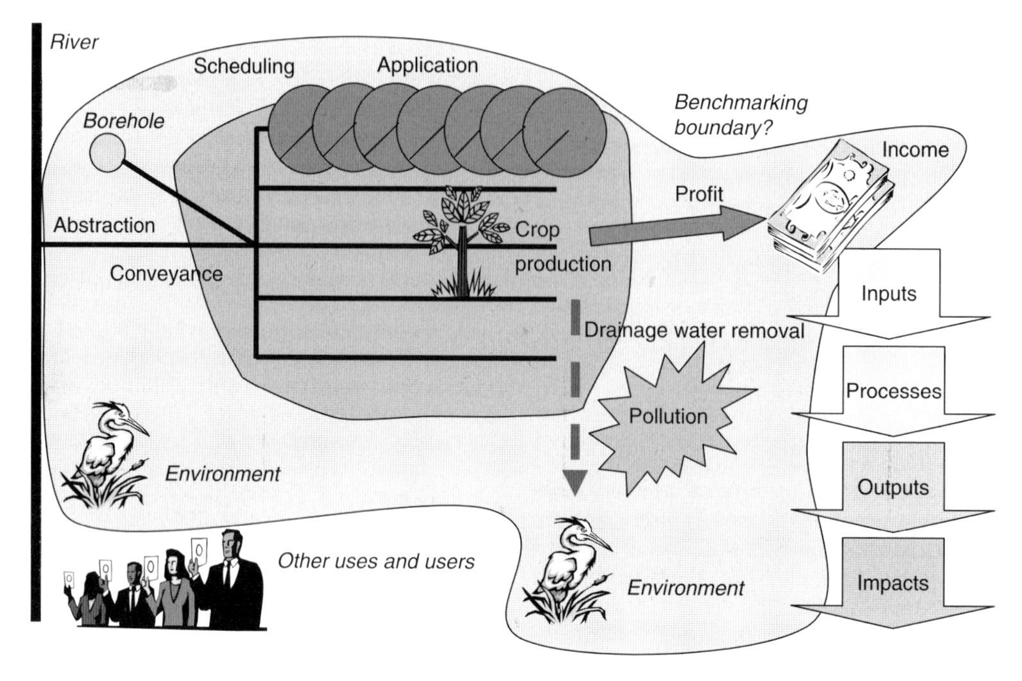

Fig. 9.9. Identification of key processes.

- Business processes:
  - cash flow (investment versus returns);
  - total annual income;
  - annual profit.
- Environmental impact:
  - wastewater quality (biological/chemical content);
  - minimum flow levels in river.

Stage 2: Data collection

Data are collected and the value of performance indicators determined. The data collection programme will identify what data are to be collected, by whom, how frequently, where, and how accurate the data need to be. These data are for the system under review and the benchmark system(s), and will include input, process, output, outcome and impact performance indicators. Additional data may have to be collected for the benchmarking exercise beyond those already collected for day-to-day system management, operation and maintenance.

Stage 3: Analysis

Data are analysed and the performance gap(s) identified in the key processes (Fig. 9.10). The analysis also identifies the cause of the performance gap, and the action(s) to close the gap. Recommendations are formulated from the options available, and then reviewed and refined. Further data collection may be required for diagnostic analysis where additional information and understanding are required to identify root causes of the performance gap. This can be either the beneficial causes of the better performing system(s) or the constraining causes of the less well performing systems.

Stage 4: Integration

The action plan developed from the analysis phase must be integrated into the operational processes and procedures of the organization in order to bring about the desired change. It is crucial that those responsible for benchmarking

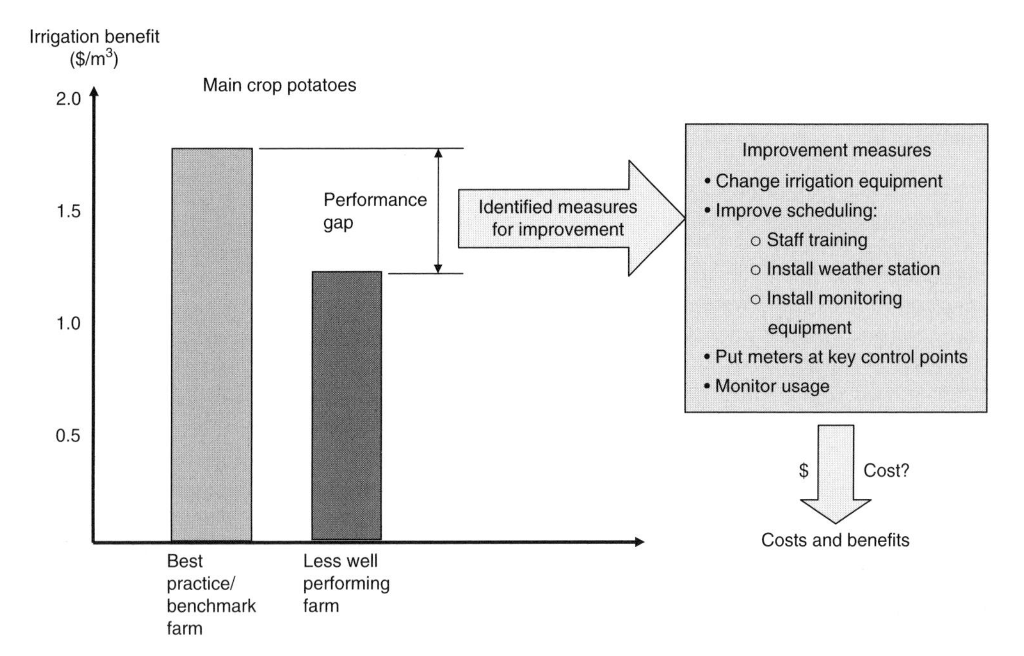

Fig. 9.10. Identification and costing of measures to close the performance gap.

have the power within the organization to bring about change. Benchmarking programmes often fail at this stage, leaving those involved disillusioned with the process, and with the performance of the organization.

The process of gaining adoption of the new processes and procedures is often termed 'internal marketing', and leads to the development of a sense of ownership and support by key personnel for the benchmarking process. Training is a key element of this process.

Stage 5: Action

Once acceptance of the new processes and procedures has been gained they can be put into place to bring about the desired change. Leadership by senior management plays a key role in ensuring that the action plan is implemented successfully. Careful monitoring of the process is required at this stage to ensure that desired targets are being achieved, and that corrective action, where necessary, is taken in time.

Stage 6: Monitoring and evaluation

The success of benchmarking is marked by the continuing measurement of the organization's performance against the target norms and standards established during the analysis and integration stages. These targets are, however, changing over time, and continual updating and revision of the targets is necessary to maintain best practices and relative performance.

Figure 9.8 shows a cyclical programme of activities, though there may be a break of some years between one benchmarking exercise and another. During this period the lessons learned from the benchmarking programme are implemented, monitored and evaluated, with refinements being made as experience is gained with implementing the new processes and procedures. As mentioned previously it has been the case with some organizations that they have so improved their performance that they have become the benchmark.

Example of benchmarking in Australia

The Australian National Committee of Irrigation and Drainage (ANCID) was one of the first organizations to implement a benchmarking programme in the irrigation and drainage sector. It began in 1998 with 33 schemes managed by irrigation service providers and now has over 40 schemes in the programme, covering some 75% of the irrigation water provider business in Australia. The total business distributes 18,000 Gl of water annually, providing water for some 2 million ha and generating an annual business turnover of AU$200 million (US$162 million) from a production base of some AU$7 billion (US$5.7 billion) (Alexander and Potter, 2004). The crops grown include rice, maize, grape vines, cotton, sugarcane, pasture, citrus and vegetables.

The benchmarking programme used 65 performance indicators:

- system operation (n = 12);
- business processes (n = 25);
- financial management (n = 14);
- environmental management (n = 14).

These indicators have been formulated to fit with the 'triple bottom line' approach adopted by the industry, measuring performance in economic, environmental and social dimensions.

A key feature of the Australian benchmarking programme is the 'three tier' reporting of data to protect commercial confidentiality. Tier 1 collects data on general irrigation water provision ('Who we are'), Tier 2 collects data on performance ('How we interact') and Tier 3 collects data on confidential internal business performance benchmarking ('How we improve'). The data are collected each year using a standard questionnaire, each contributor indicating what data can and cannot be released. The data are analysed and the report made available to all contributors, with anonymous data presented for others to compare their performance with. If a contributor wishes to obtain more information on the confidential data he/she writes to ANCID who forward his/her request on to the relevant contributor.

Figure 9.11 presents examples of the performance indicators used. As can be seen there is a wide range in the values of each of the indicators, this is due to individual differences between the systems (the crop types, method of irrigation, lined/unlined canals, etc.). This highlights the importance of using the system descriptors (Box 9.3) to categorize systems to enable comparison of like with like.

The achievements of the benchmarking programme in Australia are summarized as (Alexander and Potter, 2004):

- allowing comparison of the performance of irrigation water providers relative to each other, both at the domestic and international level;
- providing a more progressive and accountable image of the irrigation sector;
- monitoring the uptake and impact of modern technology;
- improvement in record keeping and performance analysis by service providers;
- availability of objective and reliable data across a substantial part of the irrigation industry;
- adoption by businesses of the ANCID benchmarking approach and formulation of their own inter-business benchmarking systems;
- more confident setting by business managers of targets for water delivery efficiency, operation, health and safety, and resource use.

Example of Implementing a Benchmarking Programme

This section outlines the procedures followed for implementing a benchmarking programme. Using the framework outlined in the sections above, the key components of the benchmarking programme are now detailed.

Drivers of the benchmarking process

The key driver for benchmarking is the Government's interest in institutional reform to facilitate improvements in irrigation and drainage service delivery in the public sector.

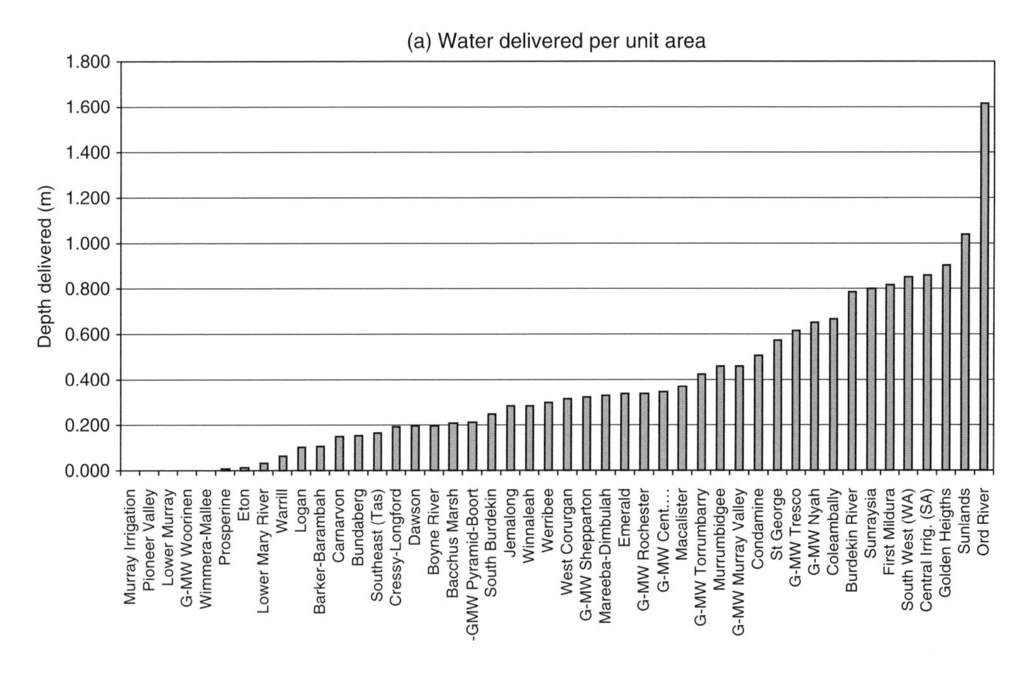

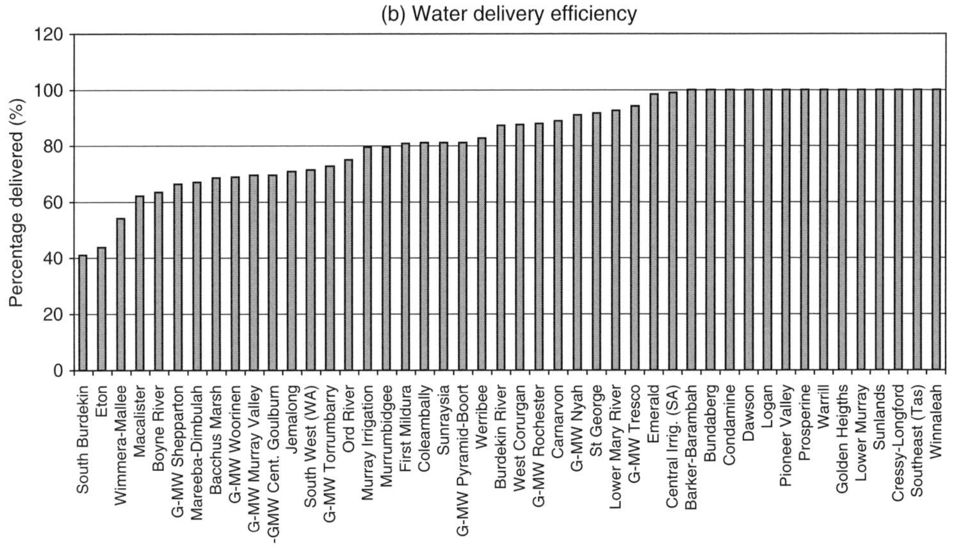

Fig. 9.11. Examples of performance data plots from the Australian benchmarking programme. (Data from ANCID, 2000.)

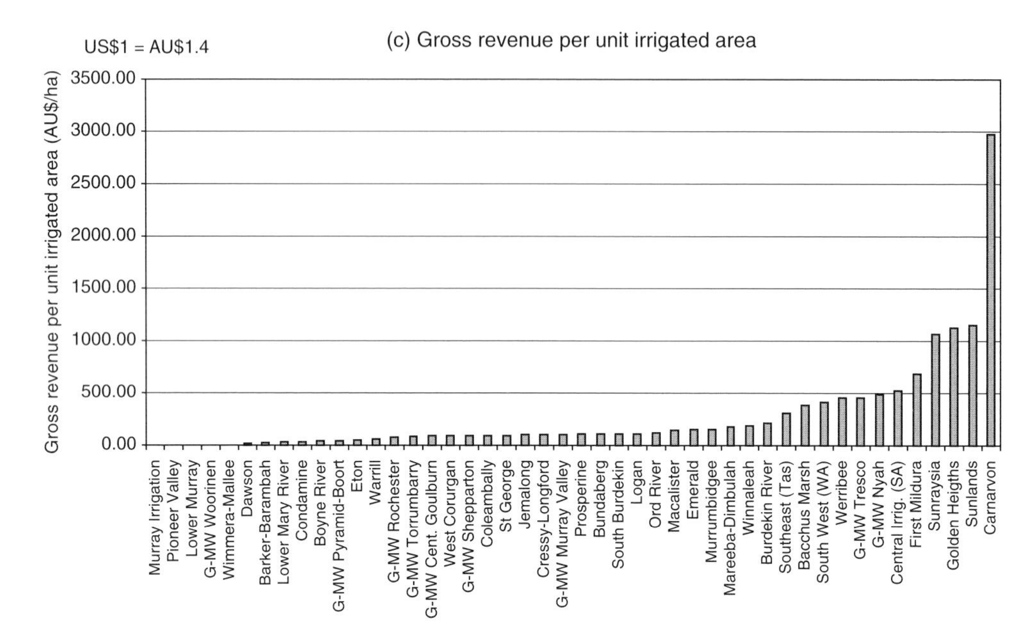

Fig. 9.11. Continued

Objectives of the benchmarking programme

The overall objective of the benchmarking programme is to sustain and increase agricultural production, while:

- improving the efficiency and productivity of water use, thus reducing the amount of water diverted for irrigation;
- minimizing the cost of irrigation water delivery and drainage water removal, consistent with providing reliable, timely and adequate irrigation water supplies and drainage water removal, and sustainable levels of system maintenance;
- sustaining soil fertility and the crop growth environment through effective drainage water removal and drainage system maintenance.

Boundaries

The main processes to benchmark are identified as:

- irrigation water delivery;
- drainage water removal;
- maintenance of infrastructure;
- environmental protection (through management of water quality).

The physical boundary has been identified as the secondary canal (Table 9.22). Secondary canals have been chosen as suitable management units to benchmark, as:

- they are the lowest management unit run by the Irrigation Agency;
- they are at the front end of service delivery to the client (the farmers);

Table 9.22. Summary details of secondary canals selected for benchmarking.

Secondary canal name	Main canal name	Command area (ha)
P1/S1	P1	5500
P1/S2	P1	2350
P1/S3	P1	5600
P2/S1	P2	5200
P2/S2	P2	5640
P2/S3	P2	3630

- they are discrete management units, with measurable inputs, outputs and processes;
- water delivery and drainage water removal processes in the secondary canal command area are strongly influenced by how these processes are managed;
- improvements in the management processes at this level can have a marked impact on crop production (output performance);
- secondary canals in a given locality have similar basic features, allowing meaningful comparisons in performance;
- data collection is feasible.

Programme

The programme established for benchmarking is summarized in Table 9.23.

Performance indicators

The indicators related to water delivery and removal, agricultural crop production and environmental protection for the four related processes are detailed in Table 9.24. There are 30 indicators identified, which require 24 sets of data.

Table 9.23. Programme for benchmarking performance in the irrigation and drainage sector.

No.	Activity	Example/explanation
1.	Identify the objectives of the total process	• Increased agricultural production • Improved efficiency and productivity of water use • Minimizing costs while maintaining adequate operation and maintenance standards • Sustain soil fertility and crop growth environment
2.	Identify the key outputs	• Irrigation water delivery • Drainage water removal • Crop production
3.	Identify performance indicators for measurement of outputs	• Crop production (in kg and MU) • Crop production (in kg and MU) per unit area • Crop production (in kg and MU) per unit water supply
4.	Collect data for output indicators and benchmark performance against comparable units	• Crop type, area, yield, input costs, market price, water supplied
5.	Quantify the gap in output performance	This may be between total crop production on secondary canals, or between total crop production within tertiary canals
6.	Identify the key processes that contribute to the output performance	• Irrigation water delivery (reliability, timeliness and adequacy) • Drainage water removal (timeliness, adequacy, soil water quality) • Maintenance of I&D system
7.	Identify performance indicators for these key processes	• Seasonal relative irrigation water supply (supply/demand) • Seasonal irrigation water supply per unit area (m^3/ha) • Main system water delivery efficiency • Pumping hours and discharge per unit area for tertiary canals in the head, middle and tail reaches • Seasonal average depth to groundwater (m) • Seasonal soil and drainage water quality • Cost of irrigation water delivery and drainage water removal

Continued

Table 9.23. Continued

No.	Activity	Example/explanation
8.	Collect data for these process indicators and assess and benchmark process performance against comparable processes	• Compare performance of the following indicators between secondary canals and tertiary canals: ○ Secondary canal water delivery efficiencies ○ Relative irrigation water supply ○ Tertiary canal pumping hours per unit area ○ Irrigation water supplies per unit area ○ Average depths to groundwater ○ Groundwater and soil quality
9.	Identify the gaps in process performance	This may be between secondary canals or between tertiary canals on a secondary canal
10.	Identify the key factors that influence this performance, and propose remedies	• Tail-end tertiary canals, for example, may be getting less water per unit area than head-end tertiary canals • Groundwater levels and soil salinity levels may be high, thus reducing crop yields
11.	Prepare an Action Plan for introduction and implementation of the proposals	The Action Plan might require senior management to take action, and/or for WUA representatives, or others, to take action. Need to specify who is involved, what resources are required (time, people, finances), and the programme for implementation
12.	Gain acceptance of the Action Plan by key stakeholders	• Agreement from senior managers within I&D agencies • Agreement between WUAs on a secondary canal
13.	Implement the Action Plan	• Disseminate the details of the Action Plan widely to explain what is being done • Leadership will be required by key stakeholders to ensure Action Plan is implemented properly • Make step-by-step improvements
14.	Monitor implementation and degree of change effected	Monitoring data fed back to all key stakeholders, including senior management and to WUA representatives
15.	Evaluate implementation and degree of change on completion	Senior management and WUA representatives to assess the change in performance as a result of implementing the Action Plan

MU, monetary unit; I&D, irrigation and drainage; WUA, water users association.

Data collection

Figure 9.12 shows the location of data collection within the secondary canal command area, and Table 9.25 summarizes where the data were collected, by whom and with what frequency.

A summary of the results of the benchmarking programme are presented in Fig. 9.13. The table shows the indicators chosen and their values for the six secondary canals. In the table the 'best' values have been highlighted in gold, while critical values are highlighted in red and areas for concern in yellow. Some of the indicators have not been given highlights as these are indicative indicators and it is not possible to judge them one against another. This is the case for example with the Total seasonal crop water demand (at field) and the Total seasonal irrigation water supply per unit command area, where the value depends on the cropping pattern within the secondary canal – there is no one 'best' figure here but the value does serve to show the relative scale of supply to, and demand by, each secondary canal. The Seasonal relative irrigation water supply is then the prime indicator linking the supply and demand.

From this example benchmarking study the following conclusions can be drawn.

- The process has identified the performance in key management units (the secondary canals). Comparing the

Table 9.24. Benchmarking processes and indicators.

Objective/ desired output	Output indicators	Processes	Process indicators	Definition
Agricultural crop production	Total seasonal area cropped per unit command area (cropping intensity)			$\frac{\text{Total area cropped seasonally}}{\text{Total command area of system}}$
	Total seasonal crop production (t)			Total seasonal crop production by crop type within command area
	Total seasonal value of crop production (MU)			Total seasonal value of agricultural crop production received by producers
	Total seasonal crop production per unit command area (crop yield, kg/ha)			$\frac{\text{Total seasonal crop production}}{\text{Total command area of system}}$
	Total seasonal value of crop production per unit command area (MU/ha)			$\frac{\text{Total seasonal value of crop production}}{\text{Total command area of system}}$
	Total seasonal crop production per unit water supply (kg/m^3)			$\frac{\text{Total seasonal crop production}}{\text{Total seasonal volume of irrigation water supply}}$
	Total seasonal value of crop production per unit water consumed (MU/m^3)			$\frac{\text{Total seasonal value of crop production}}{\text{Total seasonal volume of crop water demand } (ET_c)}$
	Total seasonal value of crop production per unit water supplied (MU/m^3)			$\frac{\text{Total seasonal value of crop production}}{\text{Total seasonal volume of irrigation water supply}}$
		Irrigation water delivery	Total seasonal volume of irrigation water supply (MCM)	Total seasonal volume of water diverted or pumped for irrigation (not including diversion of internal drainage)

Continued

Table 9.24. Continued

Objective/ desired output	Output indicators	Processes	Process indicators	Definition
			Seasonal irrigation water supply per unit command area (m^3/ha)	$\frac{\text{Total seasonal volume of irrigation water supply}}{\text{Total command area of system}}$
			Main system water delivery efficiency	$\frac{\text{Total seasonal volume of irrigation water delivery}}{\text{Total seasonal volume of irrigation water supply}}$
			Seasonal relative irrigation water supply	$\frac{\text{Total seasonal volume of irrigation water supply}}{\text{Total seasonal volume of crop water demand}}$
			Water delivery capacity	$\frac{\text{Canal capacity at head of system}}{\text{Peak irrigation water demand at head of system}}$
		Irrigation MOM finance and staffing levels	Total seasonal MOM cost for irrigation water delivery per unit command area (MU/ha)	$\frac{\text{Total seasonal MOM cost for irrigation water delivery}}{\text{Total command area of system}}$
			Total seasonal MOM cost for irrigation water delivery per unit irrigation water supply (MU/m^3)	$\frac{\text{Total seasonal MOM cost for irrigation water delivery}}{\text{Total seasonal volume of irrigation water supply}}$
			Total seasonal maintenance expenditure for irrigation water delivery per unit command area (MU/ha)	$\frac{\text{Total seasonal maintenance expenditure for irrigation water delivery}}{\text{Total command area of system}}$
			Total seasonal maintenance expenditure fraction for irrigation water delivery	$\frac{\text{Total seasonal maintenance expenditure for irrigation water delivery}}{\text{Total seasonal MOM cost for irrigation water delivery}}$

	Total cost per person employed on water delivery (MU/person)	$\frac{\text{Total cost of personnel engaged in irrigation water delivery}}{\text{Total number of personnel engaged in irrigation water delivery}}$
	Command area per unit irrigation staff (ha/person)	$\frac{\text{Total command area of system}}{\text{Total number of personnel engaged in irrigation water delivery}}$
Drainage water removal	Average depth to groundwater (m)	Average seasonal depth to groundwater calculated from water table observations over the irrigation area
Drainage MOM finance and staffing levels	Total seasonal MOM cost for drainage water removal per unit command area (MU/ha)	$\frac{\text{Total seasonal MOM cost for drainage water removal}}{\text{Total command area of system}}$
	Total seasonal MOM cost for drainage water removal per unit drainage water removal (MU/m^3)	$\frac{\text{Total seasonal MOM cost for drainage water removal}}{\text{Total seasonal volume of drainage water removed}}$
	Total seasonal maintenance expenditure for drainage water removal per unit command area (MU/ha)	$\frac{\text{Total seasonal maintenance expenditure for drainage water removal}}{\text{Total command area of system}}$
	Total seasonal maintenance expenditure fraction for drainage water removal	$\frac{\text{Total seasonal maintenance expenditure for drainage water removal}}{\text{Total seasonal MOM cost for drainage water removal}}$
	Total cost per person employed on drainage water removal (MU/person)	$\frac{\text{Total cost of personnel engaged in drainage water removal}}{\text{Total number of personnel engaged in drainage water removal}}$

Continued

Table 9.24. Continued

Objective/desired output	Output indicators	Processes	Process indicators	Definition
			Command area per unit of drainage staff (ha/person)	$\frac{\text{Total command area of system}}{\text{Total number of personnel engaged in drainage water removal}}$
		Environmental protection	Salinity of soil water (mmho/cm)	Electrical conductivity of soil water
			Soil salinity (mmho/cm)	Electrical conductivity of soil
			Salinity of water in open drain (mmho/cm)	Electrical conductivity of water in open drains
			Drainage water quality: biological (mg/l)	Biological load of drainage water expressed as biological oxygen demand (BOD)
			Drainage water quality: chemical (mg/l)	Chemical load of drainage water expressed as chemical oxygen demand (COD)

MU, monetary unit; MCM, millions of cubic metres; MOM, management, operation and maintenance.

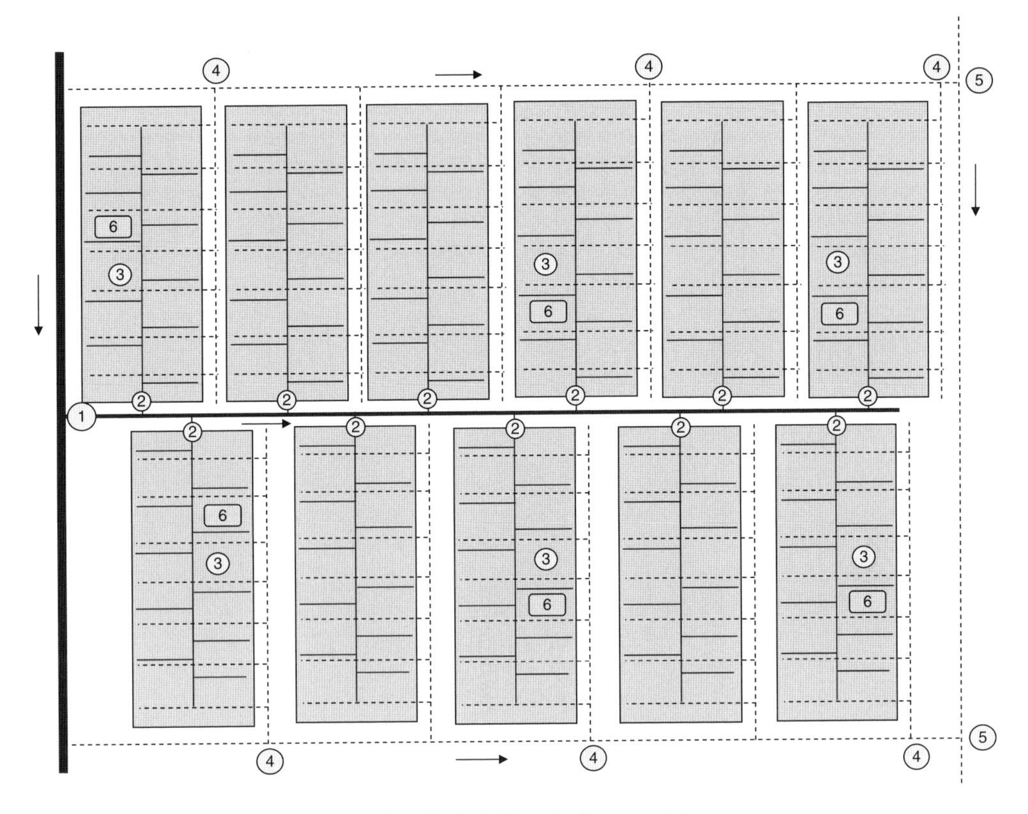

Fig. 9.12. Location of data collection (see Table 9.25 for further details).

performance of similar management units enabled best practice and suitable performance targets to be identified, identified gaps in performance and provided (some) answers to the root causes of these performance gaps.

- Diagnostic analysis is a fundamental part of benchmarking. Analysis of the initial set of performance indicators can lead on to further data collection and interviews with water users to identify the root causes of poor levels of performance.
- The value of comparative performance assessment and establishing benchmarks for selected performance indicators cannot be overemphasized; it provides real targets against which less well performing systems can be judged.
- Involvement of the water users in the process through discussions and questionnaires is an essential part of the benchmarking process.
- Due to the varying levels of performance across a range of indicators it is not always possible to identify one 'best practice' system. In some cases the irrigation water delivery performance is good, but the drainage performance is poor, and vice versa. Nevertheless, individual, achievable targets are obtained to use as benchmarks.
- If benchmarking is to be adopted on a wider scale as a management tool there should be greater involvement with the system managers (the District I&D Engineers) in the process, and the water users. These key stakeholders must be engaged in the process at the outset, and the analysis and findings shared with them at all stages.
- In future developments a Geographic Information System would be a useful tool to process, analyse and present the data.

Table 9.25. Data requirements for benchmarking.

Map location	Location	Data collected	Units	By whom collected	How collected	Frequency of collection	Period collected	Remarks
1	Secondary canal intake	Discharge entering secondary canal:		I&D agency field staff	Measurement	Daily	Season	Level data recorded daily by I&D staff
		• Flow depth	m					
		• Gate opening	m					
		• Discharge	m^3/s					
		• Duration of flow	h, min					
1	Secondary canal intake	• Water quality	mmho/cm	I&D agency field staff	Measurement	Once per month	Season	Samples collected by I&D staff and sent to laboratory
1a	Secondary canal tail escape	Discharge leaving secondary canal:		I&D agency field staff	Measurement	Daily	Season	Level data recorded daily by I&D staff
		• Flow depth	m					
		• Discharge	m^3/s					
		• Duration of flow	h, min					
2	Tertiary canal intake	Discharge delivered to tertiary canal:		WUA pump operator	Measurement	Hourly	Season	Data collected by WUA for all tertiary canals for charging and cost calculation purposes
		• Pumping hours	h					
		• Pumping head (intake, delivery)	m					
		• Fuel consumed	l					
3	Selected tertiary canals (two head, two middle, two tail)	Groundwater and soil data:		I&D agency field staff			Season	Twelve piezometers installed in each secondary canal command
		• Depth to groundwater	m			Ten to 12 times per season		
		• Salinity of groundwater (EC)	mmho/cm					
		• Soil salinity at 40 cm depth	mmho/cm			Once per season		

4	Selected tertiary canals (outfalls to selected tertiary canals two head, two middle, two tail)	Drainage water levels: • Number of days collector outlet submerged during season	 days	I&D agency field staff	Measurement	Periodically	Season	I&D agency field staff to monitor selected collector drain outfalls during the season and record the number of days they are submerged
5	Secondary drain outfall	Drainage water level and flow: • Drainage water level • Discharge • Water quality (EC)	 m m^3/s mmho/cm	I&D agency field staff	Measurement	 Daily Monthly	Season	Samples collected by I&D staff and sent to laboratory
6	Selected tertiary canals along secondary canal (ten head, ten middle, ten tail)	Command and crop areas: • Command area For a typical 10 ha sample area: • Crop type • Crop area • Crop duration • Crop production (bags) • Weight of bags (by crop type) • Crop market price • Cost of production	 ha – ha days no. kg MU MU	WUA staff I&D agency field staff	Interviews with farmers From agricultural cooperatives and Ministry of Agriculture	Once per season	Season	Simple crop data collection procedures to be tested with WUAs to ascertain if reliable crop data can be obtained for comparison between WUAs. These can be cross-checked with data collected from other sources (crop cuttings by Ministry of Agriculture, data collected by agricultural cooperatives, etc.)
6	Selected tertiary canals along secondary canal (ten head, ten middle, ten tail)	Water user satisfaction survey: • Satisfaction with water delivery • Satisfaction with drainage removal • Problems/constraints	 – – –	I&D agency field staff	Survey	Twice per season (mid-season and just after harvest)	Season	

Continued

Table 9.25. Continued.

Map location	Location	Data collected	Units	By whom collected	How collected	Frequency of collection	Period collected	Remarks
7	District I&D system	District MOM expenditure and staffing: • Total command area • Total annual MOM expenditure (salaries, office costs, operation, maintenance, etc.) • Total annual *planned* maintenance expenditure on canals and drains • Total annual *actual* maintenance expenditure on canals and drains • Total number of staff • Total cost of staff	 ha MU MU MU no. MU	District I&D engineer	Office records	Seasonally	Season	These data are available at the District Office. Historic data can also be obtained and analysed for comparative purposes and trend analysis
9	Secondary canal and tertiary canals	Complaints: • Number of complaints • Nature of complaint • Action taken	 no. no. –	District I&D engineer	Office records	Each season	Season	
9	Secondary canal collector drain and secondary drains	Complaints: • Number of complaints • Nature of complaint • Action taken	 no. no. –	District I&D engineer	Office records	Each season	Season	

I&D, irrigation and drainage; WUA, water users association; MU, monetary unit.

Description	Units	Secondary canal					
		P1/S1	P1/S2	P1/S3	P2/S1	P2/S2	P2/S3
Irrigation							
Total seasonal value of crop production per unit command area	MU/ha	2676	2684	**2935**	2419	2730	2886
Total seasonal value of crop production per unit water supply	MU/m^3	0.88	0.82	1.28	0.68	**3.23**	2.62
Total seasonal volume of crop water demand (at field)	m^3/ha	2236	2352	2226	1828	2155	2326
Total seasonal irrigation water supply per unit command area	m^3/ha	3024	3289	2286	3577	846	1110
Total seasonal irrigation water delivery per unit command area	m^3/ha	1340	2037	1339	1587	574	N/a
Main system water delivery efficiency	%	44%	**62%**	59%	44%	68%	N/a
Seasonal relative irrigation water supply	–	0.60	**0.87**	0.60	**0.87**	0.41	N/a
Total seasonal MOM costs for irrigation water delivery per unit command area	MU/ha	18.43	18.43	18.43	20.00	20.00	20.00
Total seasonal MOM costs for irrigation water delivery per unit irrigation water supply	MU/m^3	0.012	0.011	0.016	0.011	0.031	0.036
Total seasonal maintenance expenditure for irrigation water delivery per unit command area	MU/ha	13.47	13.47	13.47	10.00	10.00	10.00
Total annual maintenance expenditure fraction for irrigation water delivery	–	0.73	0.73	0.73	0.50	0.50	0.50
Total cost per person employed on water delivery	MU/person	3902	3902	3902	3750	3750	3750
Irrigation command area per unit staff	ha/person	393	393	393	375	375	375
Head:Tail tertiary canal pumping hours ratio	–	1.06	1.03	N/a	0.66	0.88	N/a
Drainage							
Groundwater level (depth to)	m	0.80	0.58	0.82	0.75	0.58	0.95
Groundwater salinity	mmhos/cm	2.2	3.0	2.1	2.0	3.2	6.1
Soil salinity	mmhos/cm	0.8	0.7	1.1	2.8	3.9	3.5
Farmer questionnaire							
Irrigation problems:							
• Very severe	counts	–	–	–	–	–	–
• Severe	counts	–	–	–	–	1	3
• Mild	counts	2	7	0	0	18	2
Drainage problems:							
• Severe	counts	–	–	–	–	–	1
• Mild	counts	–	–	–	5	–	3
• Little	counts	–	1	–	14	1	2

Legend	Gold	Best value	Red	Critical value	Yellow	Area of concern	

Fig. 9.13. Summary performance table for the irrigation and drainage system (MOM, management, operation and maintenance; MU, monetary units, US$1=5.80 MU; N/a, not available).

Endnotes

[1] Project management is time-bounded and requires that specified activities are carried out within a given timeframe to deliver specific outputs.
[2] Adapted from Chapter 2 written by the author in Bos, M.G., Burton, M.A. and Molden, D.J. (eds) *Irrigation Performance Assessment: Practical Guidelines* (Bos *et al.*, 2005).
[3] If required, the on-farm level could be further subdivided into on-farm and in-field.
[4] Note that some of these target values are specific to this example; they should be reviewed and adapted for other situations. Where monetary units are used it is important that the target figures are updated annually to allow for inflation.
[5] This section has been adapted from *A Toolkit for Monitoring and Evaluation of Agricultural Water Management Projects*, written by the author in association with Laurence Smith and Julienne Roux for the Agriculture and Rural Development Division of the World Bank (Burton *et al.*, 2008).
[6] Also referred to as the Purpose in some organizations.

References

Adams, J., Hayes, J. and Hopson, B. (1976) *Transitions – Understanding and Managing Personal Change*. Martin Robertson, Oxford, UK.

Alexander, P.J. and Potter, M.O. (2004) Benchmarking of Australian irrigation water provider businesses. In: Special Issue – Benchmarking in the Irrigation and Drainage Sector. *Irrigation and Drainage* 53(2), 165–174.

ANCID (2000) *1998/99 Australian Irrigation Water Provider – Benchmarking Report*. Australian National Committee on Irrigation and Drainage, Victoria, Australia.

Beadle, A.D., Burton, M.A., Smout, I.K. and Snell, M.J. (1988) Integration of engineering, institutional and social requirements into rehabilitation design – a case study from Nepal. *Irrigation and Drainage Systems* 2(1), 79–92.

Bloom, B.S. (1956) *Taxonomy of Educational Objectives, Handbook I: The Cognitive Domain*. David McKay Co. Inc., New York.

Bos, M.G. (1989) *Discharge Measurement Structures. ILRI Publication No. 20*. International Institute for Land Reclamation and Improvement, Wageningen, The Netherlands.

Bos, M.G. and Nugteren, J. (1974) *On Irrigation Efficiencies. ILRI Publication No. 19*. International Institute for Land Reclamation and Improvement, Wageningen, The Netherlands.

Bos, M.G., Burton, M.A. and Molden, D. (2005) *Irrigation and Drainage Performance Assessment: Practical Guidelines*. CAB International, Wallingford, UK.

Bos, M.G., Keslik, R.A.L., Allen, R.G. and Molden, D. (2009) *Water Requirements for Irrigation and the Environment*. Springer, Dordrecht, The Netherlands.

Burton, M.A. (1986) *Training Programmes for Irrigation Staff. ODI/IIMI Irrigation Management Network Paper No. 86/1d*. Overseas Development Institute, London.

Burton, M.A. (1988) *Improving Water Management in Developing Countries: A Question of Training. ODI/IIMI Irrigation Management Network Paper No. 88/1b*. Overseas Development Institute, London.

Burton, M.A. (1989a) Experiences with the Irrigation Management Game. *Irrigation and Drainage Systems* 3, 217–228.

Burton, M.A. (1989b) Putting theory into practice: simplified scheduling procedures for smallholder irrigation schemes. In: Rydzewski, J.R. and Ward, C.F. (eds) *Irrigation: Theory and Practice*. Pentech Press, London, pp. 514–526.

Burton, M.A. (1993) A simulation approach to irrigation water management. PhD thesis, University of Southampton, Southampton, UK.

Burton, M.A. (1994) The Irrigation Management Game: a role playing exercise for training in irrigation management. *Irrigation and Drainage Systems* 7, 305–318.

Burton, M.A. (2003) Irrigation management transfer: a study of change management. MBA dissertation, Henley Management College, Henley, UK.

Burton, M.A. and Molden, D. (2005) Making sound decisions: information needs for basin water management. In: Svendsen, M. (ed.) *Irrigation and River Basin Management: Options for Governance and Institutions.* CAB International, Wallingford, UK, pp. 125–144.
Burton, M.A., Kingdom, W.D. and Welch, J.W. (1996) Strategic investment planning for irrigation – the 'Asset Management' approach. *Irrigation and Drainage Systems* 10, 207–226.
Burton, M.A., Wester, P. and Scott, C. (2002) Safeguarding the needs of locally managed irrigation in the water scarce Lerma-Chapala river basin, Mexico. In: *Proceedings of the 18th International Congress on Irrigation and Drainage*, Montreal, Canada, 21–28 July (CD-ROM). International Commission on Irrigation and Drainage, New Delhi, R5.03.
Burton, M.A., Malano, H. and Makin, I. (2005) Benchmarking for improved performance in irrigation and drainage. In: *Shaping the Future of Water for Agriculture: A Sourcebook for Investment in Agricultural Water Management.* World Bank, Washington, DC, pp. 299–305.
Burton, M.A., Smith, L. and Roux, J. (2008) *Toolkit for Monitoring and Evaluation of Agricultural Water Management Projects.* Agricultural and Rural Development Division, World Bank, Washington, DC.
Carnall, C.A. (1999) *Managing Change in Organisations*, 3rd edn. FT/Prentice Hall, Harlow, UK.
Carney, D. (1998) *Changing Public and Private Roles in Agricultural Service Provision.* Overseas Development Institute, London.
Casley, D. and Kumar, K. (1987) *Project Monitoring and Evaluation in Agriculture, A Joint Study.* World Bank, International Fund for Agricultural Development and Food and Agriculture Organization of the United Nations. Johns Hopkins University Press, Baltimore, Maryland.
Chambers, R. (1988) *Managing Canal Irrigation: Practical Analysis from South Asia.* Cambridge University Press, Cambridge, UK.
Davies, A. (1993) An asset management programme for irrigation agencies in Indonesia. MSc dissertation, Institute of Irrigation Studies, University of Southampton, Southampton, UK.
de Vries, K. and Miller, D. (1984) *The Neurotic Organisation.* Jossey-Bass, New York.
Earthscan/IWMI (2007) *Water for Food, Water for Life: A Comprehensive Assessment of Water Management in Agriculture.* Earthscan, London and International Water Management Institute, Colombo.
EDI (1994) *Irrigation Training in the Public Sector; Guidelines for Preparing Strategies and Programs*, 2nd edn. Economic Development Institute, World Bank, Washington, DC.
EPI (2009) Data Center. Food and Agriculture. World Irrigated Area and Irrigated Area per Thousand People, 1950–2007. Earth Policy Institute, Washington, DC (available at http://www.earth-policy.org/index.php?/datacenter/xls/book_pb4_ch2_8.xls).
European Commission (2004) *Aid Delivery Methods – Volume 1: Project Cycle Management Guidelines.* EuropeAid Cooperation Office, European Commission, Brussels.
FAO (1977) *Crop Water Requirements. FAO Irrigation and Drainage Paper No. 24* (Doorenbos, J. and Pruitt, W.O.). Food and Agriculture Organization of the United Nations, Rome.
FAO (1978) *Effective Rainfall. FAO Irrigation and Drainage Paper No. 25* (Dastane, N.G.). Food and Agriculture Organization of the United Nations, Rome.
FAO (1979) *Yield Response to Water. FAO Irrigation and Drainage Paper No. 33* (Doorenbos, J. and Kassam, A.H.). Food and Agriculture Organization of the United Nations, Rome.
FAO (1982) *Organization, Operation and Maintenance of Irrigation Schemes. FAO Irrigation and Drainage Paper No. 40* (Sagardoy, J.A., Bottrall, A. and Uittenbogaard, G.O.). Food and Agricultural Organization of the United Nations, Rome.
FAO (1984) *Irrigation Practice and Water Management. FAO Irrigation and Drainage Paper No. 1, Revision 1* (Doneen, L.D. and Westcot, D.W.). Food and Agriculture Organization of the United Nations, Rome.
FAO (1989) *Guidelines for Designing and Evaluating Surface Irrigation Systems. FAO Irrigation and Drainage Paper No. 45* (Walker, W.). Food and Agriculture Organization of the United Nations, Rome.
FAO (1998) *Crop Evapotranspiration: Guidelines for Computing Crop Water Requirements. FAO Irrigation and Drainage Paper No. 56* (Allen, R., Pereira, L., Raes, D. and Smith, M.). Food and Agriculture Organization of the United Nations, Rome.
FAO (1999) *Transfer of Irrigation Management Services. FAO Irrigation and Drainage Paper No. 58* (Vermillion, D. and Sagardoy, J.A.). Food and Agriculture Organization of the United Nations, Rome.
FAO (2003) *Legislation on Water Users' Organizations: A Comparative Analysis. FAO Legislative Study No. 79* (Hodgson, S.). Food and Agriculture Organization of the United Nations, Rome.
FAO (2009) *Creating Legal Space for Water User Organizations: Transparency, Governance and the Law. FAO Legislative Study No. 100* (Hodgson, S.). Food and Agriculture Organization of the United Nations, Rome.

Frederiksen, H.D. and Vissia, R.I. (1998) *Considerations in Formulating the Transfer of Services in the Water Sector*. International Water Management Institute, Colombo.

Goleman, D. (1998) What makes a leader? *Harvard Business Review* 76, November/December, 93.

Horst, L. (1990) *Interactions Between Technical Infrastructure and Management. ODI/IIMI Irrigation Management Network Paper No. 90/3b*. Overseas Development Institute, London.

Huppert, W. and Urban, K. (1998) *Analysing Service Provision: Instruments for Development Cooperation Illustrated by Examples from Irrigation. GTZ Publication No. 263*. Deutsche Gesellschaft fur Technische Zusammenarbeit GmbH, Eschborn, Germany.

Hurst, D.K. (1995) *Crisis and Renewal*. Harvard Business School Press, Cambridge, Massachusetts.

ICID (1993) *ICID Environmental Checklist to Identify Environmental Effects of Irrigation, Drainage and Flood Control Projects*. International Commission on Irrigation and Drainage, New Delhi.

IIS (1995a) *Asset Management Procedures for Irrigation Schemes – Final Report*. Institute of Irrigation Studies, University of Southampton, Southampton, UK.

IIS (1995b) *Preliminary Guidelines for the Preparation of an Asset Management Plan for Irrigation Infrastructure*. Institute of Irrigation Studies, University of Southampton, Southampton, UK.

Israelson, O.W. and Hansen, V.E. (1962) *Irrigation Principles and Practices*, 3rd edn. John Wiley and Sons, New York.

IWMI (2006) Water for Food, Water for Life: Insights from a Comprehensive Assessment of Water Management in Agriculture. Presented at Stockholm World Water Week, 20–26 August 2006. International Water Management Institute, Colombo.

Jurriens, M. (1991) Rehabilitation and Management of Irrigation Projects – Short Course Lecture Notes. Institute of Irrigation Studies, University of Southampton, United Kingdom.

Kay, M. (1983) *Sprinkler Irrigation: Equipment and Practice*. Batsford Academic and Educational Ltd, London.

Kay, M. (1986) *Surface Irrigation: Systems and Practices*. Cranfield Press, Bedford, UK.

Keller, J., Keller, A. and Davids, G. (1998) River basin development phases and implications of closure. *Journal of Applied Irrigation Science* 33(2), 145–163.

Khan, M.H. (1978) Improving irrigation operation in East Java. In: *Transactions of the 10th ICID Congress, Athens*. International Commission on Irrigation and Drainage, New Delhi.

Kloezen, W. and Samad, M. (1995) *Synthesis of Issues Discussed at the International Conference on Irrigation Management Transfer. Short Report Series on Locally Managed Irrigation, Report No. 12*. International Irrigation Management Institute, Colombo.

Kotter, J.P. (1990) What leaders really do. *Harvard Business Review*; reprinted in 2001 in *Best of HBR*, December, 85–96.

Kotter, J.P. (1995) Leading change: why transformation efforts fail. *Harvard Business Review*, March/April, 59–67.

Kotter, J.P. and Schelsinger, L.A. (1979) Choosing strategies for change. *Harvard Business Review*, March/April, 106–113.

Lorange, P. and Nelson, G. (1987) *Strategic Control*. West Publications, San Francisco, California.

MacDonald, I. and Hearle, D. (1984) *Communication Skills for Rural Development*. Evans Brothers Ltd, London.

Malano, H. and Burton, M. (2001) *Guidelines for Benchmarking Performance in the Irrigation and Drainage Sector*. Food and Agriculture Organization of the United Nations, Rome and the International Programme for Technology and Research in Irrigation and Drainage, Rome.

Malano, H., Burton, M. and Makin, I. (2004) Benchmarking performance in the irrigation and drainage sector: a tool for change. In: Special Issue – Benchmarking in the Irrigation and Drainage Sector. *Irrigation and Drainage* 53(2), 119–134.

Malhotra, S.P. (1982) *The Warabandi System and Its Infrastructure. Central Board of Irrigation and Power Publication No. 157*. Central Board of Irrigation and Power, New Delhi.

Merriam, J.L. and Keller, J. (1978) *Farm Irrigation System Evaluation: A Guide to Management*. Utah State University, Logan, Utah.

Molden, D., Sakthivadivel, R. and Samad, M. (2001) Accounting for changes in water use and the need for institutional adaptation. In: Abernathy, C.L. (ed.) *Intersectoral Management of River Basins*. International Water Management Institute, Colombo, pp. 73–88.

Murray-Rust, D.H. and Snellen, W.B. (1993) *Irrigation System Performance Assessment and Diagnosis*. Joint IIMI/ILRI/IHEE Publication. International Irrigation Management Institute, Colombo.

Neville, D.F.J. (1996) Environmental Impact Assessment course notes. Institute of Irrigation and Drainage Studies, University of Southampton, Southampton, UK.

OECD (2002) *Glossary of Key Terms in Evaluation and Results-based Management*. Organisation for Economic Co-operation and Development, Paris.
OFWAT (1992) *Strategic Business Plan (AMP2) Manual*. Office of Water Services, Birmingham, UK.
OIP (2008) Project Status Report, Quarter 1, Project Implementation Unit, On-Farm Irrigation Project, Kyrgyz Republic. Prepared by Project Implementation Unit for World Bank Missions, Bishkek.
Ostrom, E. (1992) *Crafting Institutions for Self-governing Irrigation Systems*. Institute for Contemporary Studies Press, San Francisco, California.
Pasandaran, E. (1976) Water management decision making in the Pekalen Sampean Irrigation Project, East Java, Indonesia. In: Taylor, D.C. and Wickham, T.H. (eds) *Irrigation Policy and the Management of Irrigation Systems in South East Asia*. Agricultural Development Council, Bangkok, pp. 47–60.
Plant, R. (1987) *Managing Change and Making It Stick*. Fontana/Collins, London.
Replogle, J.A. and Merriam, J.L. (1980) Scheduling and management of irrigation water delivery systems. In: *Irrigation – Challenges of the 80s, Proceedings of the ASAE Second National Irrigation Symposium*, Lincoln, Nebraska, 20–23 October. American Society of Agricultural Engineers, St Joseph, Michigan, pp. 112–126.
Sadler, P. (1995) *Managing Change*. Kogan Page, London.
Shiklomanov, I. (2000) Appraisal and assessment of world water resources. *Water International* 25(1), 11–32.
Skogerboe, G.V. and Merkley, G.P. (1996) *Irrigation Maintenance and Operations Learning Process*. Water Resources Publications, Highland Ranch, Colorado.
Skogerboe, G.V., Bennet, R.S. and Walker, W.R. (1972) *Selection and Installation of Cut-throat Flumes for Measuring Irrigation and Drainage Water. Technical Bulletin No. 120*. Experiment Station, Colorado State University, Fort Collins, Colorado.
Small, L.E. and Svendsen, M. (1992) *A Framework for Assessing Irrigation Performance. IFPRI Working Papers on Irrigation Performance No. 1*. International Food Policy Research Institute, Washington, DC.
Snell, M.J. (1997) *Cost Benefit Analysis for Engineers and Planners*. Thomas Telford, London.
SRWSC (1980) *Water Bailiff's Manual*. State Rivers and Water Supply Commission, Victoria, Australia.
Svendsen, M. (ed.) (2005) *Irrigation and River Basin Management: Options for Governance and Institutions*. CAB International, Wallingford, UK.
Uphoff, N. (1990) Farmer participation in improving irrigation system management for sustainable agriculture in Asia. Paper presented at FAO Regional Workshop on Improved Irrigation System Performance for Sustainable Agriculture, Bangkok, 22–26 October. Food and Agriculture Organization of the United Nations, Rome, 38 pp.
Walker, W.R. (1993) *SIRMOD: Surface Irrigation Simulation Software, Version 2.1*. Utah State University, Logan, Utah.
Wester, P., Scott, C.A. and Burton, M.A. (2005) River basin closure and institutional change in Mexico's Lerma-Chapala basin. In: Svendsen, M. (ed.) *Irrigation and River Basin Management: Options for Governance and Institutions*. CAB International, Wallingford, UK, pp. 125–144.
Withers, B. and Vipond, S. (1974) *Irrigation: Design and Practice*. B.T. Batsford Ltd, London.
World Bank (2005) *Shaping the Future of Water for Agriculture: A Sourcebook for Investment in Agricultural Water Management*. Agricultural and Rural Development Division, World Bank, Washington, DC.

Appendix 1

Scheduling Irrigation Water Exercise

Overview

This exercise builds on the theory outlined in Chapter 5 to outline the procedures for scheduling irrigation water within the tertiary unit (on-farm) command area. The exercise uses an example of scheduling for five fields within the command area over a 16-day period and shows how to:

- determine the irrigation demands for each field;
- determine when each field needs to be irrigated;
- organize an irrigation schedule to irrigate the fields in turn;
- organize an irrigation schedule to match the inflow available from the secondary canal.

Background Information

The irrigation command area is shown in Fig. A1.1. The command area comprises five blocks of land, each fed by a quaternary channel. Each block is divided into irrigated fields belonging to individual farmers, with an average field size of 1.3 ha. There are three soil textures in the command area, sandy loam, clay loam and clay, each with different moisture-holding characteristics.

The key data for each block of land are summarized in Table A1.1, while Table A1.2 provides the key data for Block F2. The calculations below show the procedures for scheduling the irrigation supplies to the five fields in this block for a 15-day period in July. Table A1.3 provides data on the effective rainfall, the reference crop potential evapotranspiration (ET_0) and the crop evapotranspiration (ET_c) for the two crop coefficient (K_c) values at this growth stage of 0.95 and 1.05. Table A1.4 gives the soil moisture deficit in the Block F2 fields at the start of the scheduling time period.

Procedure for Determining the Irrigation Schedule for Block F-2

The procedure for determining the irrigation schedule for Block F-2 is set out below.

Step 1: Calculate crop evapotranspiration

- Calculate ET_c by multiplying ET_0 by the K_c for this growth stage given in Table A1.2. In this example there are only two sets of ET_c figures as the mid-season K_c figures are similar for cotton and maize (1.05) and for beans and cabbage (0.95).
  - Multiply Table A1.2, col. (7) by Table A1.3, col. (2) to produce Table A1.3, cols (3) and (4).

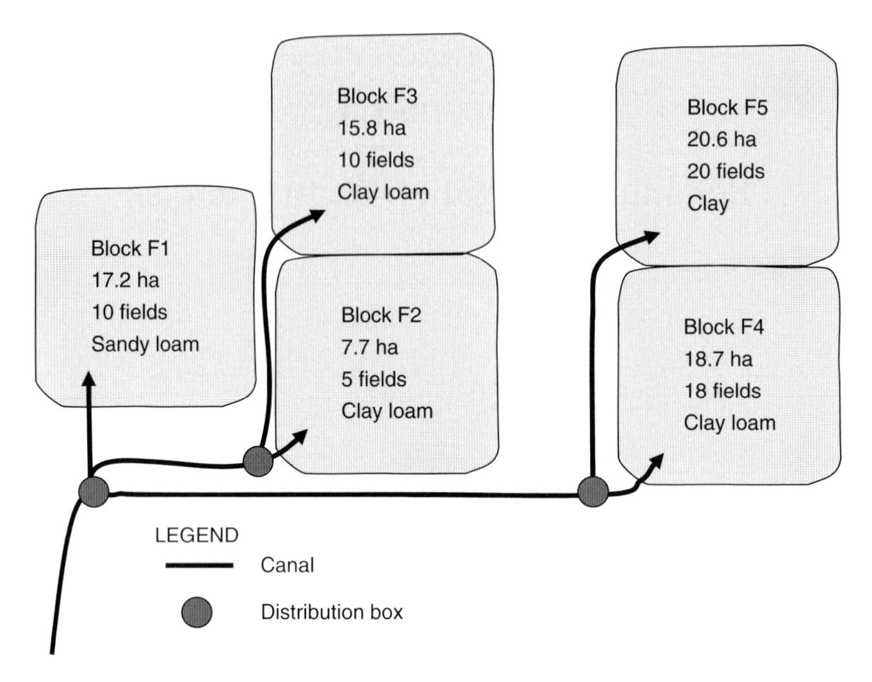

Fig. A1.1. Layout of the irrigation command area.

Table A1.1. Key data for the tertiary unit command area.

Block name	Block area (ha)	Soil texture	Total available soil water (mm/m)	Terminal infiltration rate (mm/h)	Number of individual fields in the block	Distribution efficiency
(1)	(2)	(3)	(4)	(5)	(6)	(7)
F1	17.2	Sandy loam	120	40	10	0.7
F2	7.7	Clay loam	190	20	5	0.7
F3	15.8	Clay loam	190	20	10	0.7
F4	18.7	Clay loam	190	20	18	0.7
F5	20.6	Clay	230	6	20	0.7
Total	**80**				63	

Step 2: Calculate total water available

- Calculate the total available water in the root zone for each crop (Table A1.2) by multiplying the total available soil water (TAW) in mm/m by the crop root depth.
    - Table A1.2, multiply col. (5) by col. (8) to produce col. (9).

Step 3: Determine the easily available water fraction

- Determine the '*p*' fraction for the crop from table in Chapter 5 using ET_c and the Crop Group number.
    - Use Table 5.20 and Table A1.3 cols (3) and (4) to obtain Table A1.2, col. (10).

Step 4: Calculate the easily available water limit (EAWL)

- Calculate the EAWL by multiplying the TAW by the '*p*' fraction.
    - Table A1.2, multiply col. (10) by col. (9) to obtain col. (11).

Step 5: Prepare water balance sheet for each field/crop

- Draw up a water balance sheet for each field as shown in Fig. A1.2, adding the effective rainfall and ET_c (the crop demand) for each crop. Enter the initial soil moisture deficit on 16 July from the data given in Table A1.4. For each field in turn accumulate the soil moisture deficit

Table A1.2. Data for fields in Block F2.

Field name	Area (ha)	Crop	Crop group	Maximum rooting depth (m)	Growth stage	K_c value	Total available soil water (mm/m)	Total available soil water (mm)	'p' fraction	Easily available water limit (mm)	Application efficiency
(1)	(2)	(3)	(4)	(5)	(6)	(7)	(8)	(9)	(10)	(11)	(12)
F2-1	1.6	Cotton	4	1.2	Mid-season	1.05	190	228	0.52	119	0.6
F2-2	1.4	Cabbage	2	0.4	Mid-season	0.95	190	76	0.33	25	0.6
F2-3	1.7	Beans	3	0.6	Mid-season	0.95	190	114	0.45	51	0.6
F2-4	1.4	Maize	4	1.4	Mid-season	1.05	190	266	0.52	138	0.6
F2-5	1.6	Maize	4	1.4	Mid-season	1.05	190	266	0.52	138	0.6

Table A1.3. Climatic data and potential evapotranspiration values.

Date	Reference crop potential evapotranspiration, ET_0 (mm)	Crop potential evapotranspiration, ET_c (mm) at K_c=0.95	Crop potential evapotranspiration, ET_c (mm) at K_c=1.05	Effective rainfall (mm)
(1)	(2)	(3)	(4)	(5)
16 Jul	6.0	5.70	6.30	–
17 Jul	6.0	5.70	6.30	–
18 Jul	6.0	5.70	6.30	–
19 Jul	6.0	5.70	6.30	–
20 Jul	6.0	5.70	6.30	–
21 Jul	6.5	6.18	6.83	3.0
22 Jul	6.5	6.18	6.83	–
23 Jul	6.5	6.18	6.83	–
24 Jul	6.0	5.70	6.30	–
25 Jul	6.0	5.70	6.30	–
26 Jul	6.0	5.70	6.30	–
27 Jul	6.0	5.70	6.30	–
28 Jul	6.0	5.70	6.30	5.0
29 Jul	6.0	5.70	6.30	–
30 Jul	6.0	5.70	6.30	–
31 Jul	6.0	5.70	6.30	–

Table A1.4. Soil moisture deficit in each field at start of scheduling period (16 July).

Field name	Area (ha)	Crop	Soil moisture deficit on 16 July (mm)
(1)	(2)	(3)	(4)
F2-1	1.6	Cotton	70
F2-2	1.4	Cabbage	10
F2-3	1.7	Beans	30
F2-4	1.4	Maize	100
F2-5	1.6	Maize	50

each day by adding the crop demand to the soil moisture deficit from the previous day, and subtracting any rainfall. When the EAWL is reached, apply sufficient water to return the soil close to zero water deficit. In this example, standard application volumes of 25, 50 and 125 mm depth have been used. Following the irrigation continue with accumulating the soil moisture deficit until the end of the scheduling period.

- ◦ Enter effective rainfall from Table A1.3, col. (5) into col. (2), Fig. A1.2.
- ◦ Enter crop ET_c from Table A1.3, cols (3) and (4) into crop demand columns in Fig. A1.2.
- ◦ Enter initial soil moisture deficit from Table A1.4 into soil moisture deficit columns in Fig. A1.2.

Step 6: Calculate the discharge required in litres per second per hectare

- Calculate the irrigation demand for each field in l/s/ha by multiplying the irrigation depth (mm) taken from the water balance sheet by the field area and the relevant conversion factor in Table A1.5. In this

Date	Effective rainfall (mm)	F2-1			F2-2			F2-3			F2-4			F2-5		
		Cotton			Cabbage			Beans			Maize			Maize		
		EAWL = 119			EAWL = 25			EAWL = 51			EAWL = 138			EAWL = 138		
		Crop demand, ET_c (mm)	Irrigation (mm)	Soil moisture deficit (mm)	Crop demand, ET_c (mm)	Irrigation (mm)	Soil moisture deficit (mm)	Crop demand, ET_c (mm)	Irrigation (mm)	Soil moisture deficit (mm)	Crop demand, ET_c (mm)	Irrigation (mm)	Soil moisture deficit (mm)	Crop demand, ET_c (mm)	Irrigation (mm)	Soil moisture deficit (mm)
(1)	(2)	(3)	(4)	(5)	(6)	(7)	(8)	(9)	(10)	(11)	(12)	(13)	(14)	(15)	(16)	(17)
16 Jul	0.0	6.30		70.00	5.70		10.00	5.70		30.00	6.30		100.00	6.30		50.00
17 Jul	0.0	6.30		76.30	5.70		15.70	5.70		35.70	6.30		106.30	6.30		56.30
18 Jul	0.0	6.30		82.60	5.70		21.40	5.70		41.40	6.30		112.60	6.30		62.60
19 Jul	0.0	6.30		88.90	5.70	25	2.10	5.70		47.10	6.30		118.90	6.30		68.90
20 Jul	0.0	6.30		95.20	5.70		7.80	5.70	50	2.80	6.30	125	0.20	6.30		75.20
21 Jul	3.0	6.83		99.03	6.18		10.98	6.18		5.98	6.83		4.02	6.83		79.03
22 Jul	0.0	6.83		105.85	6.18		17.15	6.18		12.15	6.83		10.85	6.83		85.85
23 Jul	0.0	6.83		112.68	6.18		23.33	6.18		18.33	6.83		17.68	6.83		92.68
24 Jul	0.0	6.30		118.98	5.70	25	4.03	5.70		24.03	6.30		23.98	6.30		98.98
25 Jul	0.0	6.30	125	0.27	5.70		9.73	5.70		29.73	6.30		30.28	6.30		105.28
26 Jul	0.0	6.30		6.57	5.70		15.43	5.70		35.43	6.30		36.58	6.30		111.58
27 Jul	0.0	6.30		12.88	5.70		21.13	5.70		41.13	6.30		42.88	6.30		117.88
28 Jul	5.0	6.30		14.18	5.70		21.83	5.70		41.83	6.30		44.18	6.30		119.18
29 Jul	0.0	6.30		20.48	5.70	25	2.53	5.70		47.53	6.30		50.48	6.30	125	0.47
30 Jul	0.0	6.30		26.78	5.70		8.22	5.70	50	3.23	6.30		56.78	6.30		6.77
31 Jul	0.0	6.30		33.08	5.70		13.93	5.70		8.93	6.30		63.08	6.30		13.08

Fig. A1.2. Water balance sheets for each plot.

example it is assumed that each field is irrigated in one 16-hour day, thus the conversion factor used is 0.1736. Divide the figure obtained by the application efficiency (given in Table A1.2) and enter the discharge required for each field in Table A1.6. Add the daily figures for each field to get the total daily discharge required for the block, and divide by the distribution efficiency to obtain the discharge required at the intake to the tertiary unit system.

Table A1.5. Table to convert irrigation depth (mm/day) into discharge (l/s/ha) for different daily irrigation durations.

Irrigation depth (mm)	Irrigation volume (m³/ha)	Time per day irrigating		Conversion factor, mm depth to l/s/ha
		Hours	Seconds	
(1)	(2)	(3)	(4)	(5)
1	10	24	86,400	0.1157
1	10	20	72,000	0.1389
1	10	16	57,600	0.1736
1	10	14	50,400	0.1984
1	10	12	43,200	0.2315
1	10	10	36,000	0.2778
1	10	8	28,800	0.3472
1	10	6	21,600	0.4630
1	10	4	14,400	0.6944
1	10	2	72,00	1.3889
1	10	1	3,600	2.7778
10	100	1	3,600	27.78
100	1000	1	3,600	277.8

Table A1.6. Irrigation water demands.

	Irrigation water demand (l/s)						
	Including application efficiency						Incl. distribution efficiency
Date	F2-1	F2-2	F2-3	F2-4	F2-5	Total	
(1)	(2)	(3)	(4)	(5)	(6)	(7)	(8)
16 Jul	–	–	–	–	–	–	–
17 Jul	–	–	–	–	–	–	–
18 Jul	–	–	–	–	–	–	–
19 Jul	–	10.1	–	–	–	10.1	14.5
20 Jul	–	–	24.6	50.6	–	75.2	107.5
21 Jul	–	–	–	–	–	–	–
22 Jul	–	–	–	–	–	–	–
23 Jul	–	–	–	–	–	–	–
24 Jul	–	10.1	–	–	–	10.1	14.5
25 Jul	57.9	–	–	–	–	57.9	82.7
26 Jul	–	–	–	–	–	–	–
27 Jul	–	–	–	–	–	–	–
28 Jul	–	–	–	–	–	–	–
29 Jul	–	10.1	–	–	57.9	68.0	97.1
30 Jul	–	–	24.6	–	–	24.6	35.1
31 Jul	–	–	–	–	–	–	–

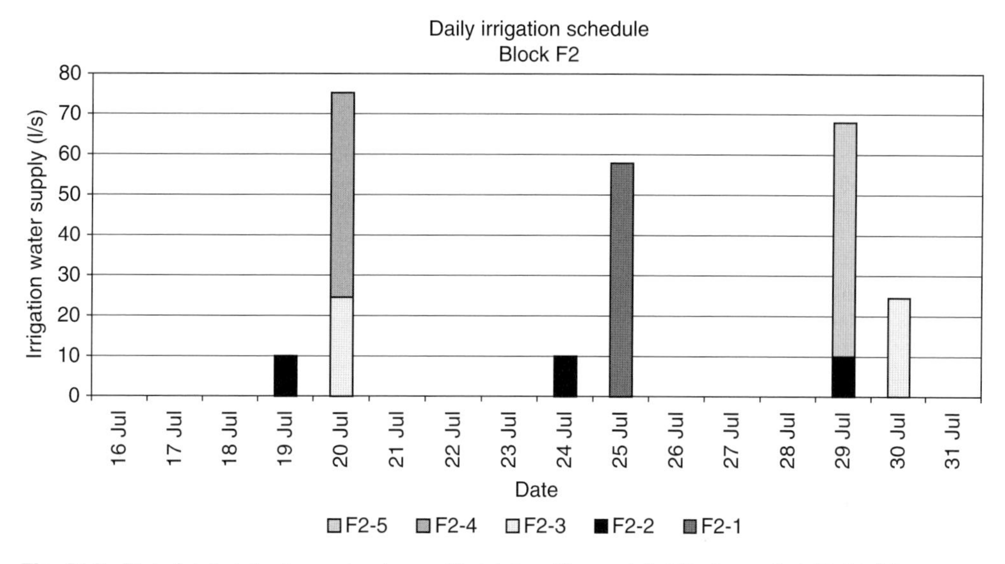

Fig. A1.3. Plot of daily irrigation water demand in total and for each field in the period 16–31 July.

- For F2-1: Table A1.6, col. (2) = Fig. A1.2, col. (4) (mm depth) multiplied by Table A1.2, col. (2) (area) multiplied by Table A1.5, col. (5) (conversion factor) divided by Table A1.2, col. (12) (application efficiency).
- Add Table A1.6, cols (2) to (6) to obtain col. (7). Divide col. (7) by Table A1.1, col. (7) to obtain col. (8).

Step 7: Plot the daily irrigation demand

- Plot the daily irrigation demand as shown in Fig. A1.3.

Summary

1. In the above example irrigation is only required for the fields in Block F2 for six out of the 16 days. On the other days the flow to the quaternary channel is closed off and the water used in other blocks.

2. Care has to be taken when working out the schedule that the required discharge does not exceed the capacity of the quaternary canal. In this example on the 20th and 29th two fields are irrigated on the same day, pushing up the discharge. The discharge can be reduced, and the daily discharge evened out, by irrigating one of the fields a day earlier or a day later.

3. The flow rates required in the quaternary canal range from 14.5 to 107.5 l/s, or 14.5 to 57.9 l/s if no two fields are irrigated on the same day. This is quite a range.

4. The above calculations are time-consuming and complicated for systems with a large number of smallholdings. Some suggestions for simplifying these procedures include the following.

- Calculations such as those above can be calculated using climatic data from previous years and rules developed for irrigation schedules within each block based on the block's soil type and the crop grown. Using the example above the rule would be to irrigate cabbage every 5 days with 25 mm depth (250 m^3/ha) during mid-season. With a flow rate of 25 l/s (90 m^3/h) it would take 4 h to irrigate a field of 1.4 ha (250×1.4/90). The difficulty will be in not applying too much water; this can be overcome by cultivating the cabbages with furrows in small basins.
- Having done these calculations there are some further options.
  - Develop irrigation schedules which irrigate with a fixed volume of water within the EAWL. Thus for the cotton and maize crop

a rule could be to irrigate each 30 days with 180 mm (1800 m^3/ha). This would replenish the water lost by evapotranspiration (at about 6 mm/day) and would ensure that the crop does not reach the EAWL. With a flow rate of 50 l/s (180 m^3/h) it would take 10 h to irrigate a 1 ha field (or 16 hours for fields F2-1 and F2-5, and 14 hours for F2-4, in the example).

◦ Devise a rotation schedule which uses a constant inflow to the block. With a total evapotranspiration demand of about 6.2 mm/day and a total area of 7.7 ha for Block F-2 the daily water demand is 477.4 m^3/day (0.0062×10,000×7.7), or 796 m^3/day if the application efficiency (0.6) is taken into account. Irrigating for 14 hours (50,400 s) per day would thus require a flow rate of 16 l/s (796/50,400), or 23 l/s including the distribution efficiency (0.7).

5. The calculations shown above can thus be used as a guide to developing a workable irrigation schedule that fits with the farming practices and yet which is based on calculated demands based on the climatic conditions, the soil type and the rooting depths of the crops.

6. The exercise serves to demonstrate how complicated it is to schedule irrigation water to match crop needs where there is a large number of landholdings and that simplified procedures need to be developed based on the procedures outlined in this example.

Appendix 2

Checklist for Assessing the Performance of Water Users Associations or Federations of Water Users Associations

Introduction

This checklist summarizes the characteristics of successful water users associations (WUAs) or federations of WUAs. The checklist can be used to make a rapid assessment of the performance of an Association or Federation. It is anticipated that the assessment will be more qualitative than quantitative, with data acquired through short interviews with WUA and Federation staff. The checklist is provided as Table A2.1 and the suggested standards for each measure as Table A2.2, following the discussion below.

Categories

The performance of a WUA or Federation can be assessed in the following areas:

- general management;
- finance;
- water management;
- asset management and maintenance;
- level of service provision.

General management

General management relates to the effectiveness of the WUA or Federation Chairman and staff in managing the WUA or Federation. Performance relates to the personal skills of the WUA/Federation managers, and to the institutional arrangements that have been established within the WUA/Federation. The key indicators of performance are:

- level of membership;
- General Assembly attendance;
- regular weekly meetings;
- management style;
- liaison and communication with water users;
- level of facilities;
- level of democracy.

Level of membership

The number of water users within the command area who are members of the association is an indicator of the viability and success of the WUA/Federation. If the level of service provided and the degree of control exercised by the WUA/Federation are high, then one can expect the membership level to be high.

In some areas the farmers do not want to join the Association but do pay for water. In these locations the fee recovery rate may be a better indicator of performance.

General/Representative Assembly attendance

The number of water users or representatives attending the General or Representative Assembly is a crucial indicator of the viability of the WUA/Federation. The level of discussion and issues raised within the Assembly are also indicators of success/viability.

Regular weekly meetings

The regularity, attendance and content of discussions at WUA/Federation weekly meetings are indicators of success. Well-attended, regular meetings with vibrant discussion at which important management decisions are made are indicators of success/viability.

Management style

The management style of the WUA/Federation Chairman and the committee influences the success/viability of the organization. A strong leader who gains consensus can be good; a dominating leader who refuses to allow discussion and debate can be bad.

Liaison and communication with water users

The WUA/Federation belongs to the water users; it is essential that the WUA/Federation management maintains close links with the water users and keeps them informed. The degree of communication/liaison can be assessed through speaking with the water users to see how informed they are about WUA/Federation activities.

Level of facilities

The level of facilities owned by the WUA/Federation will influence performance. The facilities required include:

- office accommodation;
- tables and chairs;
- motorcycles;
- radios, mobile phones;
- maps;
- records of command areas and assets within the irrigation and drainage system;
- computers and printer, or typewriter;
- filing cabinets;
- records (water indents, fees paid, membership register, asset register, etc.).

Level of democracy/participation

A valuable indicator of the performance of the Association is the level of democracy and participation exhibited within the WUA/Federation. Information needs to be obtained by questioning farmers on their level of knowledge of, involvement and participation in the affairs of the WUA/Federation, and the degree to which the WUA/Federation management liaises with them and allows them to participate.

Finance

The financial situation and financial management of the WUA/Federation are central to the long-term sustainability of the WUA/Federation. The indicators of performance in respect of financial matters include:

- setting of the irrigation service fee (ISF);
- level of fee recovery – numbers contributing;
- level of fee recovery – amount contributed;
- the audit report.

Setting of the irrigation service fee

The ISF should be set by the WUA/Federation in relation to the identified needs of the WUA/Federation. These will include:

- salaries for WUA/Federation staff;
- costs of maintaining facilities (offices, electricity, rental, stationery, etc.);
- costs of maintaining equipment and plant (motorcycles, maintenance plant, etc.);
- operation costs (pumping costs, motorcycle costs, etc.);
- maintenance costs.

Initially the WUA/Federation may set the fee based on recommendations made by the WUA Support Unit, or external body. Over time this should change to a more needs-based assessment based on the above items. At the start of each year the total operational needs of the WUA/Federation should be determined and divided by one of the following:

1. The total area irrigated, to obtain the ISF per hectare;
2. The total number of water users, to obtain the ISF per irrigator;
3. The total number of estimated irrigations (based on total area times the average number of irrigations), to obtain the ISF per irrigation.

Thus if the total estimated cost is $3000 for a 300 ha command area with 200 water users and an average number of 1.5 irrigations per hectare, the cost per water user will be $10/ha, $15/person or $6.67/irrigation.

Level of fee recovery – numbers contributing

The number of members paying their ISF is a good indicator of success. Records should be kept of water users who take water but do not pay.

Level of fee recovery – amount contributed

The total ISF collected in relation to the estimated annual requirement for management, operation and maintenance of the system is a good indicator of long-term viability of the WUA/Federation.

The audit report

The audit report summarizes the financial performance of the Association. The audit covers checks of: the annual budget, the crop book, the membership book, the membership receipt forms, the irrigation book, the irrigation invoices/receipts, the payroll, the maintenance expenditure statements, the cash book and the bank statements.

Water management

The sustainability of the WUA/Federation rests in the long term on the effectiveness of the water management practices – if these are poor then water users will not get a good level of service and will eventually stop paying their ISF. The indicators for water management performance are:

- pre-season irrigation planning;
- pre-season irrigation agreements;
- in-season irrigation planning;
- in-season irrigation implementation;
- in-season irrigation monitoring and evaluation;
- discharge measurement;
- end-of season performance evaluation.

Pre-season irrigation planning

The WUA/Federation should work with the water users to plan the irrigation season. This is especially important where the irrigation area is supplied from a reservoir or river where water is in short supply. The degree of pre-season planning and the seriousness with which it is carried out are important indicators of WUA/Federation viability.

Pre-season irrigation agreements

Arising from the pre-season plans the WUA/Federation should discuss the anticipated water supply situation relative to the planned cropped area and agree policies and strategies for the coming season. These should include:

- what to do in times of water scarcity (rotation plans, etc.);
- what to do when it rains;
- rules on priority allocations;
- agreements to cooperate between WUAs.

In-season irrigation planning

The extent and adequacy of weekly in-season irrigation planning is a strong indicator of the viability of a WUA/Federation. Irrigation indents should be made each week by the farmers to the WUA Water Master and passed up the system to the WUA/Federation where allocations should be made based on the available water supplies.

In-season irrigation implementation

The implementation of the weekly irrigation plan is central to the whole irrigation management process, and is the most important indicator for assessing water management. In far too many cases a plan is made on paper but the actual implementation is never carried out. Thus the management performance ratio (ratio of actual supply divided by planned supply) is the prime water management performance indicator.

In-season irrigation monitoring and evaluation

The level of monitoring and evaluation carried out by the WUA/Federation of the in-season irrigation allocation, scheduling and actual supply is a useful indicator of WUA/Federation performance. The area irrigated in relation to the total command area (the cropping intensity) should also be used.

Discharge measurement

The accuracy and extent of discharge monitoring, data processing and analysis indicate the management performance of the WUA/Federation.

End-of season performance evaluation

At the end of the season the WUA/Federation should carry out an evaluation of the performance of the irrigation system, and its management of that system. The extent to which this is done, and the extent to which the lessons learnt are applied, is an indicator of management competence.

Asset management and maintenance

The productivity of the irrigation system is dependent on the ability of the irrigation infrastructure to deliver the right amount of water at the right place and the right time, in accordance with the irrigation plan. Knowledge of the asset base, its type, extent and condition and maintenance requirement is an essential part of the management process. The following are therefore important:

- asset database;
- pre-season maintenance plan;
- annual maintenance identification, costing, planning, implementation and recording;
- total maintenance expenditure;
- degree of control and measurement;
- overall infrastructure condition.

Asset database

In order to manage the water the WUA/Federation must know what assets it has, and the condition of those assets. Without operational control and measurement structures the WUA/Federation cannot distribute water according to the irrigation plan.

Pre-season maintenance plan

An important management task is to carry out a pre-season maintenance inspection to document the maintenance requirement prior to the start of the irrigation season. Following this inspection a maintenance plan must be drawn up and costed, and the maintenance work implemented before the start of the irrigation season.

Annual maintenance identification, costing, planning, implementation and recording

An annual maintenance plan must be drawn up by the WUA/Federation. This plan will be made from experience during the irrigation season and a full inspection while the irrigation system is operational, plus a full inspection at the end of the season when the canals are dry. The maintenance requirement will be identified, costed, prioritized, selected (in relation to the available budget) and implemented (either through direct labour or through tendered contracts). The extent to which such work is done is an indication of the management capability of the WUA/Federation.

Total maintenance expenditure

The total expenditure on maintenance in relation to (i) the irrigable command area, (ii) the collected revenue and (iii) the total estimated management, operation and maintenance (cost) need, provides an indication of the sustainability of the irrigation system, and the management's ability to organize maintenance work.

Degree of control and measurement

The capability to control and measure the irrigation water supplies is an important factor in the management of the available water supplies. The greater the level of control and measurement, the higher the level of service achievable.

Overall infrastructure condition

Allied to the control and measurement capability is the ability of the physical infrastructure to convey the available water supply and dispose of excess drainage water. Systems in poor physical condition will be less manageable and have higher losses. Heavy vegetation in channels is of particular concern in this respect, and a useful indicator of the WUA/Federation's capabilities.

Level of service provision

As stated previously, the WUA/Federation is an organization that is there to provide water users with service – the delivery of irrigation water and removal of surplus drainage water. The concept of level of service provision is central to the success of the WUA/Federation. Measures of the level of service provided can be gained from assessing:

- knowledge by water users of WUA/Federation activities;
- satisfaction with level of service provided;
- number of registered complaints during season.

Knowledge by water users of Association/Federation activities

An indication of the level of communication and the level of service provided to water users can be gained from questioning water users to ascertain their knowledge of WUA/Federation activities. If the knowledge levels are low then the level of service provision may also be low.

Satisfaction with level of service provided

Questioning water users on the level of service provision that they receive in relation to obtaining irrigation water when they want and in the quantity that they require is an important part of the evaluation of water management. Measurement of the farmers' perception of the value for money of the level of service that they receive is also useful.

Number of registered complaints during season

The WUA/Federation should have a system for registering and dealing with complaints from water users. The number and nature of the complaints, and the action taken to address these complaints, provides an indicator of the level of service being provided.

Table A2.1. Checklist for assessment of Water Users Associations (WUAs) or Federations of WUAs.

Survey for: WUA ❑ or Federation ❑ (Tick)
Name: ....
Location: District: ....
Irrigation system: ....

Performance criterion	Score		
	Good	Medium	Poor
1. General management			
1.1 Level of membership			
1.2 General/Representative Assembly attendance			
1.3 Regular weekly management meetings held			
1.4 Management style			
1.5 Liaison and communication with water users			
1.6 Level of facilities			
1.7 Level of democracy/participation			
2. Finances			
2.1 Setting of irrigation service fee			
2.2 Level of fee recovery – numbers contributing			
2.3 Level of fee recovery – amount contributed			
2.4 Audit report			
3. Water management			
3.1 Pre-season irrigation planning			
3.2 Pre-season irrigation agreements			
3.3 In-season irrigation planning			
3.4 In-season irrigation implementation (scheduling, etc.)			
3.5 In-season irrigation monitoring and discussion			
3.6 Discharge measurement			
3.7 End-of-season performance evaluation			
4. Asset management and maintenance			
4.1 Asset database			
4.2 Pre-season maintenance plan			
4.3 Annual maintenance identification, costing, planning, implementation and recording			
4.4 Total maintenance expenditure			
4.5 Degree of control and measurement			
4.6 Overall infrastructure condition			
5. Level of service provision			
5.1 Knowledge by water users of WUA/Federation activities			
5.2 Satisfaction with level of service provided			
5.3 Number of registered complaints during season			

Table A2.2. Performance standards to be achieved by Water Users Associations (WUAs) or Federations of WUAs.

Performance indicator	Desired standard		
	Good	Medium	Poor
1. General management			
1.1 Level of membership	>80% of all potential members	50–79% of all potential members	<50% of all potential members
1.2 General/ Representative Assembly attendance	>60% of all potential members	40–59% of all potential members	<40% of all potential members
1.3 Regular weekly management meetings held	Average attendance >80%	Average attendance 50–79%	Average attendance <49%
	Meetings held/weeks in season >0.9	Meetings held/weeks in season >0.7	Meetings held/weeks in season <0.69
	Average key decisions made >5	Average key decisions made=2–4	Average key decisions made <2
1.4 Management style	Firm	Moderately firm	Weak
	Clear vision	Some vision	Little or no vision
	Clear understanding of duties and responsibilities	Some understanding of duties and responsibilities	Little or no understanding of duties and responsibilities
	Well organized	Moderately well organized	Badly organized
	Participative	Some participation	Egocentric
	Consensus driven	Some consensus	Authoritarian Non-participative
1.5 Liaison and communication with water users	Good, water users fully aware of own duties and responsibilities, and those of WUA/ Federation	Reasonable, water users aware of own duties and responsibilities, and those of WUA/ Federation	Poor, water users not aware of own duties and responsibilities, and those of WUA/ Federation
	Water users fully aware of and up-to-date with WUA/ Federation activities	Water users partially aware of and up-to-date with WUA/Federation activities	Water users not aware of and not up-to-date with WUA/Federation activities
1.6 Level of facilities	Fully equipped with motorbikes, office facilities and equipment, maps and records	Partially equipped with motorbikes, office facilities and equipment, maps and records	Poorly equipped with motorbikes, office facilities and equipment, maps and records
		Missing some key items	Missing many key items
1.7 Level of democracy/ participation	Full participation by members	Moderate participation by members	Poor participation by members
	All members know what is happening in the WUA/ Federation	Moderate levels of involvement, moderate levels of knowledge of WUA/ Federation activities	Low level of involvement, low level of knowledge of WUA/Federation activities

Continued

Table A2.2. Continued

Performance indicator	Desired standard		
	Good	Medium	Poor
2. Finances			
2.1 Setting of the ISF	ISF determined from full needs based assessment	ISF partially determined from needs based assessment	ISF not determined from needs based assessment
2.2 Level of fee recovery – numbers contributing	>90% of members paying full ISF	60–89% of members paying full ISF	<59% of members paying full ISF
2.3 Level of fee recovery – amount contributed	>90% of estimated annual requirement collected	60–89% of estimated annual requirement collected	<59% of estimated annual requirement collected
2.4 Audit report	Satisfactory audit report, no problems reported	Only moderately satisfactory audit report, some problems identified and reported	Unsatisfactory audit report, serious problems identified and reported
3. Water management			
3.1 Pre-season irrigation planning	Irrigation plan prepared and used	Irrigation plan partially prepared and used	Irrigation plan not prepared or used
3.2 Pre-season irrigation agreements	Full set of agreements reached and formally documented	Partial set of agreements reached and formally documented	Few or no agreements reached
3.3 In-season irrigation planning	Weekly irrigation schedules prepared and agreed	Weekly irrigation schedules prepared and agreed in part	No weekly irrigation schedules prepared or agreed
3.4 In-season irrigation implementation	Average management performance ratio (MPR) >0.80	Average MPR = 0.50–0.79	Average MPR<0.49
3.5 In-season irrigation monitoring and discussion	Full weekly monitoring and feedback to WUA/Federation	Partial weekly monitoring and feedback to WUA/Federation	No weekly monitoring and feedback to WUA/Federation
3.6 Discharge measurement	Discharges measured daily at all control points	Most discharges measured daily at most control points	Discharges not measured daily, or only at some control points
3.7 End-of-season performance evaluation	Full evaluation carried out of season's performance	Partial evaluation carried out of season's performance	No evaluation carried out of season's performance
4. Asset extent, management and maintenance			
4.1 Asset database	Full and up-to-date asset database available – type, extent, condition	Partial asset database available	Limited or no asset database available
4.2 Pre-season maintenance plan	Pre-season maintenance plan done and acted upon	Partial maintenance plan prepared and acted upon	No maintenance plan prepared or acted upon
4.3 Annual maintenance identification, costing, planning, implementation and recording	Maintenance requirements identified, costed, prioritized and acted upon	Maintenance requirements partially identified, costed, prioritized and acted upon	No maintenance requirements identified, costed, prioritized or acted upon

Continued

Table A2.2. Continued

Performance indicator	Desired standard		
	Good	Medium	Poor
4.4 Total maintenance expenditure	Total maintenance expenditure >$20/ha and/or >60% of collected revenue, and/or >60% of MOM costs	Total maintenance expenditure $10–19/ha and/or 40–59% of collected revenue, and/or 40–59% of MOM costs	Total maintenance expenditure <$10/ha and/or <39% of collected revenue, and/or <39% of MOM costs
4.5 Degree of control and measurement	Control and measurement possible, and functioning, at all control points	Control and measurement at control points constrained, either through lack of structures or poor condition	Control and/or measurement at control points either non-existent or not functioning
4.6 Overall infrastructure condition	Infrastructure in good condition, with no constraints to operation	Operation constrained by infrastructure condition, some restriction on system functionality	Infrastructure in poor condition, with serious constraints for operation
5. Level of service provision			
5.1 Knowledge by water users of WUA/Federation activities	Water users fully aware of WUA/Federation activities	Water users partially aware of WUA/Federation activities	Water users not aware of WUA/Federation activities
5.2 Satisfaction with level of service provided	>80% of water users 100% satisfied with level of service provided	50–79% of water users 100% satisfied with level of service provided	<49% of water users 100% satisfied with level of service provided
5.3 Number of registered complaints during season	<5 serious complaints per 100 water users during season	6–20 serious complaints per 100 water users during season	>20 serious complaints per 100 water users during season

ISF, irrigation service fee; MOM, management, operation and maintenance.

Index

Note: page numbers in **bold** refer to figures, tables and boxes